METHODS IN MOLECULAR BIOLOGY

Series Editor
John M. Walker
School of Life and Medical Sciences
University of Hertfordshire
Hatfield, Hertfordshire, AL10 9AB, UK

For further volumes:
http://www.springer.com/series/7651

Single Cell Protein Analysis

Methods and Protocols

Edited by

Anup K. Singh and Aarthi Chandrasekaran

Sandia National Laboratories, Livermore, CA, USA

Joint BioEnergy Institute, Emeryville, CA, USA

Editors
Anup K. Singh
Sandia National Laboratories
Livermore, CA, USA

Joint BioEnergy Institute
Emeryville, CA, USA

Aarthi Chandrasekaran
Sandia National Laboratories
Livermore, CA, USA

Joint BioEnergy Institute
Emeryville, CA, USA

ISSN 1064-3745 ISSN 1940-6029 (electronic)
Methods in Molecular Biology
ISBN 978-1-4939-2986-3 ISBN 978-1-4939-2987-0 (eBook)
DOI 10.1007/978-1-4939-2987-0

Library of Congress Control Number: 2015952453

Springer New York Heidelberg Dordrecht London

Printed on acid-free paper

Humana Press is a brand of Springer
Springer Science+Business Media LLC New York is part of Springer Science+Business Media (www.springer.com)

Preface

Proteins are one of the most important classes of molecules in living cells and tissues. Proteins and their complexes are involved in virtually all cellular processes including catalysis of biochemical reactions, transport of molecules across membranes, cell growth and division, cell adhesion and migration, and providing structural support. Characterizing the quantity, association, and activity of proteins is therefore critical for understanding the molecular mechanisms of cellular processes involved in cell differentiation and fate, cell signaling, disease progression, and for discovery and development of novel therapeutics, vaccines, and diagnostics. Measuring DNA and RNA can provide qualitative information on gene products (proteins) but cannot provide information on protein concentration, activity, location, posttranslational modifications, or interactions with other proteins. Therefore, we need tools and assays to directly measure proteins, their interactions and activities, and modifications. Numerous analytical methods have been developed to analyze proteins such as gel electrophoresis, immunoassays, enzyme assays, chromatography, and mass spectrometry. At the conventional scale however, these methods require a large number of cells for analysis, resulting in a population-averaged measurement. Cells are heterogeneous in nature and even genetically identical cells exhibit heterogeneous behavior. This heterogeneity may have important biological consequences for both the individual cells and the population. Population-averaged data, which assumes that all the cells are identical, can be misleading and hence more desirable data in many instances is data from analyses of single cells. One well-known example is the response of bacteria to antibiotics where at certain doses many cells die but some survive and develop resistance. Similarly, one of the unanswered questions in cancer therapy has been why essentially identical cells respond differently to a drug. Single-cell level measurement of proteins (and other molecules) has provided valuable insight into mechanisms that dictate heterogeneity in cellular response to drugs and other internal and external stimuli. Usefulness of single-cell measurements is also obvious for stem cell research as decisions in individual cells determine their fate.

Despite the need, tools that allow quantitative measurements of proteins in single cells have not been easily available. The biggest challenge to measuring proteins in single cells is the exceedingly low amount present in a cell. The complexity and large concentration range (fM to high nM) of the proteome add additional challenges. Significant efforts are currently being made to overcome these challenges and achieve selective and sensitive analysis of the proteome in individual cells. This volume highlights recent developments in flow cytometry, affinity assays, imaging, mass spectrometry, microfluidics, and other technologies that have enabled analysis of the proteins at the single-cell level. We also include chapters covering a suite of biochemical and biophysical methods capable of making the entire gamut of proteomic measurements including analysis of protein abundance or expression, protein interaction networks, posttranslational modifications, translocation, and enzymatic activity. The book should be useful to researchers and students in biological and biomedical sciences who have an interest in proteomic measurements in cells.

Livermore, CA, USA — *Anup K. Singh*
Livermore, CA, USA — *Aarthi Chandrasekaran*

Contents

Preface . . . . *v*
Contributors . . . . *ix*

1 Single-Cell Western Blotting 1
Elly Sinkala and Amy E. Herr

2 A Microfluidic Device for Immunoassay-Based Protein Analysis of Single *E. coli* Bacteria 11
Simone Stratz and Petra S. Dittrich

3 Enzyme-Linked ImmunoSpot (ELISpot) for Single-Cell Analysis 27
Sylvia Janetzki and Rachel Rabin

4 Photocleavable DNA Barcoding Antibodies for Multiplexed Protein Analysis in Single Cells 47
Adeeti V. Ullal and Ralph Weissleder

5 Genome-Wide Analysis of Protein and mRNA Copy Numbers in Single *Escherichia coli* Cells with Single-Molecule Sensitivity 55
Yuichi Taniguchi

6 Microfluidic Flow Cytometry for Single Cell Protein Analysis 69
Meiye Wu and Anup K. Singh

7 Microfluidic Image Cytometry for Single-Cell Phenotyping of Human Pluripotent Stem Cells 85
Yasumasa Mashimo and Ken-ichiro Kamei

8 Characterizing Phenotypes and Signaling Networks of Single Human Cells by Mass Cytometry 99
Nalin Leelatian, Kirsten E. Diggins, and Jonathan M. Irish

9 Multiplexed Peptide-MHC Tetramer Staining with Mass Cytometry 115
Mei Ling Leong and Evan W. Newell

10 Imaging and Mapping of Tissue Constituents at the Single-Cell Level Using MALDI MSI and Quantitative Laser Scanning Cytometry 133
Catherine M. Rawlins, Joseph P. Salisbury, Daniel R. Feldman, Sinan Isim, Nathalie Y.R. Agar, Ed Luther, and Jeffery N. Agar

11 SPLIFF: A Single-Cell Method to Map Protein-Protein Interactions in Time and Space 151
Alexander Dünkler, Reinhild Rösler, Hans A. Kestler, Daniel Moreno-Andrés, and Nils Johnsson

12 Microfluidic Proximity Ligation Assay for Profiling Signaling Networks with Single-Cell Resolution 169
Matthias Blazek, Günter Roth, Roland Zengerle, and Matthias Meier

13 Dynamics and Interactions of Individual Proteins in the Membrane of Single, Living Cells 185
Stephen Anthony, Amanda Carroll-Portillo, and Jerilyn Timlin

14 Microfluidics-Enabled Enzyme Activity Measurement in Single Cells. 209
Cinzia Tesauro, Rikke Frøhlich, Magnus Stougaard, Yi-Ping Ho, and Birgitta R. Knudsen

15 Microfluidic Chemical Cytometry for Enzyme Assays of Single Cells 221
Livia Shehaj, Lorena Lazo de la Vega, and Michelle L. Kovarik

16 Quantitative Detection of Nucleocytoplasmic Transport of Native Proteins in Single Cells 239
Zhenning Cao and Chang Lu

Index 253

Contributors

JEFFERY N. AGAR • *Barnett Institute of Chemical and Biological Analysis, Northeastern University, Boston, MA, USA; Department of Pharmaceutical Sciences, Northeastern University, Boston, MA, USA*

NATHALIE Y.R. AGAR • *Department of Neurosurgery, Brigham and Women's Hospital, Harvard Medical School, Boston, MA, USA*

STEPHEN ANTHONY • *Sandia National Laboratories, Bioenergy and Defense Technologies, Albuquerque, NM, USA*

MATTHIAS BLAZEK • *Microfluidic and Biological Engineering, IMTEK — Department of Microsystems Engineering, University of Freiburg, Freiburg, Germany; BIOSS — Centre for Biological Signalling Studies, University of Freiburg, Freiburg, Germany*

ZHENNING CAO • *School of Biomedical Engineering and Sciences, Virginia Tech, Blacksburg, VA, USA*

AMANDA CARROLL-PORTILLO • *Sandia National Laboratories, Bioenergy and Defense Technologies, Albuquerque, NM, USA*

AARTHI CHANDRASEKARAN • *Sandia National Laboratories, Livermore, CA, USA; Joint BioEnergy Institute, Emeryville, CA, USA*

KIRSTEN E. DIGGINS • *Cancer Biology, Vanderbilt University School of Medicine, Nashville, TN, USA*

PETRA S. DITTRICH • *Department of Biosystems Science and Engineering, ETH Zurich, Zurich, Switzerland*

ALEXANDER DÜNKLER • *Institute of Molecular Genetics and Cell Biology, Department of Biology, Ulm University, Ulm, Germany*

DANIEL R. FELDMAN • *Department of Neurosurgery, Brigham and Women's Hospital, Harvard Medical School, Boston, MA, USA*

RIKKE FRØHLICH • *Department of Molecular Biology and Genetics, Aarhus University, Aarhus C, Denmark*

AMY E. HERR • *Department of Bioengineering, University of California, Berkeley, Berkeley, CA, USA*

YI-PING HO • *Interdisciplinary Nanoscience Center (iNANO), Aarhus University, Aarhus, Denmark*

JONATHAN M. IRISH • *Cancer Biology, Vanderbilt University School of Medicine, Nashville, TN, USA; Pathology, Microbiology and Immunology, Vanderbilt University School of Medicine, Nashville, TN, USA*

SINAN ISIM • *Life Sciences Department, Brandeis University, Waltham, MA, USA*

SYLVIA JANETZKI • *ZellNet Consulting, Inc., Fort Lee, NJ, USA*

NILS JOHNSSON • *Institute of Molecular Genetics and Cell Biology, Department of Biology, Ulm University, Ulm, Germany*

KEN-ICHIRO KAMEI • *Institute for Integrated Cell-Material Sciences (WPI-iCeMS), Kyoto University, Kyoto, Japan*

HANS A. KESTLER • *Research Group for Bioinformatics and Systems Biology, Institute of Neural Information Processing, Ulm University, Ulm, Germany*

BIRGITTA R. KNUDSEN • *Department of Molecular Biology and Genetics, Aarhus University, Aarhus C, Denmark*
MICHELLE L. KOVARIK • *Department of Chemistry, Trinity College, Hartford, CT, USA*
LORENA LAZO DE LA VEGA • *Department of Chemistry, Trinity College, Hartford, CT, USA*
NALIN LEELATIAN • *Cancer Biology, Vanderbilt University School of Medicine, Nashville, TN, USA*
MEI LING LEONG • *School of Biological Sciences, Nanyang Technological University, Singapore, Singapore*
CHANG LU • *Department of Chemical Engineering, Virginia Tech, Blacksburg, VA, USA*
ED LUTHER • *Department of Pharmaceutical Sciences, Northeastern University, Boston, MA, USA*
YASUMASA MASHIMO • *Institute for Integrated Cell-Material Sciences (WPI-iCeMS), Kyoto University, Kyoto, Japan*
MATTHIAS MEIER • *Microfluidic and Biological Engineering, IMTEK — Department of Microsystems Engineering, University of Freiburg, Freiburg, Germany; BIOSS — Centre for Biological Signalling Studies, University of Freiburg, Freiburg, Germany*
DANIEL MORENO-ANDRÉS • *Friedrich Miescher Laboratory of the Max Planck Society, Tübingen, Germany*
EVAN W. NEWELL • *Singapore Immunology Network (SIgN), Agency for Science Technology and Research (A-STAR), Singapore, Singapore*
RACHEL RABIN • *ZellNet Consulting, Inc., Fort Lee, NJ, USA*
CATHERINE M. RAWLINS • *Barnett Institute of Chemical and Biological Analysis, Northeastern University, Boston, MA, USA*
REINHILD RÖSLER • *Institute of Molecular Genetics and Cell Biology, Department of Biology, Ulm University, Ulm, Germany*
GÜNTER ROTH • *BIOSS — Centre for Biological Signalling Studies, University of Freiburg, Freiburg, Germany; Center for Biological Systems Analysis (ZBSA), University of Freiburg, Freiburg, Germany*
JOSEPH P. SALISBURY • *Barnett Institute of Chemical and Biological Analysis, Northeastern University, Boston, MA, USA*
LIVIA SHEHAJ • *Department of Chemistry, Trinity College, Hartford, CT, USA*
ANUP K. SINGH • *Biological Science and Technology, Sandia National Laboratories, Livermore, CA, USA; Joint BioEnergy Institute, Emeryville, CA, USA*
ELLY SINKALA • *Department of Bioengineering, University of California, Berkeley, Berkeley, CA, USA*
MAGNUS STOUGAARD • *Department of Pathology, Aarhus University Hospital, Aarhus, Denmark*
SIMONE STRATZ • *Department of Biosystems Science and Engineering, ETH Zurich, Zurich, Switzerland*
YUICHI TANIGUCHI • *Laboratory for Single Cell Gene Dynamics, Quantitative Biology Center, RIKEN, Suita, Osaka, Japan*
CINZIA TESAURO • *Department of Molecular Biology and Genetics, Aarhus University, Aarhus C, Denmark*
JERILYN TIMLIN • *Sandia National Laboratories, Bioenergy and Defense Technologies, Albuquerque, NM, USA*

ADEETI V. ULLAL • *Center for Systems Biology, Massachusetts General Hospital, Boston, MA, USA*

LORENA LAZO DE LA VEGA • *Department of Chemistry, Trinity College, Hartford, CT, USA*

RALPH WEISSLEDER • *Center for Systems Biology, Massachusetts General Hospital, Boston, MA, USA*

MEIYE WU • *Biotechnology and Bioengineering, Sandia National Laboratories, Livermore, CA, USA*

ROLAND ZENGERLE • *BIOSS — Centre for Biological Signalling Studies, University of Freiburg, Freiburg, Germany; Laboratory for MEMS Applications, IMTEK — Department of Microsystems Engineering, University of Freiburg, Freiburg, Germany*

Chapter 1

Single-Cell Western Blotting

Elly Sinkala and Amy E. Herr

Abstract

Little headway has been made in single cell protein analysis, aside from tools that rely solely on antibody-probe based detection (i.e., flow cytometry, immunocytochemistry), which are limited by low specificity and multiplexing capabilities. To address these protein analysis gaps, we have introduced a single-cell western blot (scWestern). The protein assay is capable of highly specific analysis by coupling antibody-based detection with a polyacrylamide gel electrophoresis (PAGE) protein separation. Cells are settled via gravity into polyacrylamide (PA) microwells, chemically lysed in the wells, and then subjected to PAGE through the walls of the microwells and into the surrounding PA gel. Over a thousand single-cell separations are performed simultaneously, and multiple protein targets of interest are investigated. After PAGE separation, photo-immobilization of all proteins to the gel allows for antibody probing and lends to the archival quality of the scWestern assay where new proteins targets can be investigated months after the initial separations are performed.

Key words Immunoblot, Single cell, Immunoassay, Proteomics, Lab-on-a-chip

1 Introduction

Proteins are responsible for a vast majority of cellular processes. Collectively, changes in protein state or concentration can modulate overall cellular behavior. Often, distinct cell sub-populations are drivers of unique cellular processes, which include stem cell differentiation [1], cancer [2, 3], and immune response [4]. Single-cell experiments often reveal different outcomes as compared to similar studies performed at the population level [5, 6]. The inherent heterogeneity within these cell populations requires a move away from population-based protein analysis [7, 8] (i.e., Western blots) to single-cell proteomic assays to understand behavior at the single cell level. Current single-cell proteomic assays, including flow cytometry [4] and immunocytochemistry (ICC) [9], are limited by the specificity of the antibody-antigen interaction. Cross-reactivity is also a concern in enzyme-linked immunosorbent assays (ELISAs) and protein microarrays [10]. To address these current limitations in single-cell protein analysis, we developed the

Anup K. Singh and Aarthi Chandrasekaran (eds.), *Single Cell Protein Analysis: Methods and Protocols*, Methods in Molecular Biology, vol. 1346, DOI 10.1007/978-1-4939-2987-0_1,

scWestern assay [11]. The scWestern can perform ~10^3 concurrent single-cell protein separations on a standard microscope slide. This multiplex, high-throughput assay is possible through (1) high-density microwell arrays [12], (2) short separation distances (0.5 or 1 mm) for electrophoresis, (3) UV-immobilized proteins for multiple probings and (4) thin 30–60 μm polyacrylamide gel compatible with diffusive antibody transport. Additionally, the scWestern provides direct quantitative protein analysis.

The scWestern consists of a thin photoactive polyacrylamide (PA) gel micropatterned with microwell arrays. The microwell diameter is tuned to the size of the analyzed cells (15–50 μm diameter), which are passively settled onto the microwell patterned PA gel. Trapped cells are lysed with a combined denaturing radioimmunoprecipitation assay (RIPA) and electrophoresis buffer. An electric field is applied across the scWestern slide, and proteins migrate through the microwell into the adjacent PA gel. Separated proteins are immobilized with UV light where upon exposure the benzophenone methacrylamide co-monomer cross-linked into the PA gel [13] forms covalent bonds with the migrated proteins. Primary and secondary antibodies are incubated against the scWestern gel to detect proteins targets of interest. The scWestern gels can be stripped and reprobed repeatedly to detect different protein targets.

The scWestern PAGE performance is comparable to the conventional Western blot where proteins with ~50 % difference in molecular mass are resolvable in a 500 μm separation length and 30-s separation time. Altogether, the %T of the PA gel, well diameter, separation length and time are optimized to enhance the overall separations. As with SDS-PAGE, we observe protein stacking at microwell interface, and a log-linear relationship between molecular mass and protein migration was confirmed [11]. Antibodies used in a conventional Western are compatible with the scWestern, and fluorescently labeled antibodies with a microarray scanner provide the readout for protein quantification. The protein immobilization with the gel provides long-term storage and reanalysis of protein targets. We have investigated up to 12 targets on a single cell and have the potential include more. The ability to investigate many targets in a single cell allows us to investigate questions related to signal pathways that are difficult to test otherwise.

2 Materials

2.1 Microwell scWestern SU-8 Mold Components

1. Mechanical grade 100 mm diameter silicon wafers.
2. SU8-8 2025 photoresist (Y111069, MicroChem).
3. SU8 developer solution (Y020100, MicroChem).
4. Mylar mask with 20–30 μm features at 20,000 d.p.i.

5. Silane dichlorodimethylsilane.
6. Isopropanol.
7. Acetone.
8. Photoresist spinner.
9. Mask Aligner.

2.2 Polyacrylamide Gel Precursor Components

Reagents are prepared in deionized water unless otherwise noted. APS and TEMED must be prepared the day of PA gel fabrication.

1. Deionized water.
2. 30 %T, 2.7 %C Bis-acrylamide.
3. 1.5 M Tris–HCl pH 8.8.
4. 100 mM of benzophenone methacrylate conmonomer (BPMAC) in dimethyl sulfoxide (DMSO).
5. 5 % SDS.
6. 5 % Triton X-100.
7. 10 % w/v ammonium Persulfate (APS).
8. 10 % v/v tetramethylethylenediamine (TEMED).
9. Plain glass microscope glass slide silanized in-house or pre-silanized glass slides (SuperMethacrylate 3, Arrayit).

2.3 scWestern Lysis and Electrophoresis Components

1. Microwell-patterned PA gel.
2. 1× phosphate buffer solution from a 10× stock solution (PBS).
3. 60 mm polystyrene dish.
4. Glass microscope slide.
5. Electrophoresis chamber.
6. Petroleum jelly.
7. UV mercury arc lamp (such as Lightningcure LC5, Hamamatsu).
8. Lumatec series 380 liquid light guide with inline UV filter (300–380-nm bandpass, XF1001, Omega Optical).
9. UV meter.
10. RIPA-like lysis/electrophoresis buffer: 0.5 % SDS, 0.1 % v/v Triton X-100, 0.25 % sodium deoxycholate in 12.5 mM Tris, 96 mM glycine, pH 8.3 (0.5× from a 10× stock). Store at 4 °C.
11. Tris buffered saline with Tween 20 (TBST): 100 mM Tris titrated to pH 7.5 with HCl, 150 mM NaCl, 0.1 % Tween 20.

2.4 scWestern Immunoblotting, Imaging, and Stripping Components

1. Primary antibody.
2. Alexa Fluor conjugated secondary antibody.
3. TBST supplemented with 2 % bovine serum albumin (BSA).
4. 75 mm × 50 mm glass microscope slides.

5. Tape.
6. DI water.
7. GenePix 4300A microarray scanner or fluorescent microscope.
8. Harsh stripping buffer: 2.5 % SDS and 1 % β-mercaptoethanol in 62.5 mM Tris titrated to pH 6.8 with HCl.

3 Methods

3.1 SU-8 Mold for PA gel Microwells Fabrication

1. Wash silicon wafer in acetone under orbital shaking for 5 min. Rinse with isopropanol and water.
2. Place silicon wafer on 150 °C hot plate to dehydrate the surface.
3. Place silicon wafer onto spinner and deposit ~5 mL of SU-8 2025 photoresist in the center of the wafer. Spin for 15 s at 500 RPM with 100 RPM/s acceleration then 40 s at 2500 RPM with 300 RPM/s acceleration (*see* **Note 1**).
4. Place on 65 °C hot plate for 3 min and a 95 °C hot plate for 6 min. Cool to room temperature.
5. Position the mylar mask over the SU-8 and expose with 365-nm UV light at ~40 mW/cm^2 for 12 s.
6. Place exposed wafer on 65 °C hot plate for 1 min and a 95 °C hot plate for 5 min. Cool to room temperature.
7. Develop the microwell features by spraying SU-8 developer over wafer. Follow with a rinse of isopropanol and water.
8. Silanize by vapor-deposition of 200 mL of DCDMS for 1 h in vacuo, wash thoroughly with deionized (DI) water and dry under nitrogen immediately before use.

3.2 scWestern Gel Fabrication

1. Silanize glass slides according to standard protocols [14] (*see* **Note 2**).
2. Place a silanized glass slide, treated side down, onto the fabricated SU-8 mold.
3. Prepare a 7 %T polyacrylamide gel by mixing 315 μL of water, 25 μL of Tris–HCl, 117 μL of acrylamide, and 15 μL of BPMAC into a 1.5 mL tube. Gently vortex. Puncture tube cap with needle and degas the precursor (*see* **Note 3**).
4. Add 10 μL of 5 % SDS, 10 μL of 5 % Triton X-100, 4 μL of APS, and 4 μL of TEMED to the degassed precursor solution. Briefly vortex to mix ensuring not to introduce bubbles into the solution (*see* **Note 4**).
5. Immediately inject the gel solution in the gap between the glass slide and silicon wafer. Allow ~15–30 s for the solution to

wick to the opposite edge of the slide. Wait 20 min for the PA gel to polymerize (*see* **Note 5**).

6. Pipet 2 mL of PBS around the edge of the slide. Use a sharp razor blade to lift the slide from the silicon wafer (*see* **Note 6**).
7. Confirm the wells are clearly formed under a microscope. Place the gel into a dish filled with PBS gel side up. The slide can be stored in PBS at 4 °C until experiment.

3.3 Cell Settling on scWestern Gel

1. Disassociate cells with trypsin. Add media with serum to stop the reaction after cells are detached. Spin down cells, remove all media and resuspend cells in PBS to achieve a concentration of 1–3 × 10^6 cells/mL (*see* **Note 7**).
2. Remove gel from PBS and remove any excess solution by placing a absorbent tissue at the corner of the slide. Place slide in a petri dish and deposit 1–1.5 mL of cell solution. Allow 10–60 min for the cells to settle in the wells (*see* **Note 8**).
3. Rinse the untrapped cells by tilting the slide 10–20° and pipetting 1 mL of PBS at the corner of the slide. Continue until there is minimal debris. Confirm with microscope.
4. Gently pipet 1 mL of PBS onto the surface of the gel. Once the entire slide is wetted, place a clean glass slide onto the gel. Use an absorbent tissue to remove any excess solution.
5. Determine the well occupancy by imaging the entire slide at 4× magnification with 1× 1 binning (*see* **Note 9**).

3.4 Lysis and Separation

Details further described in Fig. 1.

1. Remove the top glass slide and place the gel into an electrophoresis chamber using petroleum jelly to secure and immobilize the slide during electrophoresis (*see* **Note 10**).

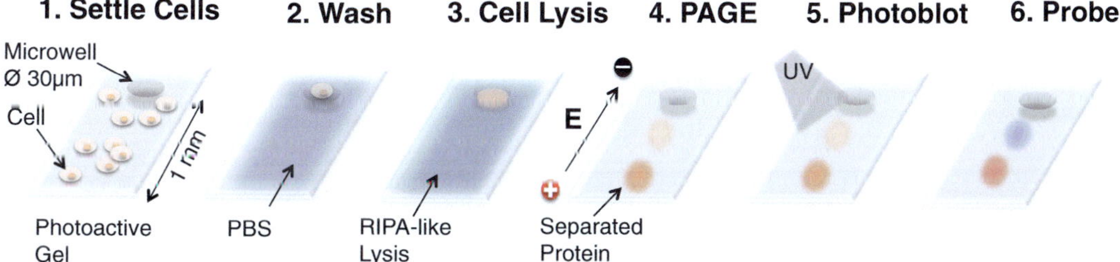

Fig. 1 The scWestern is a 4 h assay that produces ~10^3 protein separations on a single gel. The assay contains six separate steps. (1) Cells are harvested and settled onto the PA gel, (2) PBS rinsing removes uncaptured cells from the gel leaving cells isolated in the microwells. (3) A RIPA-like lysis buffer lyses the cells, and proteins are denatured with SDS. (4) The RIPA-like lysis buffer has sufficient conductivity to perform PAGE and separate proteins into the gel. (5) The gel is exposed to UV to immobilized separated proteins. (6) Proteins of interest are probed with primary and secondary antibodies. The scWestern PA gel is compatible with antibody stripping, important for investigating multiple protein targets of interest

2. Gently pour 10 mL of RIPA-like lysis/electrophoresis buffer on one corner of the electrophoresis chamber (*see* **Note 11**). Allow cells to lyse for 10–20 s. The RIPA-like lysis buffer may also be heated to enhance protein denaturation (*see* **Note 12**).
3. Immediately apply 200–250 V (E=40 V/cm^2) to perform protein separation for 20–30 s (*see* **Note 13**).
4. Immediately stop the applied voltage and expose the gel slide with UV light using a light guide. Hold the light guide ~10 cm above the gel and expose for 45 s at a power of ~40 mW/cm^2 (*see* **Note 14**).
5. Remove slide from electrophoresis chamber and store in PBS at 4 °C until probing.

3.5 Immunoblotting

1. Use a 75 × 50 mm microscope glass slide. Use standard lab tape to create shims to create a gap between the gel slide and the microscope slide for antibody probing. Place the slide, gel side down, onto the tape.
2. Create a 1:10 to 1:20 fold dilution of antibody in TBST and 2 % BSA. A gel on a 75 × 25 mm will require ~100 μL of antibody cocktail. Spin the antibodies at 10,000 × *g* for 5 min to remove aggregates.
3. Pipet antibody cocktail between the gel and microscope slide. Incubate for 2 h for primary antibodies (*see* **Note 15**).
4. Wash gel slides with TBST for 1 h under gentle orbital shaking. Exchange with fresh TBST solution every 15 min.
5. Repeat **steps 3** and **4** for secondary antibodies. 1 h incubation is sufficient.
6. Rinse gel slide with DI water and completely dry PA gel under nitrogen.
7. Image gel slide with fluorescence microscope or a microarray scanner for scWestern fluorescent readout (*see* **Note 16**).

3.6 Antibody Stripping

Details further described in Fig. 2.

1. Rehydrate dried gel slide in PBS for 10 min before stripping. Remove excess PBS and place the gel into a slide holder filled with harsh stripping buffer. Place container into a 55 °C water bath and incubate at least 3 h to overnight.
2. Remove the gel from container and place in a clean petri dish with TBST. Thoroughly wash the gel slide for 1 h changing the buffer every 15 min (*see* **Note 17**).
3. Store the stripped slide in TBST at 4 °C until probing.

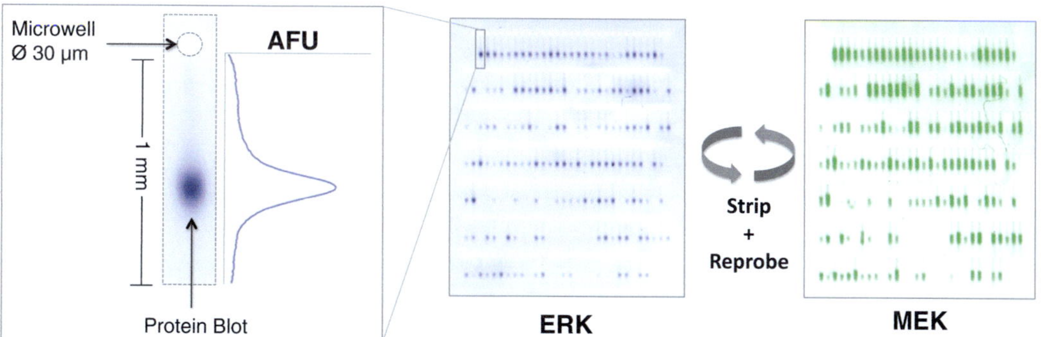

Fig. 2 Breast cancer cells (MDA-MB-231s) were settled into 30 μm diameter wells and PAGE was performed over a 1 mm separation distance. Protein quantification is achieved by measuring the area under the curve of the scWestern fluorescent readout. *AFU* arbitrary fluorescent units. Additionally, protein peak location provides further discrimination between many different proteins of interest. Proteins ERK and MEK are key downstream targets for several different pathways associated with breast cancer development and progression. Here, the gel was initially probed for ERK and subsequently stripped and reprobed for MEK. The immobilized proteins on the PA gel archives the separations permitting the investigation of multiple protein targets on a single cell

4 Notes

1. The SU-8 protocol produces ~30 μm tall posts for ~30 μm deep PA gel microwells. Refer to Microchem SU-8 protocol to create posts with different heights. The microwell depth can influence how well a cell is captured and remains in the microwell during handling. It is important to optimize the microwell diameter to cell size to reduce the number of microwells with multiple cells or to ensure larger cells are trapped. When handling a new cell line, a silicon master with multiple microwell diameters sizes (15–50 μm) can identify the optimal diameter by counting the microwell diameter that captured the most cells while keeping the trapping close to 1 cell per microwell.
2. Microscope glass slides are silanized with a self-assembled surface monolayer of methacrylate functional groups to bond the polyacrylamide gel to the slide. The glass slides can be silanized in house or purchased pre-salinized.
3. Acrylamide is a known carcinogen. Use gloves when handling acrylamide or any objects that come in direct contact with the solution. Handle the scWestern slides with gloves since the gel can contain acrylamide even after polymerization. The ratio of the acrylamide/bis-acrylamide solution can be adapted for specific experimental procedures.
4. The acrylamide gel solution must be used immediately after the addition of APS and TEMED to prevent gel polymerization

outside the glass slide master construct. Both APS and TEMED initiate gel polymerization. The gel precursor solution must be thoroughly degassed to limit the introduction of oxygen into the gel precursor, which can affect the length of polymerization.

5. The remaining gel precursor in the tube can be used to monitor the state of polymerization. As on the silanized slide, the gel in the tube will also be polymerized after 20 min. The gel is polymerized when it is solidified. If bubbles are formed in the gel when pipeting onto the SU-8 wafer, the glass slide can be gently moved to release the bubbles. However, do not disturbed the PA gel 30s after introduction. The polymerization is already underway and disturbing the gel may produce malformations in the gel.
6. To lift the gel without disturbing the microwell features, use one hand to firmly hold down the wafer and slide the razor blade slightly underneath the gel slide. Slowly lift the razor blade directly up to release the slide. Rinse the master wafer with water before storage to remove any remaining PA gel. The silanization is robust, and the master does not require re-silanization between gel fabrications.
7. Use a cell strainer tube to isolate large clusters of aggregated cells for a single cell suspension. Refrain from using complete media to settle cells onto the gel. The serum in the media can partition into the wells and obscure signal from the cell lysate. It is best to run the scWestern within 1 h of harvesting the cells. Place the harvested cells onto ice if not processed immediately. Non-adherent cells do not require removal from plate/dish and are simply resuspended in PBS before cell settling on gel.
8. Gently agitate the slide occasionally to redistribute the cells on the gel surface increasing probability for cell capture in microwell. To prevent the gel from drying, ensure there is sufficient solution placed onto the gel. The settling rate can vary between cell types.
9. A single microwell can capture multiple cells. Imaging the slide is important to determine the number of cells per microwell. Fluorescent images of stained cells (e.g., Hoechst 3342) can be obtained during this step. The imaging of multiple fluorescent channels or at a higher magnification will increase the total imaging duration. Ensure the slide has sufficient PBS and the secondary glass slide remains on the gel to reduce evaporation.
10. Work quickly after removing the glass cover slide to prevent the cells and slide from drying.
11. The electrophoresis chamber is constructed with two parallel platinum wires that run along a container sufficient to house a 75 × 25 mm slide. The distance between the wires should be large enough to accommodate the slide. The platinum wire

should be straight to produce the most uniform electric field. Loop the ends of the platinum wires for secure connections with the alligator clips of the power supply. To view the lysis and separation, use a chamber with clear glass or plastic to use with microscope.

12. Lysis and separation time and RIPA-like lysis buffer temperature all control the degree of separations. For larger proteins of interest, it is helpful to increase the separation time. Heating the lysis buffer, up to 55 °C, can also improve separations.
13. Lysis and separation time is optimized for specific protein targets of interest.
14. Wear proper UV blocking glasses and facemask during the immobilization step.
15. We found that 2 h and 1 h are the optimal incubation times for primary and secondary antibodies respectively. However, the antibody may be incubated for longer if needed.
16. We quantify the fluorescent protein content by measuring the area under the curve for single blot.
17. It is important to remove and thoroughly wash all harsh strip ping buffer from gel. Any remaining solution can interfere with subsequent probings.

References

1. Chang HH, Hemberg M, Barahona M et al (2008) Transcriptome-wide noise controls lineage choice in mammalian progenitor cells. Nature 453(7194):544–547
2. Dalerba P, Kalisky T, Sahoo D et al (2011) Single-cell dissection of transcriptional heterogeneity in human colon tumors. Nat Biotechnol 29(12):1120–1127
3. Balic M, Williams A, Lin H et al (2013) Circulating tumor cells: from bench to bedside. Annu Rev Med 64:31–44
4. Bendall SC, Simonds EF, Qiu P et al (2011) Single-cell mass cytometry of differential immune and drug responses across a human hematopoietic continuum. Science 332(6030):687–696
5. Tay S, Hughey JJ, Lee TK et al (2010) Single-cell NF-[kgr] B dynamics reveal digital activation and analogue information processing. Nature 466(7303):267–271
6. Altschuler SJ, Wu LF (2010) Cellular heterogeneity: do differences make a difference? Cell 141(4):559–563
7. Ciaccio MF, Wagner JP, Chuu C et al (2010) Systems analysis of EGF receptor signaling dynamics with microwestern arrays. Nat Methods 7(2):148–155
8. Hughes AJ, Herr AE (2012) Microfluidic western blotting. Proc Natl Acad Sci 109(52):21450–21455
9. Ashton RS, Conway A, Pangarkar C et al (2012) Astrocytes regulate adult hippocampal neurogenesis through ephrin-B signaling. Nat Neurosci 15(10):1399–1406
10. Sevecka M, MacBeath G (2006) State-based discovery: a multidimensional screen for small-molecule modulators of EGF signaling. Nat Methods 3(10):825–831
11. Hughes AJ, Spelke DP, Xu Z et al (2014) Single-cell western blotting. Nat Methods 11(7):749–755
12. Wood DK, Weingeist DM, Bhatia SN et al (2010) Single cell trapping and DNA damage analysis using microwell arrays. Proc Natl Acad Sci 107(22):10008–10013
13. Cao X, Shoichet MS (2002) Photoimmobilization of biomolecules within a 3-dimensional hydrogel matrix. J Biomater Sci Polym Ed 13(6):623–636
14. Hughes AJ, Herr AE (2010) Quantitative enzyme activity determination with zeptomole sensitivity by microfluidic gradient-gel zymography. Anal Chem 82(9):3803–3811

Chapter 2

A Microfluidic Device for Immunoassay-Based Protein Analysis of Single *E. coli* Bacteria

Simone Stratz and Petra S. Dittrich

Abstract

We present a method suitable for quantitative analysis of intracellular proteins, metabolites and secondary messengers of single bacterial cells. The method integrates the concept of immunoassays on a microfluidic device that facilitates single cell trapping and isolating in a small volume of a few tens of picoliters. Combination of the benefits of microfluidic systems for single cell analysis with the high analytical selectivity and sensitivity of immunoassays enables the detection of even low abundant intracellular analytes which occur only at a few hundred copies per bacterium.

Key words Single-cell analysis, Bacteria analysis, *E. coli*, Immunoassay, Microfluidics, Enzyme activity

1 Introduction

Individual genetically identical cells subjected to the same microenvironment can differ considerably in their phenotypic properties [1–3]. This diversity originates mainly from the stochastic nature of cellular gene transcription and translation processes [4, 5]. Investigation of the underlying control mechanisms regulating the noise level in a cell population requires analysis of single cells [6, 7]. In facing the challenges of single cell analysis, microfluidic platforms offer particular advantages over conventional macro scale systems [8, 9]. Precise handling and control of small fluid volumes in the picoliter and nanoliter range allows transport and positioning of cells and cell lysates without sample dilution or loss [10]. Due to these benefits, several microfluidic devices highly suitable for single cell DNA and RNA analysis were presented in recent years [11–13]. In case of DNA and RNA analysis, the sensitivity can be greatly enhanced by amplification of the nucleic acids via polymerase chain reaction (PCR). For the detection of proteins in the low concentration range, the situation is more difficult, as there are no convenient signal amplification methods

Anup K. Singh and Aarthi Chandrasekaran (eds.), *Single Cell Protein Analysis: Methods and Protocols*, Methods in Molecular Biology, vol. 1346, DOI 10.1007/978-1-4939-2987-0_2,

available [14, 15]. Moreover, selectivity is a critical point in the choice of detection method, as there is a great variety of different proteins present in a cell. In previous work, we successfully performed enzyme-linked immunosorbent assays (ELISA) on-chip which enabled the protein analysis of single mammalian cells [16]. Due to the high sensitivity and selectivity provided by immunoassays, it was possible to quantify the amount of the housekeeping protein GAPDH in U937 cells as well as HEK293 cells and the amount of the secondary messenger cAMP in MLT cells. The above referenced projects illustrate the successful practical use of microfluidic devices for single cell analysis, but are limited with respect to the cell types. Most often, single cell analysis systems are optimized for mammalian cells that are relatively large and contain large analyte quantities compared to bacteria [17, 18]. Microfluidic platforms capable of handling small bacteria with on average a few hundred analyte copy numbers per cell are rarely found [19–21].

A better understanding of the origins of microbial phenotypic diversity and the underlying control mechanisms are of great importance for developing strategies against antibiotic or biocide resistant microorganisms. Therefore, we developed a system optimized for the analysis of single *E. coli* bacteria. For this purpose, the design of a previously used microfluidic device [16] was significantly downscaled in size. As a result, single *E. coli* bacteria could mechanically be trapped and isolated in microchambers with a volume as small as 155 pL [22]. As the analysis microchambers could be opened and closed repeatedly, different surface- and cell treatment steps could be performed successively. As a proof-of-concept study, an immunoassay for the quantification of the enzyme ß-galactosidase in individual *E. coli* bacteria was carried out. However, the approach could be extended to any other proteins for which antibodies are available. In the following protocol the major procedures of the method including bacteria cultivation, device fabrication, immunoassay performance, and data acquisition are described in greater detail.

2 Materials

2.1 Bacteria Cultivation

1. LB Agar plates: Add 250 mL deionized water to a 500 mL media storage bottle, weigh out 14 g of premixed LB-Agar powder and stir the mixture vigorously for 30 min. Add deionized water to total volume of 400 mL, label with autoclave tap and autoclave for 20 min at 120 °C and 1 bar above atmospheric pressure. Let the mixture cool to 55 °C. Meanwhile, place a Bunsen burner and 16 sterile petri dishes (petri dish square with vents) on a bench top that has been cleaned carefully with ethanol. Light the Bunsen burner, flame the lid of

the LB agar bottle with one hand (wear gloves), remove the lid of the first petri dish with the other hand and pour about 25 mL LB agar into the petri dish until it is almost half filled. After filling, immediately replace the lid and do not move the petri dish until the LB Agar is solid. Repeat the process for the remaining petri dishes. Write name, date and content on the upper edge of the lids and store the hardened LB Agar plates in plastic bags at 4 °C in the fridge (*see* **Note 1**).

2. LB Broth (Lennox) medium: Add 16 g LB Broth (Lennox) powder microbial growth medium to a 1000 mL media storage bottle filled with 800 mL deionized water and stir rigorously for 20 min. Then, autoclave the solution for 20 min at 120 °C and 1 bar above atmospheric pressure. Store the autoclaved LB Broth (Lennox) medium at room temperature.
3. 40 wt% glucose stock solution: Add 320 g glucose to a 1000 mL media storage bottle filled with 480 mL deionized water. Add glucose slowly while stirring the solution vigorously (*see* **Note 2**). Then, autoclave the solution for 20 min at 120 °C and 1 bar above atmospheric pressure. Store the autoclaved solution at room temperature.
4. M9 minimal medium: Buy or prepare the following stock solutions in deionized water: M9 minimal salts (5×), 1 M $MgSO_4$ solution, and 1 M $CaCl_2$ solution.
5. Preparation of M9 Minimal medium: To a 1000 mL media storage bottle filled with 640 mL deionized water, add 160 mL M9 salts (5×), 0.8 mL $MgSO_4$ (1 M), and 0.08 mL $CaCl_2$ (1 M). Stir the solution for 10 min. Then, autoclave the solution for 20 min at 120 °C and 1 bar above atmospheric pressure. Store the completed M9 minimal medium at room temperature.
6. 10 wt% lactose stock solution: Add 80 g lactose to a 1000 mL media storage bottle filled with 720 mL deionized water. Stir the solution for 15 min. Then autoclave the solution for 20 min at 120 °C and 1 bar above atmospheric pressure. Store the autoclaved solution at room temperature.
7. *E. coli* bacteria of the strain K-12 MG1655.
8. 1,1′-dioctadecyl-3,3,3′3′-tetramethylindocarbocyanine perchlorate (DiI) stock solution (100 μM in MeOH). Store at −20 °C.

2.2 Fabrication of the Microfluidic Device

1. Photoresist SU-8 2010.
2. Mr-Dev600 developer.
3. Silanization: 1*H*,1*H*,2*H*,2*H*-perfluorodecyl-trichlorosilane.
4. Poly(dimethylsiloxane) (PDMS) mixture: Poly(dimethylsiloxane) PDMS monomer and curing agent.

2.3 Surface Modification of the Microfluidic Device

1. Phosphate buffered saline (PBS): Buy or prepare a phosphate buffered saline solution with the following formulation of inorganic salts: KCl (200 mg/L), KH_2PO_4 (200 mg/L), NaCl (8000 mg/L), $Na_2HPO_4 \times 7H_2O$ (2160 mg/L). Dissolve all salts in deionized water. Filter the PBS sterile before use to avoid contamination of the microfluidic device with dust particles. Store at room temperature until usage.
2. Biotin-derivatized poly(L-lysine)-grafted poly(ethylene glycol) (PLL-g-PEG-biotin) solution: Mix 2.5 μL of the biotin-derivatized poly(L-lysine)-grafted poly(ethylene glycol) (PLL-g-PEG-biotin) stock solution (0.5 wt% in PBS) with 2.5 μL of the poly(L-lysine)-grafted poly(ethylene glycol) (PLL-g-PEG) stock solution (0.5 wt% in PBS) and 44 μL PBS in a 1.5 mL Eppendorf tube. Store the prepared solution at –20 °C until usage.
3. Avidin solution: Mix 5 μL of the avidin stock solution (0.5 wt% in PBS) and 45 μL PBS in a 1.5 mL Eppendorf tube. Store the prepared solution at –20 °C until usage.
4. Biotin-conjugated Protein G solution: Mix 5 μL of the biotin-conjugated Protein G stock solution (0.025 wt% in PBS) and 45 μL PBS in a 1.5 mL Eppendorf tube. Store the prepared solution at –20 °C until usage.
5. Anti-ß-galactosidase antibody solution: Mix 5 μL of the anti-ß-galactosidase antibody stock solution (ß-galactosidase mouse monoclonal IgG_1, Santa Cruz Biotechnology, USA, 50 μg/mL in PBS) and 45 μL PBS in a 1.5 mL Eppendorf tube. Store the prepared solution at –20 °C until usage.

2.4 Immunoassay

1. Lysis buffer: Buy or prepare the following stock solutions in deionized water: 10 mM Tris-HCl (pH 8) solution, 50 mM EDTA solution, and 1 mM NaCl solution.

   Mix 5 mL of the 1 M NaCl solution with 1 mL of the EDTA stock solution and 44 mL of the 10 mM Tris–HCl (pH 8) stock solution in a 100 mL media storage bottle and store at 4 °C until use for a maximum of 4 weeks. Directly before performing the immunoassay, weigh 5 mg lysozyme powder (lysozyme from chicken egg white) to a 2 mL Eppendorf tube and dissolve in 500 μL of the lysis buffer stock solution. It is important to prepare a fresh lysozyme lysis buffer solution for every experiment (*see* **Note 3**).
2. Substrate solution: Mix 1 μL of the fluorescein di-β-D-galactopyranoside (FDG) stock solution, with 399 μL PBS.

3 Methods

3.1 Bacteria Cultivation

All steps of the following protocol should be performed under sterile conditions.

1. For short-term maintenance and use, streak *E. coli* bacteria of the strain K-12 MG1655 with the lac operon on dry LB Agar plates. Incubate the plates, agar side up, at 37 °C for 24 h. Seal the plates with Parafilm to keep the moisture in the culture and store them at 4 °C for a maximum of 4 weeks (*see* **Note 4**).
2. Preparation of the preculture: Inoculate a 15 mL Falcon tube containing 3 mL of LB Broth (Lennox with additional 20 mM glucose) with a single *E. coli* colony of the previously prepared LB Agar plates. Carry out the incubation at 37 °C with constant shaking at 220 rpm for 8 h. Inoculate the preculture into a 15 mL Falcon tube containing 3 mL M9 minimal medium (with additional 20 mM lactose) at a dilution of 1–200 (*see* **Note 5**). Carry out the incubation at 37 °C with constant shaking at 220 rpm until the OD_{600} is approximately 1.

3.2 Fabrication of the Microfluidic Device

The microfluidic device is composed of two poly(dimethylsiloxane) (PDMS) layers which are bonded on a glass slide to close the fluid channel (*see* Fig. 1). The top layer (pressure layer) consists of a set

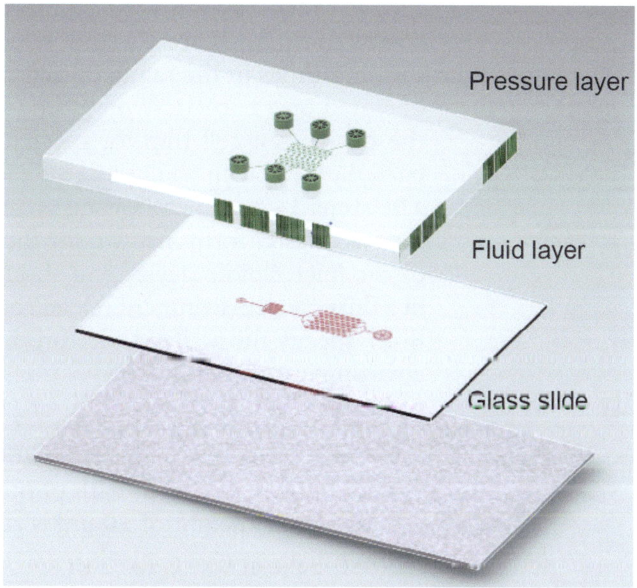

Fig. 1 Schematic drawing of the three different layers required to build the microfluidic device. Adapted from [22]

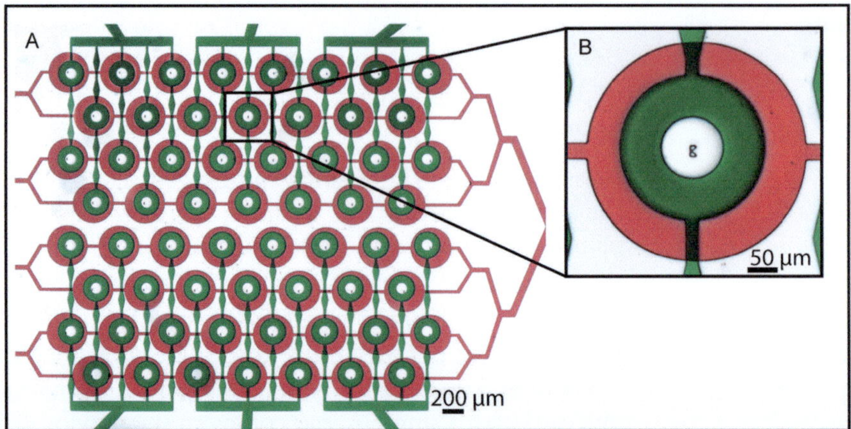

Fig. 2 (**a**) Bright-field image of the entire array. The fluid channels (height 10 μm) are filled with red food dye, the closed microchambers (volume 80 pL) are filled with colorless buffer solution and the pressure layer is filled with green food dye. (**b**) Magnified bright-field image of one microchamber. The centrally placed cell trap with a small gap size of 1 μm enables to trap individual *E. coli* bacteria

of 60 microchambers with integrated ring feature (*see* Fig. 2). The ring feature act as a valve, which when attenuated, isolates the content of the microchambers completely from the surrounding medium. The opening and closing of the valves is regulated by a homemade pressure control system (*see* **Note 6**) with nitrogen gas supply which is connected to the device by custom made metal connectors (*see* **Note** 7) and silicon tubing. To ensure precise opening times in the range of milliseconds, the pressure valves are operated by a LabVIEW program (*see* **Note 8**). The key feature of the device is a cell trap for mechanical trapping of single *E. coli* bacteria located in the center of each microchamber (*see* Fig. 2). The **steps 1**–7 of the following protocol describe the fabrication of the two master forms needed for the production of the PDMS part of the microfluidic device and should be performed preferentially in a dust-free environment (clean room) (except **step** 7). The **steps 8**–**17** specify the actual device production in a standard laboratory environment.

1. For preparation of the master forms, dehydrate two silicon wafers (100 mm diameter) at 200 °C for 10 min. Let the wafers cool down, spin-coat with photoresist at 2500 rpm for 30 s so that a final height of 10 μm is achieved.
2. Soft bake at 95 °C for 240 s.
3. Exposure with UV light (130 mJ/cm^2, measured at 365 nm) in a mask aligner and a transparency photomask.
4. Post exposure bake the photoresist at 95 °C for 240 s and develop for 3 min exposing the 10 μm high features.

5. Hard bake the wafers for 3 h at 200 °C.
6. Confirm the heights of the SU-8 features with a step profiler.
7. Silanize the master forms by storing the wafers overnight in a desiccator with 50 μL 1*H*,1*H*,2*H*,2*H*-perfluorodecyltrichlorosilane under reduced pressure of 100 mbar. After silanization, the production process of the master forms is finished.
8. For production of the poly(dimethylsiloxane) (PDMS) part of the device, mix carefully PDMS monomer and curing agent at a ratio of 10:1 (total of 60 g) in a plastic weighing dish by using a plastic spatula. Place the weighing dish in a vacuum degassing chamber and degas under reduced pressure of 50 mbar for 20 min.
9. Place the master form of the top part in the middle of the bottom part of a square petri dish and fix at the edges with tape.
10. Pour a 5 mm thick layer of PDMS mixture on the master form of the top part, place in a vacuum degassing chamber and degas under reduced pressure of 50 mbar for 20 min. Cure in the oven at 80 °C for 3 h.
11. For the bottom part containing the fluid channels, spin-coat the PDMS mixture on the master form at 2600 rpm (Spin Coater) for 60 s to build a 30 μm high PDMS coating. Afterwards cure in the oven at 80 °C for 1 h.
12. Cut the master form carefully out of the petri dish by using a scalpel. Remove the hardened PDMS from the master form (the PDMS should peel off automatically). Cut the top part to shape by a razor blade and punch holes for the pressure valves with a Biopsy puncher (1 mm diameter).
13. Place the tailored top part and the PDMS coated master form of the fluid layer in a plasma cleaner. Activate both chip parts by oxygen plasma (0.75 mbar, 18 W) for 45 s and align under a microscope. Pour PDMS around the composed device to facilitate later the peel off process.
14. Harden the composed device in the oven at 80 °C for 1 h.
15. Remove the composed device carefully from the master form, cut to shape by a scissor and punch two holes for the fluid inlet and outlet with a Biopsy puncher (1.5 mm diameter).
16. For assembling the PDMS part of the device and the glass slide (24 × 40 × 0.15 mm), place both in a plasma cleaner and activate by oxygen plasma (0.75 mbar, 18 W) for 45 s.
17. Place the PDMS part of the device on the glass slide and place the completed device on a hot plate at 50 °C for 5 min to allow bonding (*see* **Note 9**).

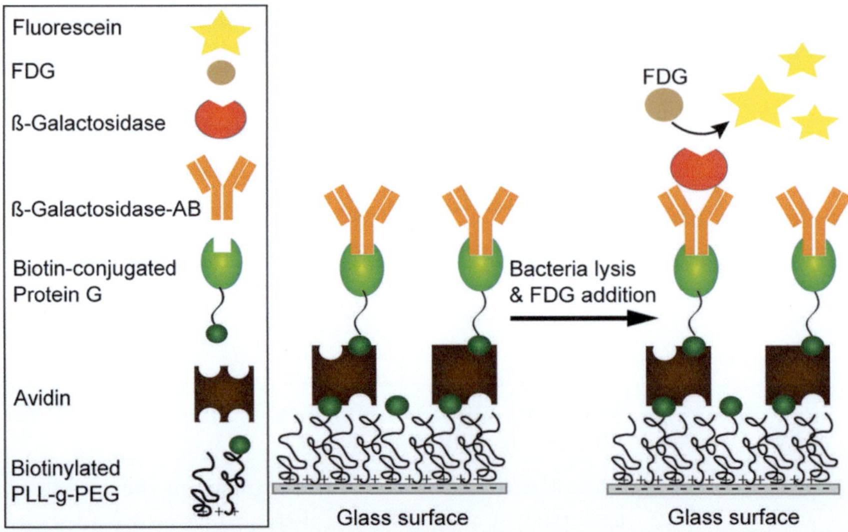

Fig. 3 Schematic drawing of the surface modification. ß-galactosidase binding antibodies are immobilized on the surface. After bacteria lysis, the released ß-Galactosidase is captured by the antibodies and converts the added substrate FDG into highly fluorescent fluorescein

3.3 Surface Modification of the Microfluidic Device (Fig. 3)

1. Centrifuge the prepared PLL-g-PEG-biotin solution into the device using a centrifuge (800 × *g*, 5 min) directly after finishing the bonding process (*see* **Note 10**). After 1 h of incubation time, flush the device with PBS using a syringe pump (*see* **Note 11**).
2. Flush the prepared avidin solution into device at a flow rate of 5 μL/min for 10 min. After a 5 min flow-free incubation time (*see* **Note 12**), continue with the washing-step and flush the device with PBS at a flow rate of 10 μL/min for 5 min.
3. Flush the prepared biotin conjugated Protein G solution into the device at a flow rate of 5 μL/min for 10 min. After a 5 min flow-free incubation time, continue with the washing step and flush the device with PBS at a flow rate of 10 μL/min for 5 min.
4. Flush the prepared anti-ß-galactosidase antibody solution into the device at a flow rate of 5 μL/min for 10 min. After a 5 min flow-free incubation time, continue with the washing step and flush the device with PBS at a flow rate of 10 μL/min for 5 min. Store the surface modified device at 4 °C until usage (*see* **Note 13**).

3.4 DiI-Staining and On-Chip Loading of E. coli Bacteria

1. Pipette 1 μL DiI stock solution into a 2 mL Eppendorf tube. Add 150 μL of the *E. coli* culture grown to an OD_{600} of 1 (*see* Subheading 3.1) and incubate at 37 °C with constant shaking at 220 rpm for 5 min in a thermomixer.
2. Add 1849 μL PBS and centrifuge for 3 min at 14.1 rcf. Discard the supernatant, resuspend the bacteria pellet with 2 mL of

fresh PBS, and repeat the centrifugation process (*see* **Note 14**). Discard again the supernatant and resuspend the bacteria cell pellet in 1.5 mL fresh PBS. Pipette 150 μL of the bacteria sample in a 2 mL Eppendorf tube, add 1350 μL PBS and mix by repeated aspirating and dispensing of the pipette tip.

3. For monitoring the trapping process of the *E. coli* bacteria and for collecting the data of the subsequently performed immunoassay, position the microfluidic device on a microscope equipped with a high-sensitivity camera (EMCCD camera).
4. Filter the *E. coli* bacteria, dilute the sample again 1 to 10 with PBS (Filter pore size of 10 μm) (*see* **Note 15**) and draw it into a 1 mm syringe. Load the *E. coli* bacteria into the device at a flow rate of 5 μL/min for 5 min (*see* **Note 16**). Afterwards, flush the device with PBS at a flow rate of 5 μL/min for 10 min (*see* **Note 17**).
5. Finally, control the number of single bacteria occupied microchambers by taking fluorescent images of each cell trap (60× water immersion objective with 1.5 zoom) and each microchamber (60× water immersion objective, exposure time 200 ms, excitation at 546 ± 12 nm, emission at 607 ± 80 nm). Close the microchambers to isolate the single trapped *E. coli* bacteria from the surrounding.

3.5 Performance of the Immunoassay

See Fig. 3 for schematic of assay. A detailed description of the preparation of the lysis buffer and the FDG solution can be found in the Subheading 2.4. Both solutions should be prepared and mixed together on the day the experiment is performed.

1. Flush the lysis buffer into the device at a flow rate of 5 μL/min for 10 min (*see* **Note 18**).
2. Open the microchambers for 0.7 s to introduce the lysis buffer (*see* **Note 19**). For precise opening of the microchambers use the previously described LabVIEW program.
3. Flush the device with PBS at a flow rate of 5 μL/min for 10 min (*see* **Note 20**).
4. Flush the FDG solution into the device at a flow rate of 5 μL/min for 5 min.
5. Open the microchambers for 0.7 s to introduce the substrate FDG (*see* **Note 21**). For precise opening of the microchambers use the LabVIEW program.

3.6 Data Acquisition and Evaluation

1. Start to monitor the reaction process directly after the substrate FDG was added. Therefore, take pictures every 30 min within a period of 3 h of all microchambers occupied by a single *E. coli* bacterium (*see* Fig. 4) (EMCCD camera, 4× objective, exposure time 300 ms, gain × 100 excitation: 470 ± 40 nm, emission: 525 ± 50 nm).

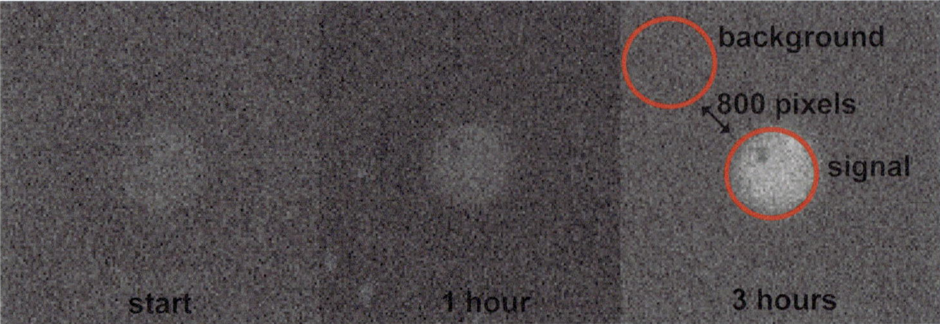

Fig. 4 Fluorescent images of the same microchamber enclosing the lysate of a single *E. coli*. The starting time is defined as the point in time the substrate FDG is added. The fluorescent signal inside the microchamber increases over time as the amount of fluorescent product of the enzymatic reaction increases. For background correction an image section directly beside the closed microchamber of identical pixel size is also analyzed and subtracted from the detected fluorescent data inside the microchamber

2. For the data analysis, analyze the micrographs with an imaging program like ImageJ (*see* **Note 22**). Measure the fluorescence signal in the microchamber by marking a circular area of around 800 pixels (*see* Fig. 4). For reliable evaluation of the fluorescence signal, it is essential to make a background correction for each data point. Therefore, measure the background signal separately for each analyzed microchamber. In this case, the background signal is the signal outside of the microchamber where no analyte is present. Generate a graphic (Excel) for each microchamber in which the background corrected fluorescence signal is plotted against time. The curve progression in case of an *E. coli* free chamber can be clearly distinguished from a chamber occupied by one or two *E. coli* (*see* Fig. 5).
3. It would be also possible to quantify the amount of enzyme present in a single *E. coli*. Therefore, perform the assay with different known enzyme concentrations (*see* as example Fig. 6) instead of with bacteria. Acquire and evaluate the data as described in **steps 1** and **2**. Perform a linear fitting (using Origin or Excel) of the plotted curve. The slope of the linear part at the beginning of data acquisition is a measure for the number of enzyme molecules present in a microchamber, as the enzymatic reaction is based on Michaelis–Menten kinetics. The Michaelis–Menten enzyme kinetics model states how the product formation speed correlates with the substrate starting concentration as well as the enzyme concentration [23]. As the substrate starting concentration is kept constant during the immunoassay, the product formation speed can be correlated directly with the ß-galactosidase concentration.

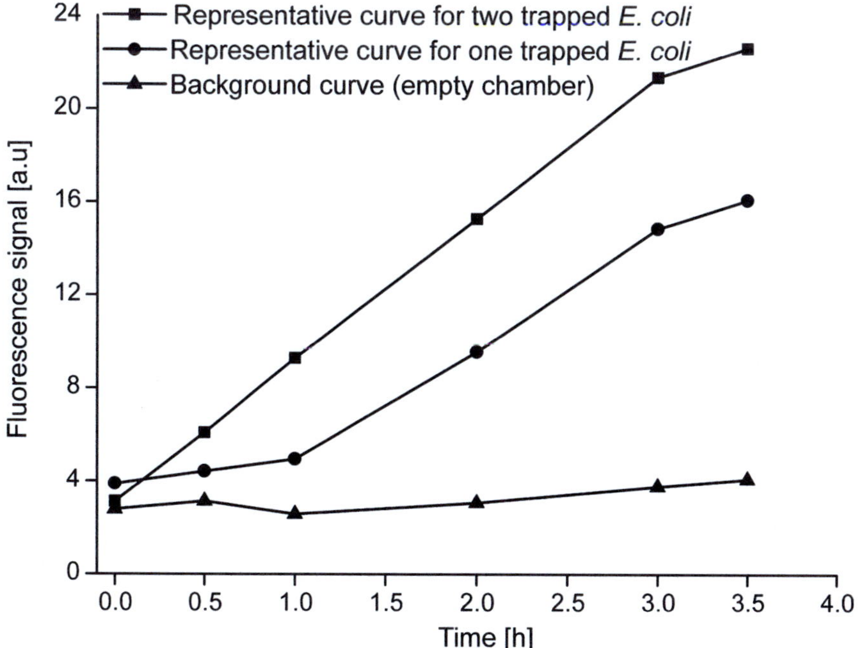

Fig. 5 Representative curves of the fluorescent signal over time progression in case of two, one, or zero lysed *E. coli* in the microchamber. The shown data originate from three different microchambers of the same microfluidic device

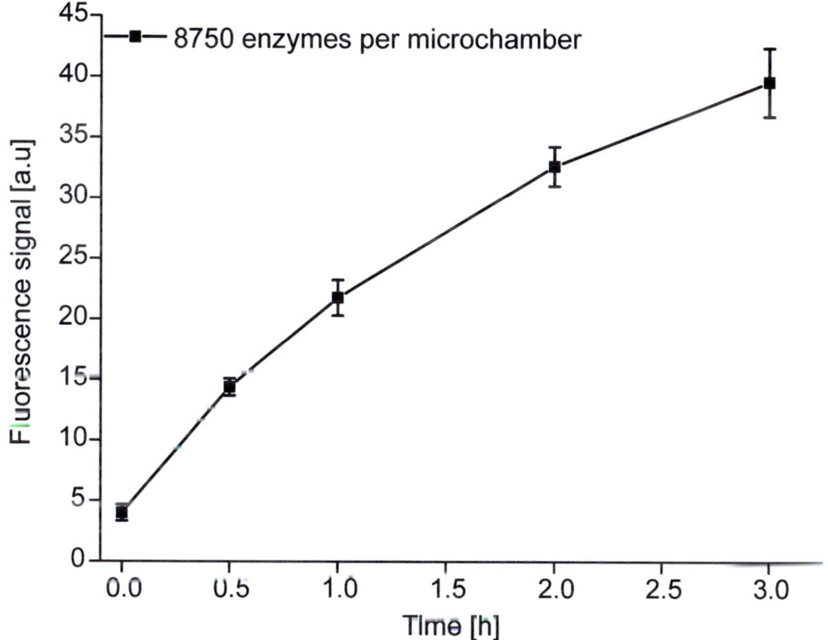

Fig. 6 Curve progression in the case where the immunoassay is performed with a known enzyme concentration of 8750 enzymes per microchamber. The slope of the curve is a measure for the number of enzymes present, and therefore, a set of curves of different known enzyme concentrations could be used to generate a calibration curve

4. To quantify the amount of enzymes present in a single *E. coli* bacterium, generate a calibration curve as described in **step 3** and insert the slope value of an actual *E. coli* experiment into the equation of the calibration curve.

4 Notes

1. Agar plates which do not contain antibiotics can generally be used as long as they are not contaminated or dried out. In order to prevent drying, wrap the lid of the petri dish with Parafilm.
2. Glucose at high concentrations can be solubilized more easily if the deionized water is gently heated.
3. Lysozymes are a type of glycoside hydrolases which disintegrate the cell wall of bacteria. They are an essential component of the *E. coli* lysis buffer. The ideal storage temperature for the enzyme is –20 °C, and fresh, defrosted enzyme powder should be added to the buffer directly before use.
4. To ensure that the *E. coli* colonies stored on LB agar plates receive an optimal supply of nutrients, streak them out on fresh LB agar plates every 4 weeks.
5. It is advisable to inoculate the *E. coli* bacteria first from LB agar plates to a complex medium and afterwards to a defined medium. By preculturing the *E. coli* bacteria in a nutrient rich complex medium, it is ensured that the bacteria are in an optimal fitness state before the change of nutrient medium and carbon source occurs.
6. Key feature of the custom-made pressure control system is a magnetic valve block (nine valves in total) which is connected to a programmable logic controller which enables to operate the valves via the LabVIEW program. For detailed information see the schematic (Fig. 7).
7. The custom-made metal connectors are made of capillary tubes (stainless steel, inner diameter of 1.35 mm) with a 90° bend.
8. The LabVIEW program controls the sequentially opening and closing of the ring-shaped valves. For opening, the pressure is released of the affected valve within a defined period of time. For closing, the valve is pressurized again. The LabVIEW program enables to operate precisely opening times in the range of milliseconds.
9. Bonding of the PDMS-part of the device to the glass slide is a highly time-critical step, as the surface activation only lasts for a few minutes.

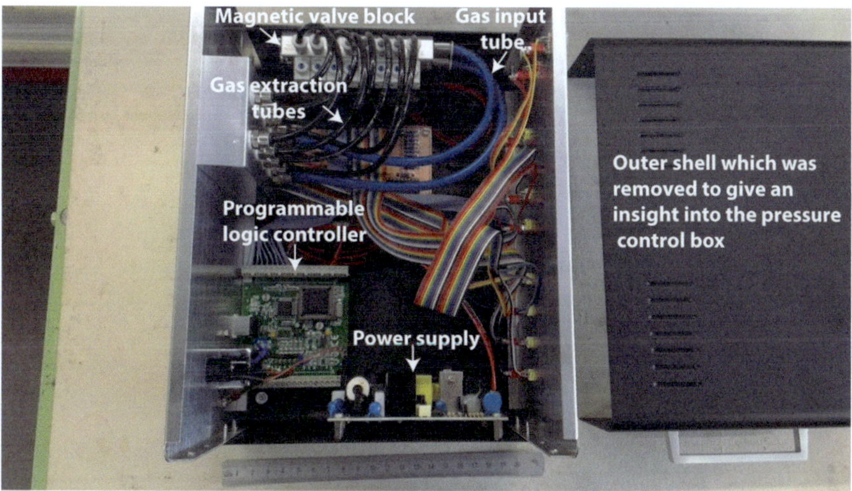

Fig. 7 Schematic of custom-made pressure control system

10. The first step of the surface modification is highly time-critical, as the glass slide of the device has to remain activated otherwise PLL-g-PEG-biotin will not attach to the glass surface. Only the first step of the surface modification protocol is time-critical, the **steps 2–4** can be performed without maintaining a strict time-frame, as long as drying-up of the surface is prevented.
11. All reagent solutions were pumped through the device by using a syringe pump guaranteeing a flow-rate accuracy of 0.5 %.
12. The flow-free incubation time is performed to ensure that the reagents flushed through the device have enough time to diffuse to the surface area of the device.
13. To prevent drying up of the modified surface, place PBS filled pipette tips in the inlet and outlet of the device. Store the device in a closed box containing a moisturized tissue at 4 °C in the fridge.
14. The DiI-labeled *E. coli* bacteria sample is washed twice to remove excess dye from the sample solution. Otherwise, the excess dye stains the PDMS when the bacteria sample is flushed into the device. As a result, the PDMS based cell trap is stained itself and it would be impossible to distinguish an empty cell trap from an occupied cell trap.
15. It is important to filter the *E. coli* bacteria sample before introducing it to the microfluidic device to prevent dust particles from the medium solution from blocking the cell traps. It is advisable to use filters with a pore size of 20 μm or 10 μm. Smaller pore sizes should be avoided as they also filter out the bacteria.

16. The bacteria trapping efficiency is higher at high flow rates. Therefore, do not reduce the flow-rate while flushing the bacteria sample into the device. Moreover, do not flush more than 25 μL of the bacteria sample into the device, as a high number of excess bacteria increase the risk of unintentional bacteria sticking to the PDMS outside the cell traps.
17. This long washing step is used to ensure that the bacteria sample is completely flushed out of the device. Thereby only bacteria physically trapped within the cell traps remain within the device.
18. As the lysis of *E. coli* bacteria can take up to 30 min, wait at least for 30 min before you insert the substrate FDG.
19. 0.7 s is the minimal opening time required to ensure a full fluid exchange within the microchamber at a flow rate of 5 μL/min. As complete lysis of *E. coli* bacteria takes up to 30 min, analyte loss by opening the chamber for 0.7 s is unlikely.
20. The lysis buffer could lead to a degradation of the substrate FDG. Therefore, it is important to ensure that the lysis buffer is flushed out completely before the substrate is introduced.
21. Opening the microchamber for introducing the substrate FDG also allows for cell lysate residues to be washed out. That means the substrate introducing step is also an integrated washing step.
22. http://imagej.nih.gov/ij/

References

1. Kaern M, Elston TC, Blake WJ et al (2005) Stochasticity in gene expression: from theories to phenotypes. Nat Rev Genet 6:451–464
2. Ito Y, Toyota H, Kaneko K et al (2009) How selection affects phenotypic fluctuation. Mol Syst Biol 5:1–7
3. Munsky B, Neuert G, Van Oudenaarden A (2012) Using gene expression noise to understand gene regulation. Science 336:183–187
4. Viney M, Reece SE (2013) Adaptive noise. Proc R Soc B 280:1–9
5. Hunt BG, Ometto L, Keller L et al (2013) Evolution at two levels in fire ants: the relationship between patterns of gene expression and protein sequence evolution. Mol Biol Evol 30:263–271
6. Taniguchi Y, Choi PJ, Li GW, Chen H et al (2010) Quantifying E. coli proteome and transcriptome with single-molecule sensitivity in single cells. Science 329:533–538
7. Mazumder A, Tummler K, Bathe M et al (2013) Single-cell analysis of ribonucleotide reductase transcriptional and translational response to DNA damage. Mol Cell Biol 33:635–642
8. Dittrich PS, Manz A (2006) Lab-on-a-chip: microfluidics in drug discovery. Nat Rev Drug Discov 5:210–218
9. Klepárník K, Foret F (2013) Recent advances in the development of single cell analysis—a review. Anal Chim Acta 800:12–21
10. Kovarik ML, Gach PC, Orno DM et al (2012) Micro total analysis systems for cell biology and biochemical assays. Anal Chem 84:516–540
11. Zhang Y, Ozdemir P (2009) Microfluidic DNA amplification—a review. Anal Chim Acta 638:115–125
12. Lounsbury JA, Karlsson A, Miranian DC et al (2013) From sample to PCR product in under 45 minutes: a polymeric integrated microdevice for clinical and forensic DNA analysis. Lab Chip 13:1384–1393
13. Chang CM, Chang WH, Wang CH et al (2013) Nucleic acid amplification using microfluidic systems. Lab Chip 13:1225–1242

14. Wu M, Singh AK (2012) Single-cell protein analysis. Curr Opin Biotechnol 23:83–88
15. Bendall SC, Simonds EF, Qiu P et al (2011) Single-cell mass cytometry of differential immune and drug responses across a human hematopoietic continuum. Science 332:687–696
16. Eyer K, Stratz S, Dittrich PS et al (2013) Implementing enzyme-linked immunosorbent assays on a microfluidic chip to quantify intracellular molecules in single cells. Anal Chem 85:3280–3287
17. Eyer K, Kuhn P, Dittrich PS et al (2012) A microchamber array for single cell isolation and analysis of intracellular biomolecules. Lab Chip 12:765–772
18. Chen Y, Zhang B, Feng H et al (2012) An automated microfluidic device for assessment of mammalian cell genetic stability. Lab Chip 12:3930–3935
19. Leung K, Zahn H, Leaver T et al (2012) A programmable droplet-based microfluidic device applied to multiparameter analysis of single microbes and microbial communities. Proc Natl Acad Sci U S A 109:7665–7670
20. Kim M, Isenberg BC, Sutin J et al (2011) Programmed trapping of individual bacteria using micrometer-size sieves. Lab Chip 11: 1089–1095
21. He M, Edgar JS, Jeffries GD et al (2005) Selective encapsulation of single cells and subcellular organelles into picoliter- and femtoliter-volume droplets. Anal Chem 77: 1539–1544
22. Stratz S, Eyer K, Kurth F, Dittrich PS (2014) On-chip enzyme quantification of single Escherichia coli bacteria by immunoassay-based analysis. Anal Chem 86: 12375–12381
23. Berg JM, Tymoczko JL, Stryer L (2002) The Michaelis-Menten model accounts for the kinetic properties of many enzymes. In: Freeman WH (ed) Biochemistry, 5th edn. Freeman W. H. and Company, New York

Chapter 3

Enzyme-Linked ImmunoSpot (ELISpot) for Single-Cell Analysis

Sylvia Janetzki and Rachel Rabin

Abstract

The ELISpot, a heterogeneous immunoassay, is widely used for detection of low abundant analytes. It is a reliable and robust assay to monitor responses of the immune system at the single-cell level by capturing secreted molecules of interest with specific, membrane-bound antibodies. Those molecules are then made visible by a cascade of ELISA-related development steps. The final results are distinct spots on the membrane as an imprint of the cell secreting the captured molecules, not only allowing their quantification but also providing insight on the kinetics and strength of secretion. This chapter describes the optimized protocol steps of the ELISpot technique, important improvements and tools available for the community, and the current expansion of the technique into polyfunctional cell analysis.

Key words ELISpot, FluoroSpot, Immune monitoring, T cell assays, B cell assays

1 Introduction

The ELISpot methodology was first described in 1983 [1]. The key advantages of ELISpot are as follows: (1) It has outstanding sensitivity, enabling the detection and analysis of cells present in even very low frequencies within cell populations, down to only a few cells including single cells [2]. (2) It is a straightforward, easy-to-adapt technology. (3) It is a functional assay. (4) It can be adapted to high throughput. (5) The assay can readily be qualified and validated. Importantly, the assay can be applied in monitoring of functions of T cells, B cells, and the cells involved in the innate immune system [3]. This easy transferability between applications is enabled by a generalizable format used as shown in the schematic setup of ELISpot (Fig. 1). Of note, the assay provides insight about the secretion of analytes on the single-cell level, not only by providing the number of analyte-secreting single cells within an investigated sample but also by giving evidence about the amount of analyte secreted and the speed at which the analyte was secreted. These parameters directly correlate with the size and staining

Anup K. Singh and Aarthi Chandrasekaran (eds.), *Single Cell Protein Analysis: Methods and Protocols*, Methods in Molecular Biology, vol. 1346, DOI 10.1007/978-1-4939-2987-0_3, © Springer Science+Business Media New York 2015

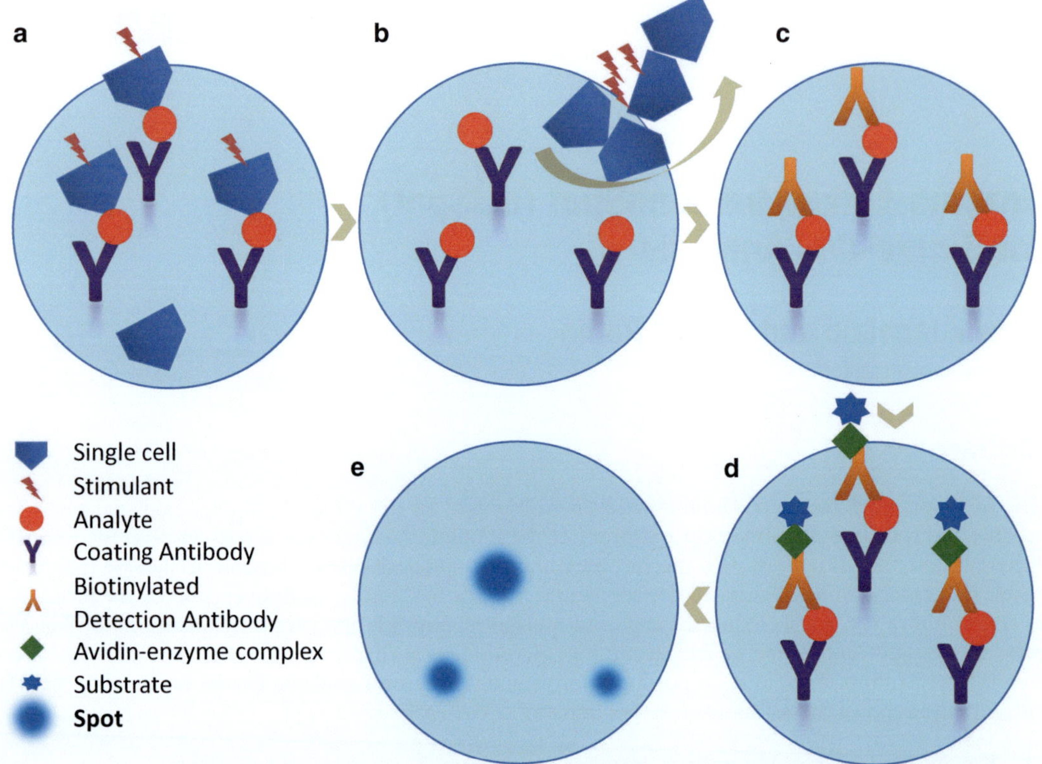

Fig. 1 Schematic setup of an ELISpot assay. (**a**) Cells and stimulants are added to a well coated with a capture antibody specific for the analyte. (**b**) The secreted analyte binds to the antibody and cells and stimulants are washed off. (**c**) An analyte-specific biotinylated detection antibody is added. (**d**) The signal is enhanced with an avidin–enzyme complex. (**e**) The added substrate precipitates at contact with the enzyme and forms spots

intensity of each spot, which is each single cell's imprint of analyte secreted over time.

ELISpot can be described briefly as follows. The bottom of a 96-well tissue culture plate (*see* **Note 1**), typically consisting of a membrane that allows high protein binding (*see* **Note 2**), is coated with an antibody directed against a cytokine, immunoglobulin or other protein of interest (from here on called "analyte" for short) that is secreted by cells under specific conditions, e.g., antigenic stimulation. Cells and stimulants such as peptides, proteins, cell lysates, tumor or infected cells, or professional antigen presenters, are added to the plate for short-term incubation, typically 16–48 h. The incubation is sought to induce secretion of the analyte of interest by the immune cells. The secreted molecule is bound by the capture antibody used for coating, cells and stimulants are washed off, and the captured protein is made visible by an ELISA-like colorimetric detection method. Detection differs from traditional microtiter plate-based ELISAs as it uses a substrate that forms an insoluble product that precipitates onto the membrane

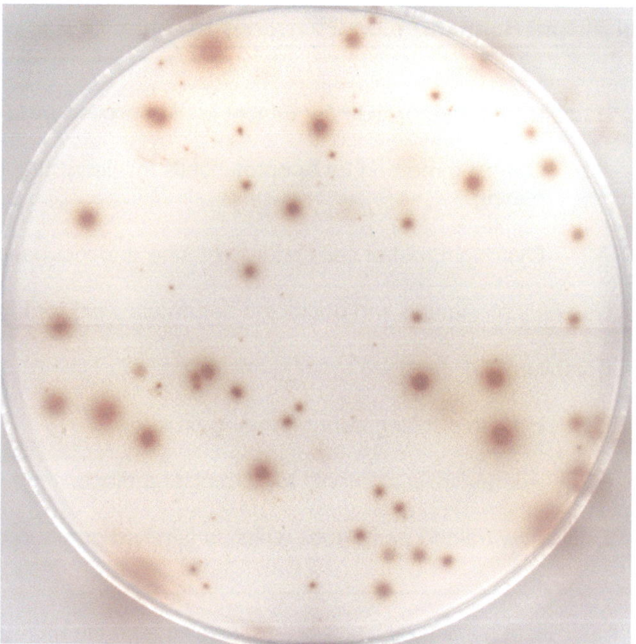

Fig. 2 Typical spot appearance with dark center (= high amount of cytokine bound in close vicinity of cells) and fading of color toward the spot periphery (= less cytokine captured further away from cell indicative of diffusion and capture kinetics of the cytokine). This spot feature helps distinguishing true spots from artifacts. The image was taken with a Zeiss ELISpot Reader (Zeiss, Thornwood, USA)

and creates a spot as an imprint of the protein-secreting cell in response to the stimulant added to the well. A typical spot is defined by a darker center with fading intensity of staining towards its periphery, caused by the immediate capture of secreted protein molecules close to the cell, and their diffusion kinetics away from the cell based on the amount and speed of secretion over time (Fig. 2). Under a given Standard Operating Procedure (SOP), the size distribution and staining intensity of spots in a sample are correlated to the activation of the secreting cells [4]. The larger the spot and the stronger its staining intensity (= more secreted molecules bound per area), the more protein has been secreted over time by the corresponding cell. Of note, spot appearance can also be strongly influenced by technical aspects of the assay, as eluded to in detail in Subheading 3, underlining the importance of standardization procedures once the assay has been optimized for its specific use [5, 6].

In addition to optimization and standardization procedures, the harmonization of ELISpot protocols has been shown to have major impact on the final ELISpot results and on reducing variability across laboratories [7]. Harmonization guidelines for ELISpot have been introduced by two nonprofit cancer

Initial Elispot Harmonization Guidelines to Optimize Assay Performance

A Establish laboratory SOP for Elispot testing procedures, including:

- **A1** Counting method for apoptotic cells for determining adequate cell dilution for plating
- **A2** Overnight rest of cells prior to plating and incubation

B Use only pre-tested and optimized serum allowing for low background : high signal ratio

C Establish SOP for plate reading, including:

- **C1** Human auditing during reading process
- **C2** Adequate adjustments for technical artifacts

D Only allow trained personnel, which is certified per laboratory SOP, to conduct assays

Fig. 3 Initial ELISpot harmonization guidelines, derived from the CIC/CRI ELISpot proficiency panel program. The figure is reproduced from Janetzki et al. [8]

immunotherapy organizations based on the results of large international ELISpot proficiency panels. The iterative testing of the same sample by many laboratories using their own SOP led to the identification of crucial protocol steps that significantly influence the assay outcome [8, 9]. The integration of such guidelines does not require the standardization toward one single SOP, an impossible undertaking, but rather focuses on general protocol steps, which will be addressed in Subheading 3 (Fig. 3).

The robustness and sensitivity of the ELISpot, together with many new developments for materials, reagents and protocol steps, have allowed the technique to also enter the polyfunctional analysis arena. The most successful approach uses fluorescent detection dyes for the simultaneous measurement of two analytes (*see* **Note 3**) within a well, allowing the identification of subgroups of cells secreting either one or both analytes (Fig. 4) [10, 11]. A specific advantage of this method is that in addition to the identification of specific subpopulations it uses only half of the cells compared to two separately run ELISpot assays analyzing one analyte each. Countless analyte combinations exist, allowing the evaluation of dual and even triple protein secretion patterns on the single-cell level with extraordinary sensitivity [11].

The protocol described in detail here is used for the analysis of cytokine-secreting immune cells. It can be easily adapted to other applications, like detection of immunoglobulin-secreting cells [12], chemokine secretors [13, 14], or cells secreting apolipoproteins [15], as well as others.

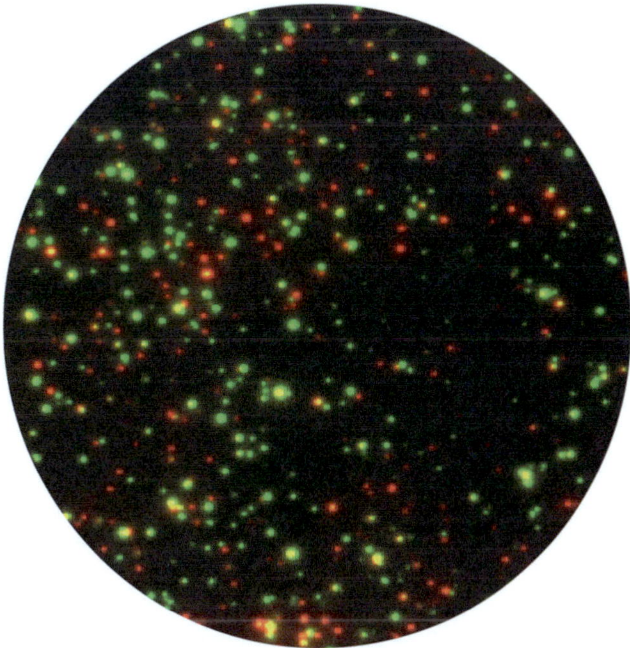

Fig. 4 Example of an IFNγ/IL-2 Fluorospot assays. Human PBMC were stimulated with PHA-L. The image was obtained with an AID iSpot ELISpot Reader (Autoimmun Diagnostika, Strassberg, Germany) using specific filter sets for FITC and Cy3 excitation and emission to acquire separate images of FITC and Cy3 spots. The two images were then overlaid by the software. *Green spots* indicate cells that secreted IFNγ only. *Red spots* indicate cells that secreted IL-2 only. *Yellow spots* indicate cells that secreted both cytokines simultaneously

2 Materials

2.1 Preparation of ELISpot Plates

1. HTS PVDF 96-well Filter plate (EMD Millipore) (*see* **Note 4**).
2. 1× sterile phosphate-buffered saline (PBS) without Calcium or Magnesium (PBS is always Ca^{2+} and Mg^{2+}-free throughout the protocol).
3. 70 % solution of pure anhydrous ethanol.
4. Coating antibody solution: Dilute capture antibody in PBS so to obtain a coating buffer solution with 10 μg/mL capture antibody (*see* **Note 5**).
5. Blocking buffer: Prepare a 1 % Bovine serum Albumin, Fraction V, in PBS.

2.2 Preparation of Cells and Stimulants

1. The source of cells to be tested in ELISpot is limitless. Cells could have been previously frozen or prepared fresh (*see* **Note 6**).
2. Similar to the multitude of cell sources, many different stimulants can be used to trigger the stimulation of the analyte of

interest, which are reviewed elsewhere [5]. One common and convenient choice is the use of overlapping peptide pools spanning an entire protein. Peptide pools consisting of 15mers overlapping by 11 amino acids have been shown to efficiently stimulate both CD4+ and CD8+ cells [16–18] (*see* **Note 7**).

3. Prepare a positive control reagent, like PHA-L (Leucoagglutinin-L), ConA (Concavalin A), PMA (phorbol 12-myristate 13-acetate) combined with ionomycin, or SEB (Staphylococcus enterotoxin) (*see* **Note 8**), by diluting the reagent stock solutions with the assay test medium to obtain the 2× working solution at the following concentration:
   (a) PHA-L at 20 μg/mL (to obtain a final assay concentration of 10 μg/mL after 1:1 dilution with cells when adding to the ELISpot plate).
   (b) ConA at 4 μg/mL (to obtain a final assay concentration of 2 μg/mL after 1:1 dilution with cells when adding to the ELISpot plate).
   (c) PMA/ionomycin at 0.4 ng/mL PMA and 0.4 μM ionomycin (to obtain a final assay concentration of 0.2 ng/mL PMA and 0.4 μM ionomycin after 1:1 dilution with cells when adding to the ELISpot plate).
   (d) SEB at 2 μg/mL (to obtain a final assay concentration of 1 μg/mL after 1:1 dilution with cells when adding to the ELISpot plate).
4. Pre-warmed appropriate assay test medium (*see* **Note 9**).
5. Benzonase, >99 % pure (EMD Millipore).

2.3 Spot Development

1. Washing buffer: non-sterile 0.05 % Tween 20 in PBS, and PBS alone.
2. Detection antibody solution: Dilute detection antibody in PBS containing 0.5 % BSA as to obtain a detection buffer solution with 1 μg/mL detection antibody.
3. Low-protein-binding syringe filter, pore size 0.2 μm.
4. Avidin–enzyme complex for use with biotinylated detection antibody (*see* **Note 10**).
5. Chromogenic substrate (*see* **Note 11**).

2.4 Equipment

1. Laminar flow hood.
2. Humidified incubator (37 °C, 5 % CO_2).
3. Centrifuge.
4. Microscope (bright-field) and hemacytometer with trypan blue (for cell counting), or

5. Automated cell counter, preferably flow-based that allows easy assessment of the degree of apoptosis in PBMC (different models available).
6. Automated ELISpot Reader (different models available).

2.5 Other Materials

1. Sterile and nonsterile reservoirs.
2. Multichannel pipettor (20 and 200 μL).
3. Single channel pipettor (20 and 200 μL).
4. Sterile and non-sterile tips for pipettor.
5. Sterile serological pipettes (1, 2, 5, 10, and 25 mL).
6. Sterile, conical polypropylene tubes (15 and 50 mL).
7. Squeeze bottle (for nonsterile washing steps) or 96-well plate washer.
8. Syringe (10 or 20 mL).
9. Aluminum foil.

3 Methods

The assay described here is a general outline of a cytokine ELISpot using frozen human PBMC. The assay can easily be adapted to other applications, for some of which further details are provided in the notes outlined in Subheading 4.

3.1 Day 1: Assay Preparation

The active protocol steps have to be carried out under sterile conditions.

1. Plan your experiment thoroughly. Attempt testing your sample in triplicates for each condition. If availability of cells allows, run your negative control (e.g., cells + medium only) in six replicates. Create a plate layout that is easy to follow when plating cells and stimulants, and that decreases the likelihood of mispipetting. Include a trending (= external) quality control (*see* **Note 12**). An optimal layout for testing three PBMC samples against three different peptide pool stimulants, a control peptide pool (CEF-E, Extended pool of 8–11mers from CMV, EBV, and influenza virus), and PHA as mitogenic control, is depicted in Fig. 5.
2. Coat your plate by adding 100 μL coating antibody (*see* **Notes 13** and **14**).
3. Wrap plate in aluminum foil and store at 4 °C overnight.
4. Transfer appropriate number of PBMC vials from liquid nitrogen to 37 °C water bath for thawing. Do not shake vials. With a little ice remaining, transfer vials into biosafety hood,

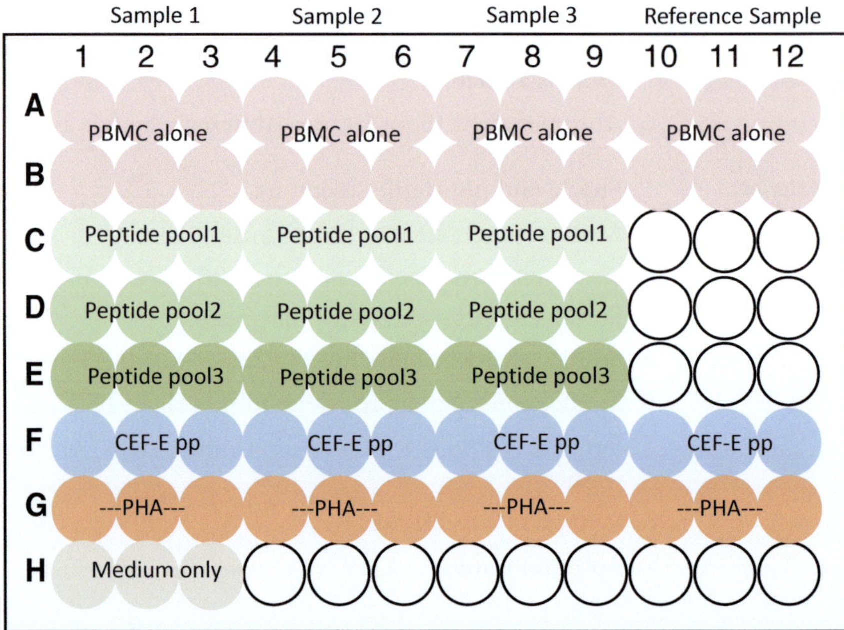

Fig. 5 Optimal plate setup for testing of three different PBMC samples against three different peptide pool stimulants, one control peptide pool, and a mitogenic control. A reference sample is also included, which only needs to be tested against the control peptide pool and mitogen for efficient trending information

clean outside of vial with ethanol wipe, and pipet the content into a 50 mL sterile conical tube.

5. Dropwise and slowly add warm assay test medium (*see* **Note 9**) supplemented with 50 U/mL benzonase to make the sample up to 10 mL (*see* **Note 15**). Shake the tube gently while adding the medium.
6. Spin down cells at 450 RCF for 10 min at room temperature.
7. Resuspend cells in test medium without benzonase. Take small sample for cell counting and repeat wash.
8. Resuspend cells at 2×10^6/mL in assay test medium without benzonase, and divide equally into 50 mL conical tubes such that no more than 10×10^6 cells (or 5 mL of cell suspension) are stored in each tube.
9. Rest cells at 37 °C in a 5 % CO_2 humidified incubator with cap slightly loosened for gas exchange for at least 18 h (*see* **Note 16**).

3.2 Day 2: Assay Setup

These protocol steps have to be carried out under sterile conditions.

1. Transfer plate into hood, dump coating buffer, and wash plate once with 200 μL PBS.
2. Add 200 μL blocking buffer. Wrap plate in aluminum foil and incubate for at least 1 h at 37 °C.

3. Take sample from the rested, previously frozen cells (Subheading 3.1) for counting (or prepare fresh cells now), wash cells once and resuspend in appropriate volume for plating into ELISpot plate (*see* **Note 17**):
   (a) Prepare 4–8 × 10^6 cells/mL test medium for stimulation with peptides, peptide pools or proteins (to plate 2–4 × 10^5 cells per well) (*see* **Note 18**).
   (b) Prepare a maximum of 3 × 10^6 cells/mL in test medium (for a maximum of 1.5 × 10^5 cells per well) for stimulation with other antigen presenting cells (e.g., pulsed dendritic cells, tumor cells, virally infected cells, others) (*see* **Note 19**).
   (c) Prepare 1 × 10^6 cells/mL test medium for stimulation with the positive control/mitogen such as PHA (for 5 × 10^4 cell per well) (*see* **Note 20**).
4. After plate blocking, dump blocking buffer, wash plate once with 200 μL test medium, and add 50 μL of cell suspension slowly to the appropriate wells. Place plate carefully back into the incubator to let cells settle down for at least 1 h.
5. Prepare stimulants:
   (a) Since stimulants will be diluted with cells at a 1:1 ratio, prepare peptides at double the stimulation concentration desired (most peptides work well at a final concentration of 1–10 μg/mL (*see* **Note 21**), but optimal working concentration should be established in optimization runs).
   (b) If stimulator cells are added, prepare cell solutions at double the desired final concentration. Number of cells to be added depends on the size of the stimulator. Enough cells need to be added so that every effector cell can potentially "see" the antigen(s) of interest. The final stimulator cell concentration ranges typically from 1 × 10^4 to 1 × 10^5 cells per well.
   (c) Prepare PHA-L solution at 20 μg/mL in test medium.
6. Carefully add 50 μL of stimulants to the appropriate wells. Be careful not to disturb cell layer. To assess the spontaneous secretion of cytokines only, add 50 μL of test medium (add DMSO to match DMSO concentration in wells with peptides, if used). If you add antigen presenting cells, you also need to test those cells alone for any cytokine release.
7. Run three wells per experiment with medium added only, without added cells or added stimulants, to assess the rate of false-positive spots in your assay (*see* Subheading 3.3, **step 2**).
8. Place plates carefully in an undisturbed incubator for 16–20 h (*see* **Note 22**).

3.3 Day 3: Spot Development

Spot development can be performed under nonsterile conditions.

1. Remove plate from incubator and rigorously wash plate 6 times with PBS/0.05 % Tween 20 buffer (*see* **Note 23**).

2. Prepare the detection antibody solution (*see* Subheading 2.3). Filter the diluted antibody with a low-protein-binding syringe filter, pore size 0.2 μm, to remove any aggregates (*see* **Note 24**).
3. Add 100 μL detection antibody solution per well. Incubate plate for at least 2 h at 37 °C.
4. During the last minutes of incubation, prepare the avidin–enzyme complex (follow the mixing instructions provided by the manufacturer).
5. Remove plate from incubator, wash three times with PBS, and add 100 μL of avidin–enzyme complex. Incubate plate for at least 1 h at room temperature.
6. During the last minutes of incubation, prepare the chromogenic substrate (following the manufacturer's instructions).
7. Wash plate three times with PBS and add 100 μL of the substrate to each well.
8. Observe spot development closely. Four to ten minutes incubation at room temperature is sufficient for most substrates. Do not overdevelop.
9. As soon as spots become visible, stop the reaction under running tap water. Remove the underdrain from the back of the ELISpot plate and rinse membranes from both sides.
10. Flick out residual water and blot back of membranes against a paper towel to absorb remaining liquid. Air-dry the plates at room temperature in the dark, preferentially overnight.

3.4 Data Acquisition (Plate Reading)

Spots are stable for a long time (months to years), especially if stored protected from light (*see* **Note 25**).

1. Use an automated ELISpot imaging system for spot counting (different automated reader systems are available with different capabilities and software features) (*see* **Note 26**).
2. Adjust reading parameters by comparing negative control wells and wells with spots. Spots from antigenic stimulation wells are preferred over positive control spots, which may have a different appearance due to different kinetics of cytokine induction and secretion. Depending on the reader and software used, settings can typically be adjusted to all or some of the parameters related to the following spot features: size, color, intensity of staining, shape, and fading of color from spot center to its periphery. Additional adjustments can be made to reading algorithms pertaining to membrane background staining, diffuseness of spots, and crowdedness of spots in a well. Artifacts of different caused and small spots, which are caused by nonspecific stimulation of cells and are clearly distinguished from spots caused by antigen-specific stimulation of cells, should be excluded from counting (*see* **Note 27**).

3. Audit plate and check results for plausibility.
4. Save all information available from the evaluation process including images and parameters applied.

3.5 Data Presentation and Analysis

1. Data sets and data presentation should always include the negative control counts (e.g., cells plus medium only). Include information about the variation (e.g., Standard Deviation).
2. It is recommended to present data as spot counts per number of cells plated. Depending on the assay setup, the assay may have only a limited linearity range, and extrapolated data may skew the overall picture and are only of limited value for comparison purposes. This specifically applies when testing cells for responses against proteins or peptides, without the addition of separate antigen presenting cells.
3. Check on the acceptability of the intra-replicate variability. The variation between three or more replicates can be expressed as the variance/(median + 1). The analysis of large data sets from multiple ELISpot proficiency panels, in which laboratories test the same samples with their own SOP, revealed that a variation of more than 2.47 is above the normal range, and a variation above 10 (using the equation given above) indicates an extremely high variation (Fig. 6) [19].

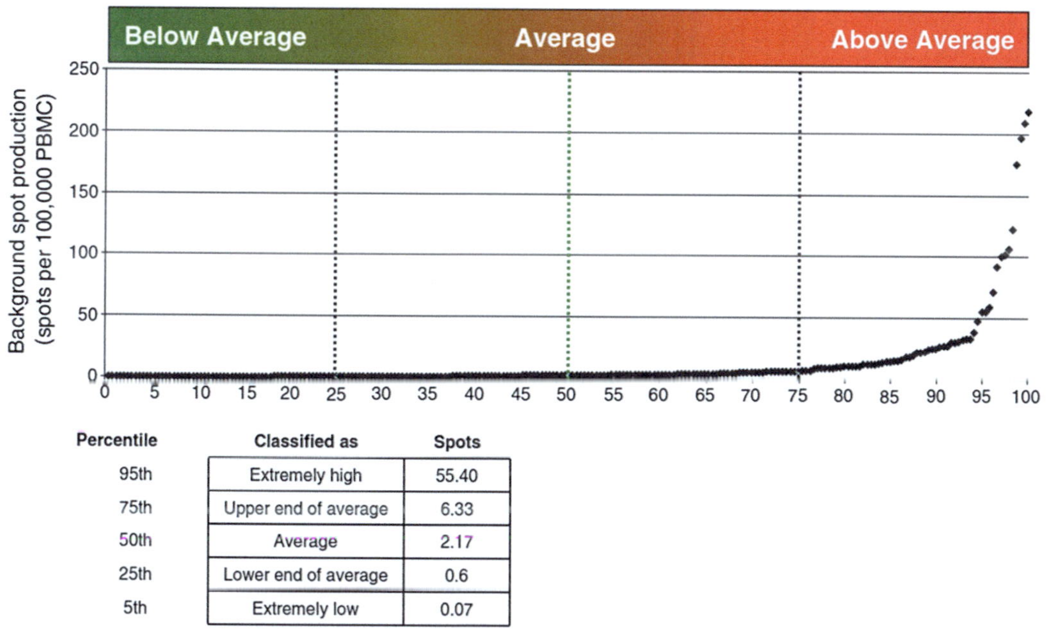

Percentile	Classified as	Spots
95th	Extremely high	55.40
75th	Upper end of average	6.33
50th	Average	2.17
25th	Lower end of average	0.6
5th	Extremely low	0.07

Fig. 6 Background spot production per 100,000 PBMC, based on 239 reported replicate sets from three phases of the CIP ELISpot proficiency panel program. Results are shown in ascending order. The *x*-axis shows the percentile rank and the *y*-axis indicates the reported mean spot number in the background control replicates (PBMC + medium alone). The figure is reproduced from Moodie et al. [19]

4. Determine the response status by either using an empiric or statistical method. An empiric response definition requires a series of experiments that demonstrates how the empiric rule keeps control of the false-positive rate. An excellent example on how to approach establishing an empiric response definition rule for ELISpot is given elsewhere [20]. For statistical testing, the Distribution-free Resampling Method (DFR) has been shown to be well suited [19]. DFR is a nonparametric test that avoids distributional assumptions. This is important since the p-value in statistical testing is typically calculated from the assumed distribution of the data. Since we deal with a low number of replicate measurements for the same condition (typically three, sometimes less, rarely more), it is impossible to predict that those few data points are normally distributed. They may, but most likely they are not, especially when keeping the inherent intra-replicate variability of ELISpot in mind (*see* **Note 28**). DFR testing offers a lenient test (DFR(eq): looking for any detectable difference) and a more stringent test (DFR(2×): the difference between negative control and stimulant conditions has to be at least twofold). DFR testing requires at least triple replicate measurements. It is recommended to run six negative control replicates. A free webtool is available: http://www.scharp.org/zoe/runDFR/.
5. Report on your assay conduct following MIATA (Minimal Information about T cell Assays) or similar reporting guidelines [21]. MIATA offers a framework for structured and transparent reporting of T cell assays in publications, to allow a good understanding and interpretation of your work. The MIATA guidelines have been established in an intense 3-year public consensus process. Multiple journals encourage MIATA-compliant reporting (*see* **Note 29**). Tools for easy implementation are available online: http://miataproject.org/.

4 Notes

1. There are now also 384-well plates available. They are of advantage if only a small number of cells is available for testing since they require less added cells per well for optimal stimulation conditions (typically around 50,000 cells per well).
2. Typical ELISpot plates contain membranes consisting of nitrocellulose or polyvinylidene fluoride (PVDF). For most ELISpot assays, PVDF plates are the preferred choice due to their optimal retention of capture antibody once plates are coated [22].
3. ELISpot kits for the simultaneous fluorescent detection of three analytes are currently in development and expected to enter the market soon. Such triple cytokine Fluorospot assays

allow the identification of seven subpopulations within one sample (three subpopulations secreting only one analyte, three subpopulations secreting a combination of two analytes, and one secreting all three analytes).

4. Filter plates come with clear or opaque frames. Certain automated ELISpot reader systems prefer a specific frame color. For Fluorospot assays use Fluorospot plates with clear frames due to the high reflection of light from opaque well walls, which leads to a wash-out effect of fluorescent signals. Clear HTS PVDF plates with a low auto-fluorescent membrane are available (EMD Millipore).

5. Manufacturers of ELISpot antibodies may give recommendation for optimal dilution of their coating and detection antibody. It is recommended to follow these recommendations closely to obtain optimally defined spots. As a rule of thumb, the total amount of 1 μg of coating antibody per well works well for most cytokine ELISpot assays. In case of B-cell assays, in which coating can be done with the antigen in order to capture antigen-specific immunoglobulin, higher concentrations might be necessary to obtain well defined spots. The required amount of detection antibody is typically lower and of lesser importance, about 0.1 μg per well.

6. In human studies, frozen cells are most commonly used, especially peripheral blood mononuclear cells (PBMC) obtained from whole blood, in order to allow for batch testing. Many critical parameters during the isolation and freezing of PBMC influence their functionality and secretion pattern of cytokines, including time frame between blood draw, PBMC isolation and freezing [23, 24], the freezing medium and procedure [25, 26] as well as the temperature at which cells are stored and the time frame of storage [27]. The detailed explanation of these steps is, however, beyond the scope of this chapter. Excellent information can be obtained from documents published by the Clinical and Laboratory Standard Institute (CLSI), and specifically from the recent guidelines on performance of single-cell immune response assays [6].

7. The size and purity of peptide pools are important considerations. Pools consisting of more than 100 peptides may give lower responses due to competition, compared to the sum of responses for each peptide contained in that pool when measured separately [18]. As DMSO is often used as solvent for peptides, and DMSO concentrations >1 % in the final assay are toxic to cells, large peptide pools may not be sufficiently dissolved. To that end it has been reported that poorly dissolvable peptides can cause false-positive spots [28]. Also, impurities can cause false-positive responses [29, 30]. For clinical monitoring applications, the purity grade of peptides should be 90 %, and for research applications at least 70 %.

8. SEB is a biological toxin and available only from specific sources. Specific requirements must be met to allow SEB use.
9. While historically cells are tested for functionality in assays using test media supplemented with L-glutamine, antibiotics and serum (most commonly used: RPMI 1640 medium supplemented with 2 mM L-glutamine, 1 % penicillin–streptomycin, and 10 % serum), it has been recently shown that serum is the leading cause for suboptimal assay performance and variability between laboratories [8]. Each serum performs uniquely, and it can potentially either suppress responses or nonspecifically stimulate cells. This applies for pooled human AB as well as for FCS/FBS sera alike. Hence, it is strongly recommended to pretest multiple serum lots before choosing one for the specific assay. An alternative and attractive choice is serum-free media, and their applicability for IFNγ ELISpot assays testing human PBMC has been demonstrated convincingly [31]. Not only do serum-free media lack the potential confounding serum-effect, but they also do not require repetitive pretesting beyond the initial test since their composition remains the same.
10. This signal amplification step is required for most ELISpot assays. (Only very few commercially available kits exist that reach a sufficient labeling of the detection antibody with enzyme.) Based on the enzyme, there are two choices for this development step: avidin–horseradish peroxidase (HRP) or avidin–alkaline phosphatase (AP) complex. The enzyme determines which substrate can be used. In case of Fluorospot, no enzyme is necessary.
11. For horseradish peroxidase, use either AEC (3-amino-9-ethylcarbazole) for red spots or TMB (tetramethylbenzidine) for bright blue spots. DAB (3,3′-diaminobenzidine) is not recommended due to its carcinogenic properties. With alkaline phosphatase use BCIP/NBT (5-bromo-4-chloro-3-indolyl phosphate/nitro blue tetrazolium) for blue-purple spots. Substrates are available as ready-to-use solutions. It is recommended to filter the solution before adding to the ELISpot plate with a low-protein-biding filter, pore size 0.45 μm, in order to remove aggregates and prevent artifacts.
12. An external control (= trending control) is a sample available in abundance that is being tested each time an assay is run under the same SOP. Ideally, such control is obtained from the same donor or animal, and frozen away in multiple aliquots. Each time an assay is performed one aliquot is tested against a standard antigen that elicits a response in this sample. A negative and nonspecific stimulation control should also be included. The assessment of such a "reference" sample allows following the precision between assays, and can prompt corrective steps if the assay performance is below expectance. An example of such trending control is given elsewhere [32].

13. It is recommended to pre-wet the PVDF membrane with 70 % ethanol to overcome its hydrophobicity. For that, add 15 μL of 70 % ethanol to each well. Tap plate to allow even wetting of the entire membrane in each well. Thorough wetting is achieved when the membrane turns from white to a blue-grayish color. Wash out the remaining ethanol by three consecutive washes with PBS. Leave the last PBS wash in the wells until the coating antibody solution is prepared. Do not use more than 15 μL 70 % ethanol since higher volumes will promote leakage. Prewetting of PVDF is not necessary for all coating antibodies. You will need to test if this step is necessary while optimizing your protocol. If spots appear fainter, more diffused, and are possibly also lower in number in none pre-treated plates compared to plates treated with ethanol, then prewetting is required for optimal binding of the coating antibody (or antigen, in case of B-cell ELISpot). *Do not* treat nitrocellulose plates with ethanol.
14. In the case of Fluorospot assays where the secretion of more than one analyte is tested, both coating antibodies are added to the plate. The membrane has a binding capacity for immunoglobulin of more than 100 μg per well. Adding two capture antibodies (=2 μg per well in total) only uses up a small percentage of that binding capacity. The exceptional binding capacity of membranes also supports the use of high amounts of antigen for coating in the case of B-cell ELISpot assays to obtain well defined spots.
15. Benzonase is a nuclease that degrades all forms of DNA and RNA that is released by dying cells, and thus prevents clumping [33].
16. Transport and storage of blood, PBMC isolation, freezing and thawing impact the functionality of PBMC and lead to programmed cell death (apoptosis). During the overnight resting period apoptotic cells eventually die and functionally impaired cells begin to recover their function [8, 34, 35]. Fresh cell preparations do not require an overnight resting step.
17. This cell count is of outmost importance. Most reliable cell counts are achieved by an automated method that also allows the assessment of the apoptotic cell fraction (e.g., Viability and Count Assay with use of Muse or Guava machines, EMD Millipore). Trypan blue-based assays generally overestimate the viability by including apoptotic cells in the living cell fraction [36, 8]. Apoptotic cells do not produce cytokines, and they may suppress responses [37].
18. A single layer of cells is typically achieved by adding $1–1.5 \times 10^5$ cells per well. Since peptides are presented by the same PBMC added to the well for testing, effective antigen presentation and co-stimulation is achieved by adding enough

cells for some pile-up. However, it is important to realize that cells in upper layers will not be detected in the assay and cause elevated background staining due to cytokine diffusion into the test medium. Hence, adding only $2–4 \times 10^5$ cells per well presents a good balance between achieving effective stimulation conditions and minimizing the loss of detection in upper PBMC layers.

19. Such single layer of PBMC can be effectively stimulated by overlaying it with a sufficient number of stimulator cells to achieve a condition in which each effector cell has contact with the antigen.
20. Adding more cells per well will lead to spot crowdedness, in which spots cannot be efficiently enumerated.
21. Peptides are typically obtained freeze-dried. DMSO can be used to efficiently dissolve most peptides. A minimum amount of DMSO should be used as to allow further dilution with PBS. Dissolved peptides need to be stored at −20 °C, or optimally at −80 °C, to avoid degradation. Repeated thaw/freeze cycles should be avoided. For optimal stability, obtain small aliquots of freeze-dried peptide, and dissolve on a need-to-use basis.
22. The incubation time is determined by the secretion kinetics of the analyte. For example, certain cytokines require a longer incubation, up to 48 h, like TNF or IL-4. Also, if whole protein is used as a stimulant, the protein requires processing by professional antigen-presenting cells, and hence, additional incubation time may be necessary.
23. This is the most important washing step. All cells need to be removed effectively, which is best achieved by using a squirt bottle with a large tip opening ensuring enough buffer flow and pressure to remove cells sticking to the membrane. An automated plate washer may be used instead. Pipetting washing buffer with a multichannel pipettor is not suitable for this washing step, and will leave cells sticking to the membrane, causing spots with white centers due to the blocking of binding sites for the secondary antibody (Fig. 7).
24. Aggregates of secondary biotinylated antibody cause false-positive spots. Removal of these aggregates before plating the detection antibody completely abolishes false spot development.
25. Even fluorescent signals from commercially available kits (e.g., Mabtech, Sweden) are stable for many months, especially Cy3 signals. FITC signals start to fade after repeated or prolonged exposure to light.
26. Obtain sufficient training for efficient operation of the automated imager.

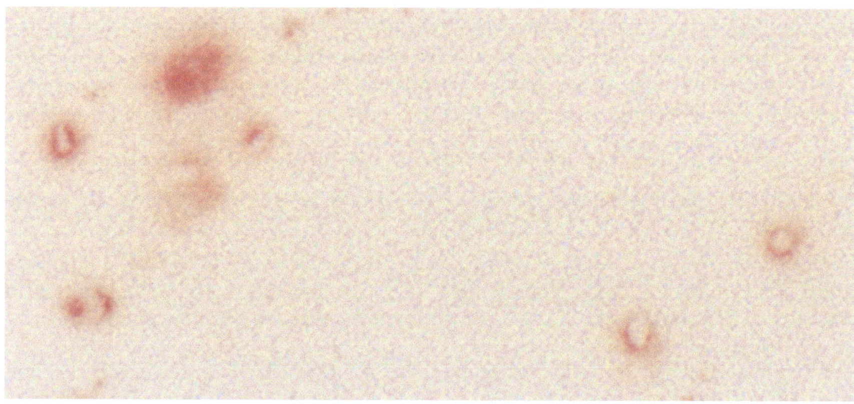

Fig. 7 White centered spots due to insufficient washing after cell incubation. Cells still sticking to the membrane prevent secondary antibody to efficiently bind to captured cytokine, leading to areas of spared spot development

27. A large international panel study has recently been conducted by the Cancer Immunotherapy Consortium (CIC) among 75 laboratories in order to obtain an overview of Elispot plate reading approaches and the degree of variability in results caused by them. Based on the results of this project and after an extensive consensus finding process, general plate evaluation guidelines were established and published [38].

28. An example is given here: For background reactivity (cells + medium only) replicates are 0, 4, and 5. The mean is 3, and the median is 4. The replicate measurements for cells plus stimulant are 4, 56, and 6. The mean is 22, while the median is 6 (Note the difference between the mean and the median!). If statistical testing is performed using a test that is based on the assumption of normal data distribution (e.g., the *t*-test), then the mean of 3 and 22 are compared, and the test will be likely to report a positive response. On the other hand, nonparametric testing, as the DFR testing, does not assume that normal distribution. In case of the DFR, it shuffles all replicate measurements (of negative control and stimulation wells), and then checks on how all different permutations change the final test statistics. Importantly, this approach takes the inherent data variability into account and does not carry the danger of invalid test results due to an improper assumption of data distribution.

29. Guidelines and various tools for easy implementation including Easy Reporting tools for ELISpot and a checklist for MIATA compliance can be found on the project's website: www.miata-project.org. With the Easy Reporting tool, specific Materials and Methods sections can be created in time-conserving fashion.

The tool contains preformatted Materials and Methods section documents which can be filled with the author's specific assay information. MIATA-compliant papers are listed in a Hall of Fame with direct link back to the published article, offering increased visibility of the published work.

References

1. Czerkinsky CC, Nilsson LA, Nygren H, Ouchterlony O, Tarkowski A (1983) A solid-phase enzyme-linked immunospot (ELISPOT) assay for enumeration of specific antibody-secreting cells. J Immunol Methods 65(1-2): 109–121. doi:10.1016/0022-1759(83)90308-3
2. Helms T, Boehm BO, Asaad RJ, Trezza RP, Lehmann PV, Tary-Lehmann M (2000) Direct visualization of cytokine-producing recall antigen-specific CD4 memory T cells in healthy individuals and HIV patients. J Immunol 164(7):3723–3732. doi:10.4049/jimmunol.164.7.3723
3. Janetzki S, Romero P, Roederer M, Bolton D, Jandus C (2012) Immune monitoring design within the developmental pipeline for an immunotherapeutic or preventive vaccine. Morrow WJW, Sheikh NA, Schmidt CS and Davies DH (ed) Vaccinology: Principles and Practices. Blackwell Publishing Ltd, Chichester, UK, pp 417–440
4. Hesse MD, Karulin AY, Boehm BO, Lehmann PV, Tary-Lehmann M (2001) A T cell clone's avidity is a function of its activation state. J Immunol 167(3):1353–1361. doi:10.4049/jimmunol.167.3.1353
5. Janetzki S, Cox JH, Oden N, Ferrari G (2005) Standardization and validation issues of the ELISPOT assay. Methods Mol Biol 302:51–86. doi:10.1385/1-59259-903-6:051
6. CLSI. Performance of Single Cell Immune Response Assays: Approved Guideline - Second Edition. CLSI document I/LA26-A2. Wayne, PA: Clinical and Laboratory Standards Institute; 2013
7. Janetzki S, Britten CM (2012) The impact of harmonization on ELISPOT assay performance. Methods Mol Biol 792:25–36. doi:10.1007/978-1-61779-325-7_2
8. Janetzki S, Panageas KS, Ben-Porat L, Boyer J, Britten CM, Clay TM, Kalos M, Maecker HT, Romero P, Yuan J, Kast WM, Hoos A (2008) Results and harmonization guidelines from two large-scale international Elispot proficiency panels conducted by the Cancer Vaccine Consortium (CVC/SVI). Cancer Immunol Immunother 57(3):303–315. doi:10.1007/s00262-007-0380-6
9. Britten CM, Gouttefangeas C, Welters MJ, Pawelec G, Koch S, Ottensmeier C, Mander A, Walter S, Paschen A, Muller-Berghaus J, Haas I, Mackensen A, Kollgaard T, thor Straten P, Schmitt M, Giannopoulos K, Maier R, Veelken H, Bertinetti C, Konur A, Huber C, Stevanovic S, Wolfel T, van der Burg SH (2008) The CIMT-monitoring panel: a two-step approach to harmonize the enumeration of antigen-specific CD8+ T lymphocytes by structural and functional assays. Cancer Immunol Immunother 57(3):289–302. doi:10.1007/s00262-007-0378-0
10. Janetzki S, Rueger M, Dillenbeck T (2014). Stepping up ELISpot: Multi-Level Analysis in FluoroSpot Assays. Cells 27 (3):1102–15. doi:10.3390
11. Ahlborg N, Axelsson B (2012) Dual- and triple-color fluorospot. Methods Mol Biol 792:77–85. doi:10.1007/978-1-61779-325-7_6
12. Walsh PN, Friedrich DP, Williams JA, Smith RJ, Stewart TL, Carter DK, Liao HX, McElrath MJ, Frahm N (2013) Optimization and qualification of a memory B-cell ELISpot for the detection of vaccine-induced memory responses in HIV vaccine trials. J Immunol Methods 394(1-2):84–93. doi:10.1016/j.jim.2013.05.007
13. Hagen J, Houchins JP, Kalyuzhny AE (2012) ELISPOT assay as a tool to study oxidative stress in lymphocytes. Methods Mol Biol 792:87–96. doi:10.1007/978-1-61779-325-7_7
14. Smedman C, Gardlund B, Nihlmark K, Gille-Johnson P, Andersson J, Paulie S (2009) ELISPOT analysis of LPS-stimulated leukocytes: human granulocytes selectively secrete IL-8, MIP-1beta and TNF-alpha. J Immunol Methods 346(1-2):1–8. doi:10.1016/j.jim.2009.04.001
15. Braesch-Andersen S, Paulie S, Smedman C, Mia S, Kumagai-Braesch M (2013) ApoE production in human monocytes and its regulation by inflammatory cytokines. PLoS One 8(11), e79908. doi:10.1371/journal.pone.0079908
16. Kern F, Faulhaber N, Frommel C, Khatamzas E, Prosch S, Schonemann C, Kretzschmar I, Volkmer-Engert R, Volk HD, Reinke P (2000) Analysis of CD8 T cell reactivity to cytomega-

lovirus using protein-spanning pools of overlapping pentadecapeptides. Eur J Immunol 30(6):1676–1682. doi:10.1002/1521-4141(200006)30:6<1676::AID-IMMU1676>3.0.CO;2-V

17. Kiecker F, Streitz M, Ay B, Cherepnev G, Volk HD, Volkmer-Engert R, Kern F (2004) Analysis of antigen-specific T-cell responses with synthetic peptides—what kind of peptide for which purpose? Hum Immunol 65(5):523–536. doi:10.1016/j.humimm.2004.02.017

18. Russell ND, Hudgens MG, Ha R, Havenar-Daughton C, McElrath MJ (2003) Moving to human immunodeficiency virus type 1 vaccine efficacy trials: defining T cell responses as potential correlates of immunity. J Infect Dis 187(2):226–242. doi:10.1086/367702

19. Moodie Z, Price L, Gouttefangeas C, Mander A, Janetzki S, Lower M, Welters MJ, Ottensmeier C, van der Burg SH, Britten CM (2010) Response definition criteria for ELISPOT assays revisited. Cancer Immunol Immunother 59(10):1489–1501. doi:10.1007/s00262-010-0875-4

20. Dubey S, Clair J, Fu TM, Guan L, Long R, Mogg R, Anderson K, Collins KB, Gaunt C, Fernandez VR, Zhu L, Kierstead L, Thaler S, Gupta SB, Straus W, Mehrotra D, Tobery TW, Casimiro DR, Shiver JW (2007) Detection of HIV vaccine-induced cell-mediated immunity in HIV-seronegative clinical trial participants using an optimized and validated enzyme-linked immunospot assay. J Acquir Immune Defic Syndr 45(1):20–27. doi:10.1097/QAI.0b013e3180377b5b

21. Britten CM, Janetzki S, Butterfield LH, Ferrari G, Gouttefangeas C, Huber C, Kalos M, Levitsky HI, Maecker HT, Melief CJ, O'Donnell-Tormey J, Odunsi K, Old LJ, Ottenhoff TH, Ottensmeier C, Pawelec G, Roederer M, Roep BO, Romero P, van der Burg SH, Walter S, Hoos A, Davis MM (2012) T cell assays and MIATA: The essential minimum for maximum impact. Immunity 37(1):1–2. doi:10.1016/j.immuni.2012.07.010

22. Weiss AJ (2012) Overview of membranes and membrane plates used in research and diagnostic ELISPOT assays. Methods Mol Biol 792:243–256. doi:10.1007/978-1-61779-325-7_19

23. Bull M, Lee D, Stucky J, Chiu YL, Rubin A, Horton H, McElrath MJ (2007) Defining blood processing parameters for optimal detection of cryopreserved antigen-specific responses for HIV vaccine trials. J Immunol Methods 322(1-2):57–69, doi:10.1016/j.jim.2007.02.003

24. Kierstead LS, Dubey S, Meyer B, Tobery TW, Mogg R, Fernandez VR, Long R, Guan L, Gaunt C, Collins K, Sykes KJ, Mehrotra DV, Chirmule N, Shiver JW, Casimiro DR (2007) Enhanced rates and magnitude of immune responses detected against an HIV vaccine: effect of using an optimized process for isolating PBMC. AIDS Res Hum Retroviruses 23(1):86–92. doi:10.1089/aid.2006.0129

25. Filbert H, Attig S, Bidmon N, Renard BY, Janetzki S, Sahin U, Welters MJ, Ottensmeier C, van der Burg SH, Gouttefangeas C, Britten CM (2013) Serum-free freezing media support high cell quality and excellent ELISPOT assay performance across a wide variety of different assay protocols. Cancer Immunol Immunother 62:615–627. doi:10.1007/s00262-012-1359-5

26. Yokoyama WM, Thompson ML, Ehrhardt RO (2012) Cryopreservation and thawing of cells. Curr Protoc Immunol Appendix 3:3G. doi:10.1002/0471142735.ima03gs99

27. Weinberg A, Song LY, Wilkening CL, Fenton T, Hural J, Louzao R, Ferrari G, Etter PE, Berrong M, Canniff JD, Carter D, Defawe OD, Garcia A, Garrelts TL, Gelman R, Lambrecht LK, Pahwa S, Pilakka-Kanthikeel S, Shugarts DL, Tustin NB (2010) Optimization of storage and shipment of cryopreserved peripheral blood mononuclear cells from HIV-infected and uninfected individuals for ELISPOT assays. J Immunol Methods 363(1):42–50. doi:10.1016/j.jim.2010.09.032

28. Karlsson RK, Jennes W, Page-Shafer K, Nixon DF, Shacklett BL (2004) Poorly soluble peptides can mimic authentic ELISPOT responses. J Immunol Methods 285(1):89–92. doi:10.1016/j.jim.2003.11.013

29. Currier JR, Galley LM, Wenschuh H, Morafo V, Ratto-Kim S, Gray CM, Maboko L, Hoelscher M, Marovich MA, Cox JH (2008) Peptide impurities in commercial synthetic peptides and their implications for vaccine trial assessment. Clin Vaccine Immunol 15(2):267–276. doi:10.1128/CVI.00284-07

30. de Beukelaar JW, Gratama JW, Smitt PA, Verjans GM, Kraan J, Luider TM, Burgers PC (2007) The impact of impurities in synthetic peptides on the outcome of T-cell stimulation assays. Rapid Commun Mass Spectrom 21(7):1282–1288. doi:10.1002/rcm.2958

31. Janetzki S, Price L, Britten CM, van der Burg SH, Caterini J, Currier JR, Ferrari G, Gouttefangeas C, Hayes P, Kaempgen E, Lennerz V, Nihlmark K, Souza V, Hoos A (2010) Performance of serum-supplemented and serum-free media in IFNgamma ELISPOT assays for human T cells. Cancer Immunol Immunother 59(4):609–618. doi:10.1007/s00262-009-0788-2

32. Cox JH, Ferrari G, Janetzki S (2006) Measurement of cytokine release at the single cell level using the ELISPOT assay. Methods 38(4):274–282, doi:10.1016/j.ymeth.2005.11.006
33. Smith JG, Liu X, Kaufhold RM, Clair J, Caulfield MJ (2001) Development and validation of a gamma interferon ELISPOT assay for quantitation of cellular immune responses to varicella-zoster virus. Clin Diagn Lab Immunol 8(5):871–879. doi:10.1128/CDLI.8.5.871-879.2001
34. Romer PS, Berr S, Avota E, Na SY, Battaglia M, ten Berge I, Einsele H, Hunig T (2011) Preculture of PBMCs at high cell density increases sensitivity of T-cell responses, revealing cytokine release by CD28 superagonist TGN1412. Blood 118(26):6772–6782. doi:10.1182/blood-2010-12-319780
35. Kutscher S, Dembek CJ, Deckert S, Russo C, Korber N, Bogner JR, Geisler F, Umgelter A, Neuenhahn M, Albrecht J, Cosma A, Protzer U, Bauer T (2013) Overnight resting of PBMC changes functional signatures of antigen specific T-cell responses: impact for immune monitoring within clinical trials. PLoS One 8(10), e76215. doi:10.1371/journal.pone.0076215
36. Mascotti K, McCullough J, Burger SR (2000) HPC viability measurement: trypan blue versus acridine orange and propidium iodide. Transfusion 40(6):693–696. doi: 10.1046/j.1537-2995.2000.40060693.x
37. Lenders K, Ogunjimi B, Beutels P, Hens N, Van Damme P, Berneman ZN, Van Tendeloo VF, Smits EL (2010) The effect of apoptotic cells on virus-specific immune responses detected using IFN-gamma ELISPOT. J Immunol Methods 357(1-2):51–54. doi:10.1016/j.jim.2010.03.001
38. Janetzki S, Price L, Schroeder H, Britten CM, Welters MJ, Hoos A (2015) Guidelines for the automated evaluation of Elispot assays. Nat Protoc 10(7):1098–115. doi:10.1016/j.jim.2010.03.001

Chapter 4

Photocleavable DNA Barcoding Antibodies for Multiplexed Protein Analysis in Single Cells

Adeeti V. Ullal and Ralph Weissleder

Abstract

We describe a DNA-barcoded antibody sensing technique for single cell protein analysis in which the barcodes are photocleaved and digitally detected without amplification steps (Ullal et al., Sci Transl Med 6:219, 2014). After photocleaving the unique ~70mer DNA barcodes we use a fluorescent hybridization technology for detection, similar to what is commonly done for nucleic acid readouts. This protocol offers a simple method for multiplexed protein detection using 100+ antibodies and can be performed on clinical samples as well as single cells.

Key words Multiplexing, Proteomics, Single cell analysis, Pathway analysis, Fine needle aspirates, DNA barcoding, Antibody conjugation

1 Introduction

A number of experimental proteomic and protein profiling techniques are being developed for single cell analyses and scant clinical materials. However, current tools for profiling key proteins in clinical samples remain limited. Standard practice employs immunocytology, which precludes broad protein analysis on scant samples and lacks automation, relying on trained pathologists [1]. Proteomic analyses by mass spectrometry is promising, but costly and technically challenging, particularly in the clinical setting [2].

DNA-barcoding techniques have previously been used to increase the throughput of protein analysis [3, 4]. Short DNA fragments are attached to an antibody of interest and subsequently measured by a variety of DNA quantification techniques, including qPCR and DNA sequencing. Unfortunately, these readouts introduce bias during amplification steps, require long processing times, or are not cost-effective. Additionally, multiplexed qPCR currently measures a maximum of 5 markers at a time. Other methods have sought to increase multiplexing using spatially encoded arrays on

Anup K. Singh and Aarthi Chandrasekaran (eds.), *Single Cell Protein Analysis: Methods and Protocols*, Methods in Molecular Biology, vol. 1346, DOI 10.1007/978-1-4939-2987-0_4,

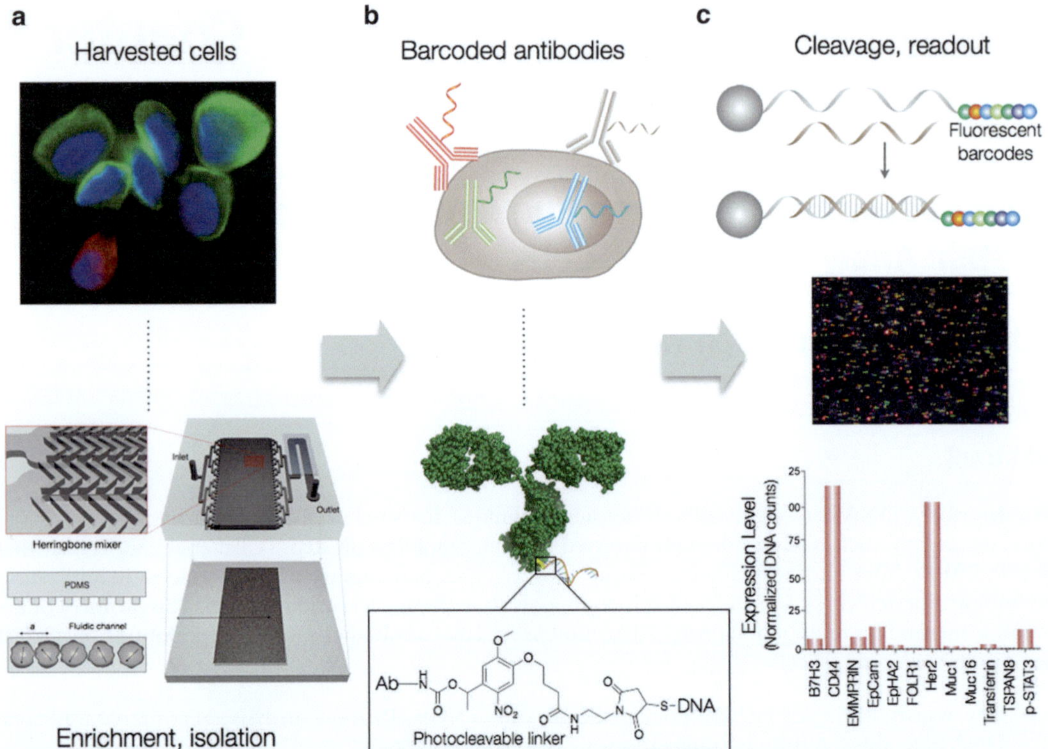

Fig. 1 Overview of ABCD method. (**a**) Patient samples are enriched and harvested for cells of interest. For fine needle aspirates or other low-cellularity samples, use of microfluidic devices will help to increase sample yield and recovery. For details on the optimized microfluidic device built for these studies please *see* [10]. (**b**) Antibodies are barcoded with specific DNA tags and are attached via a photocleavable linker. The cocktail of antibodies can be applied to the patient sample to target cell surface, intracellular, and even nuclear markers with light, reversible membrane permeabilization. (**c**) To measure marker expression, the DNA tags are cleaved with light, bound to complementary fluorescent tags, imaged and quantified

microfluidics chips [5], but this requires using and validating an additional capture antibody on the microfluidic platform (Fig. 1).

Here we describe a simple DNA-antibody barcoding strategy to perform multiplexed protein measurements in cells with an antibody panel, thereby enabling system-wide profiling of single cells or limited amounts of clinical sample material [6]. The approach interrogates cells by tagging each antibody in the cocktail with a small (~70mer) unique DNA barcode using a stable photocleavable linker [7]. The photocleavable linker efficiently releases the unique DNA barcode which can then be detected by a variety of means. In this protocol, we advance multiplexed quantitation by using the fluorescent hybridization technology from NanoString Technologies. This platform can measure up to 16,384 barcodes and has been validated for detecting as low as femtomolar concentrations of DNA and RNA [8, 9]. Our protocol describes how to extend this technique to measure proteins within cells and/or clinical samples with single cell sensitivity [6].

2 Materials

Prepare all solutions using ultrapure RNAse-free water, and take care to do all processing with RNase/DNAse-free tips. Prepare and store all reagents at room temperature (unless otherwise stated). Diligently follow all waste disposal regulations when discarding clinical waste materials. We do not add sodium azide to any of the reagents. Antibody–DNA conjugates should be aliquoted and frozen at −20 °C or −80 °C as quickly as possible.

1. Size Separating Filters (0.5 ml, 3 kDa, and 100 kDa, Amicon Ultra Filter).
2. Small molecule filtration Columns (2 ml, 7 kDa Zeba columns).
3. Sephadex columns (Illustra Nap 5, made of Sephadex G-5; GE Healthcare).
4. NanoString capture/reporter probe alien sequence kit. Kit can be ordered commercially through NanoString using their “alien” potato sequences designed with similar melting temperatures appropriate for the NanoString assay.
5. Thiolated DNA sequences (75mer sequences), designed in accordance with NanoString “alien” sequences.
6. Antibodies in carrier-free PBS (no azide, BSA, gelatin, etc.).
7. Dithiothreitol (DTT); No-weigh format (Pierce, cat # 20291).
8. Protein A/G magnetic beads (Pierce).
9. Photocleavable Linker (PCL) (For details on synthesizing this compound *see* previous work [7]).
10. Sodium bicarbonate.
11. Biology-grade DMF/DMSO.
12. Micro-BCA assay kit for protein quantitation.
13. RNAse/DNAse-free tubes.
14. 1.7 ml microcentrifuge tubes.
15. 0.5 ml or 0.3 ml tubes (optional: DNA Lo-Bind).
16. ATL lysis buffer.
17. Proteinase K.
18. Reducing buffer: 100 mM dithiothreitol (DTT) in PBS (1 mM EDTA, pH 8.0). Make fresh each time. PBS/1 mM EDTA, pH 8.0 buffer can be stored at RT.

19. Antibody conjugation buffer: PBS containing 5 % DMF and 10 % 0.1 M $NaHCO_3$. Make fresh each time.
20. Wash buffer (TBS + 0.1 % Tween). Stored at room temperature.
21. Antibody separation buffer (TBS + 0.1 %Tween 20 + 0.05 mg/ml cysteine). Make fresh each time.
22. Perm/blocking buffer: 10 % v/v Rabbit serum, 2 % BSA, 1 mg/ml SS salmon sperm DNA, 0.2 mg/ml Cysteine (Sigma Aldrich), 20× Perm/Wash Buffer diluted to 2× (BD Bioscience) (alternatively, can use 0.4 % saponin), and 0.1 % Tween 20.
23. Fixation buffer: 0.5× FB1 PhosFlow Fix Buffer (BD Biosciences) diluted in distilled, deionized water; alternatively 2 % formaldehyde can be used.
24. NanoString platform (nCounter analysis system?).

3 Methods

3.1 DNA Modification

1. Begin this step 30 min before setting up the Antibody–Linker reaction (*see* Subheading 3.2).
2. Reduce thiol-modified DNA 75-mer sequences (*see* Subheading 2) using Reducing Buffer for 2 h at room temperature (*see* **Note 1**). Calculate a molar ratio that is tenfold in excess of the antibody.
3. After 2 h, collect purified, reduced DNA in fractions from Sephadex columns. After the initial 400 μl, collect fractions drop by drop (25–50 μl).
4. Test for fractions containing free DTT by adding a 25 μl aliquot of the micro BCA assay mix (50:48:2 reagents MA–MB–MC) to each fraction, starting with the last one collected. After mixing, discard any aliquots that are clear or purple in color. Stop adding micro BCA assay mix when fraction is a very light green, which indicates the absence of DTT (*see* **Note 2**).
5. Pool all remaining DTT-free fractions together and concentrate the mix using a 3K MWCO filter (Amicon) (*see* **Note 3**). To quantify the yield, measure the volume of the fractions collected (via pipette) and the DNA concentration (via NanoDrop).

3.2 Antibody–Linker Conjugation

1. Incubate 100 μg of the carrier-free antibody (reaction concentration at 1 mg/ml or higher, *see* **Note 4**) with photocleavable bifunctional linker in Antibody Conjugation Buffer at room temperature for 1.5 h. Reactions occur with 50 M excess (*see* **Note 5**) of the photocleavable linker to the antibody in 200 μl for 1.5 h. Reactions should be covered in foil, and antibody concentration should ideally be as high as possible, or at a minimum of 1 mg/ml.

2. Remove excess linker from maleimide-activated antibodies using a desalting spin column (e.g., Zeba, 7K MWCO column, eluent: PBS) (*see* **Notes 6** and 7).

3.3 Antibody–DNA Conjugation and Purification

1. Link antibody to DNA by directly pooling the reduced DNA and purified antibody–linker conjugates into a tube. Place on a rocker covered in foil at 4 °C for 12 h.
2. Concentrate conjugates and partially remove excess DNA via size filtration with 100K MWCO centrifugal filter followed by three washes with TBS.
3. Wash protein A/G magnetic beads with TBST (Tris buffered saline and Tween 20) and then incubate with antibody conjugate for 60 min on rocker with Antibody Separation Buffer (*see* **Note 8**). Following incubation, wash three times with TBST.
4. Following binding, elute with a high-salt buffer (e.g., Gentle Elution Buffer). Elute three times with 75 µl, pipetting gently to mix thoroughly.
5. Collect eluted portions and remove high salts and exchange buffer to TBS (*see* **Note 9**) using a desalting spin column (e.g., Zeba, 7K MWCO).

3.4 Storing and Charactering Antibody–DNA Conjugates

1. Measure protein concentration using the micro-BCA assay (*see* **Note 10**).
2. Measure DNA concentration of all antibodies of interest on the NanoString platform. To do so, add antibody cocktail under two conditions: (1) "Control:" antibodies added in their native forms with DNA still attached and (2) "Released DNA:" antibodies treated with proteinase K and photocleaved. Add each antibody at the same protein concentration (2.5×10^{-5} mg/ml for a 5 µl sample). The difference in DNA readings between the two measurements reveals the relative number of DNA per antibody. This difference was divided by the isotype control measurement to account for possible inherent experimental error that may occur during determination of protein concentration (*see* **Note 11**).
3. Aliquot and store antibodies at concentrations of 0.25 mg/ml in PBS with BSA (0.15 mg/ml) at −20 °C, with adequate usage for at least 12 experimental runs (the number of runs on each NanoString cartridge) to avoid freeze–thaw cycles.

3.5 Cell Profiling

1. Block 1.7 ml centrifuge tubes with perm/blocking buffer.
2. Fix cells with fixation buffer for 15 min at room temperature, vortexing intermittently.
3. Permeabilize and block cells by incubating cells for 1 h with perm/blocking buffer at 37 °C (*see* **Note 12** about cell counts).

4. Pool all desired antibody conjugates in TBS, 0.1 % Tween, and 0.05 mg/ml cysteine.
5. Add the antibody cocktail to the fixed and permeabilized cells in an equal volume to the perm/block buffer.
6. Follow incubation with four 1.5 ml washes using wash buffer.
7. Count labeled cells and lyse desired amount in DNA Lo-Bind tubes (10 μL cell mix) with cell lysis buffer (e.g., 34.2 μl ATL lysis buffer) and proteinase K (5.8 μl). Let lysis proceed at 56 °C for a minimum of 30 min.
8. For *single cell analysis*, isolate single cells (via micromanipulation) and place each one directly into a NanoString PCR tube. In this case, 0.5 μl of proteinase K and 4.5 μl of ATL lysis buffer are used and no subsequent dilution is required.
9. Enhance release of DNA probes with photocleavage by placing tubes on handheld long wave UV lamp for 15 min.
10. Spin samples for 10 min at 14,000 rcf.
11. Collect supernatant. Dilute samples in nuclease-free water as appropriate for the NanoString Probe Kit (*see* **Note 13**). For single cells, no dilution is required, but additional blocking DNA can be added to prevent DNA loss.
12. Follow all instructions from NanoString for hybridization and detection of DNA from the lysis samples.

3.6 Protein Expression Analysis

1. Normalize raw DNA counts via the mean of the internal NanoString positive controls, which accounts for hybridization efficiency onto the cartridge. This normalization can be performed automatically using NanoString nCounter software. Only normalize over the positive controls that fall in a linear range (*see* **Note 14**).
2. To convert normalized counts to antibody expression values, divide by the relative DNA/Ab counts as quantified in **step 2**, Subheading 3.4. This step is not necessary when looking at the log fold or percent change in counts between two cell conditions.
3. Threshold expression values against nonspecific binding by removing signals lower than the appropriate isotype IgG control antibody. For stricter thresholding, also remove signals lower than one standard deviation above the isotype IgG control.
4. Significant differences can be determined by pairwise *t* testing using false discovery rates (FDR = 0.2) for post-testing or by bootstrap analysis.
5. Optional: Normalization by housekeeping genes can account for differences in cell count in case cells were not counted prior to cell lysis.

4 Notes

1. 75-mer oligos have been optimized for ideal binding with the NanoString platform. Though smaller oligos can reduce size hindrance during antibody staining, oligos below 50mer did not bind with the NanoString probes.
2. The purple color is produced by chelation of two molecules of bicinchoninic acid, which occurs in the presence of DTT.
3. Proceed to **step 1**, Subheading 3.3 as soon as possible to ensure that antibodies react with the DNA while in their reduced state.
4. Higher concentrations of antibody during the reaction resulted in higher yield.
5. Carrier-free antibodies are critical as BSA or other components can compete with the NHS ester chemistry of the conjugation reaction.
6. The molecular weight of the antibody was approximated to be 150 kDa.
7. In addition to antibody of interest, a control IgG antibody should be used for nonspecific binding (e.g., Cell Signaling). Housekeeping gene antibodies can also be used to approximate cell counts (GAPDH, tubulin, etc.).
8. Use beads in accordance with product specifications, e.g., for beads that bind 55 μg IgG antibody per mg bead, use ~2 mg of beads.
9. Do not use PBS as it will result in precipitates when mixed with the high-salt elution buffer.
10. A280 absorbance to quantify protein concentration is no longer accurate, as the absorbance will be affected by the attached DNA.
11. Antibodies can be validated both by dot blot for target binding and by comparing their target expression to the native antibodies on flow cytometry
12. Standard cell counts of 200,000 cells/tube can be used for cell culture. For clinical samples, cell counts as low as 1000 or less can be used. Note that cell loss will increase with fewer cells. For lower cell counts, increase blocking times.
13. To determine the appropriate input, begin by creating a DNA titration curve with varying concentrations of DNA vs. the total counts on NanoString. Select amounts in the linear counting range of the NanoString Analyzer. For approximately 100 antibody–DNA conjugates, 50–100 cells are typically within range. For low input amounts, nonspecific DNA blocking can also be used.

14. If nonspecific binding and higher counts are seen in the lower positive controls, clean the NanoString prep station thoroughly and reduce the amount of time samples spend at room temperature before the NanoString prep occurs.

References

1. Hsi ED (2001) A practical approach for evaluating new antibodies in the clinical immunohistochemistry laboratory. Arch Pathol Lab Med 125:289–294
2. Lanni EJ, Rubakhin SS, Sweedler JV (2012) Mass spectrometry imaging and profiling of single cells. J Proteomics 75:5036–5051
3. Gu L, et al. (2014) Multiplex single-molecule interaction profiling of DNA-barcoded proteins. Nature 515, 554–557
4. Fan R., et al. (2008) Integrated barcode chips for rapid, multiplexed analysis of proteins in microliter quantities of blood. Nat Biotech 26, 1373–1378
5. Shi Q, Qin L, Wei W et al (2012) Single-cell proteomic chip for profiling intracellular signaling pathways in single tumor cells. Proc Natl Acad Sci 109:419–424
6. Ullal AV, Peterson V, Agasti S et al (2014) Cancer cell profiling by barcoding allows multiplexed protein analysis in fine-needle aspirates. Sci Transl Med 6:219ra9
7. Agasti SS, Liong M, Peterson VM et al (2012) Photocleavable DNA barcode-antibody conjugates allow sensitive and multiplexed protein analysis in single cells. J Am Chem Soc 134: 18499–18502
8. Geiss GK, Bumgarner RE, Birditt B et al (2008) Direct multiplexed measurement of gene expression with color-coded probe pairs. Nat Biotechnol 26:317–325
9. Fortina P, Surrey S (2008) Digital mRNA profiling. Nat Biotechnol 26:293–294, PMID: 18327237
10. Chung J, Issadore D, Ullal A, Lee K, Weissleder R, Lee H (2012) Rare cell isolation and profiling on a hybrid magnetic/size-sorting chip. Biomicrofluidics 7(5):54107

Chapter 5

Genome-Wide Analysis of Protein and mRNA Copy Numbers in Single *Escherichia coli* Cells with Single-Molecule Sensitivity

Yuichi Taniguchi

Abstract

Single-cell proteomic and transcriptomic analysis is an emerging approach for providing quantitative and comprehensive characterization of gene functions in individual cells. This analysis, however, is often hampered by insufficient sensitivity for detecting low copy gene expression products such as transcription factors and regulators. Here I describe a method for the quantitative genome-wide analysis of single-cell protein and mRNA copy numbers with single molecule sensitivity for the model organism *Escherichia coli*.

Key words Stochastic gene expression, Single molecule detection, Single-cell proteomic and transcriptomic analysis, Microfluidics, Fluorescent protein fusion library, High-throughput platform, *Escherichia coli*

1 Introduction

Gene expression is often stochastic, because genes exist at single or low copy numbers in a cell. Owing to this property, protein and mRNA copy numbers vary from cell to cell in isogenic cell populations, resulting in variations in cellular properties. Single-cell proteomic and transcriptomic analysis [1–5] is an approach to comprehensively study mRNA and protein expressions in individual cells. This analysis has already been realized in many methods such as mRNA sequencing [1], cDNA microarray [2], mass spectroscopy [3], flow cytometry [4], and fluorescence microscopy [5].

Superior sensitivity is an important requisite of the analysis, since protein and mRNA molecules also exist in low copy numbers in a cell. Here, I describe a method for the genome-wide characterization of protein and mRNA copy numbers with single-molecule sensitivity in single *Escherichia coli* cells, which we have published recently [6]. Single molecule sensitivity allows us to quantify gene

Anup K. Singh and Aarthi Chandrasekaran (eds.), *Single Cell Protein Analysis: Methods and Protocols*, Methods in Molecular Biology, vol. 1346, DOI 10.1007/978-1-4939-2987-0_5,

expression stochasticity in individual cells [7], in addition to achieving universal analysis of the gene expression at any copy number.

The method described in this chapter is based on a chromosomal yellow fluorescent protein (YFP)-protein fusion library and uses a high-throughput single molecule imaging platform for analysis. The library is a collection of *E. coli* strains in which a particular gene is chromosomally tagged with the coding sequence of a YFP variant, Venus [8]. Protein expression levels can be analyzed by detecting Venus fluorescence. In addition, mRNA expression levels can be analyzed with fluorescence in situ hybridization (FISH) [9, 10] by hybridizing red fluorescently labeled oligonucleotide probes targeting *venus* mRNA in the library. The high-throughput single molecule imaging platform provides information of protein and mRNA expression with single molecule sensitivity for all library strains. For the analysis, nearly one hundred library strains are aligned in one custom-designed microfluidic device and then scanned systematically under single molecule fluorescence microscopy. With this system, 96 samples can be analyzed within 30 min, and an average of ~4000 cells can be analyzed per strain.

2 Materials

2.1 Microfluidic Chip Fabrication

1. Photomask. Design the pattern of the microfluidic device using CAD software (Fig. 1) and photo-plot a photomask with this design (*see* **Note 1**).

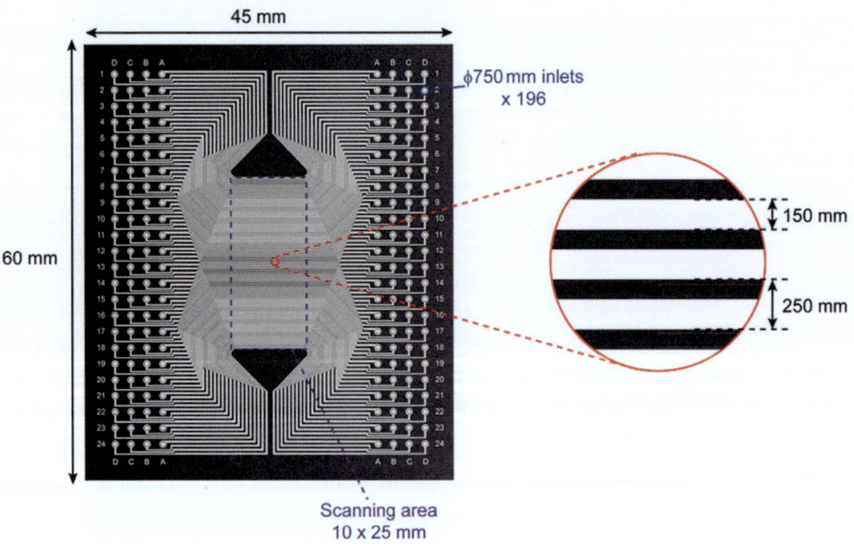

Fig. 1 Microfluidic chip design. The microfluidic chip integrates 96 independent microfluidic channels, in each of which different library strains can be imaged with single molecule fluorescence microscopy. The total area is 45×60 mm^2. A 25-μm layer is formed through the transparent part of the design on a silicon wafer. ϕ0.75 mm holes are punched through the PDMS replicas at the *circles on the right and left sides* to create the inlets and outlets of the channels. The central 10×25 mm^2 area is scanned under the microscope. Adapted from ref. [6]

2. UV-curable photoresist: SU8-2025 (Micro-Chem, Newton, MA, USA).
3. Poly-dimethylsiloxane (PDMS).
4. Large coverslip: 0.17 mm thick, 48×60 mm^2.
5. 4-in. test grade silicon wafer.
6. Developer solution: Propylene glycol monomethyl ether acetate.
7. Isopropyl alcohol.
8. Plasma cleaner.
9. Oven.
10. Hot plate.
11. Punching tool (Harris Uni-Core, Ted Pella, Inc., Redding, CA, USA).
12. 0.1 % poly-L-lysine.
13. 20 μL multichannel pipette.

2.2 Cell Culture and Sample Preparation for Microscopy

1. Venus protein fusion strain library. Obtain the library from the Coli Genetic Stock Center (New Haven, CT, USA; *see* **Note 2**).
2. Supplemented LB medium: 10 g/L tryptone, 5.0 g/L yeast extract, 10 g/L NaCl, and 20 μg/mL chloramphenicol.
3. Supplemented M9 minimal medium: 12.8 g/L $Na_2HPO_4 \times 7H_2O$, 3.0 g/L KH_2PO_4, 0.5 g/L NaCl, 1.0 g/L NH_4Cl, 2 mM $MgSO_4$, 0.1 mM $CaCl_2$, 0.4 % glucose, 1× MEM amino acids (Corning, Manassas, VA, USA), 1× MEM vitamins (Corning), and 20 μg/mL chloramphenicol.
4. 0.85 % NaCl.
5. Deep 2 mL 96-well plate and indented mat lid.
6. 10 μL, 200 μL, and 1 mL multichannel pipette.
7. Multi-dispensing pipettor.
8. 96-well transparent flat culture dish.
9. Reagent reservoir.
10. Scotch tape.

2.3 FISH Experiment

1. FISH probe: Prepare a 20-mer oligodeoxynucleotide (5′-TCCTCGATGTTGTGGCGGAT-3′) with a covalently linked red dye molecule (Atto 594) on the 5′ end (*see* **Note 3**).
2. 2× fixation solution: 7.4 % formaldehyde and 2× RNase-free PBS in DEPC-treated water.
3. RNase-free PBS solution.
4. 70 % ethanol.
5. 0.2 mg/ml BSA.

6. Wash buffer: 25 % formamide and 2× SSC in RNase-free water.
7. Hybridization buffer: 25 % formamide, 2× SSC, 10 % dextran sulfate, and 0.1 % *E. coli* tRNA.
8. 140 mM 2-mercaptoethanol.

2.4 Automated Microscope System

1. Inverted microscope (IX71, Olympus, Shinjuku, Tokyo, Japan).
2. 514 nm laser (Innova 300, Coherent, Santa Clara, CA, USA) for Venus excitation.
3. 580 nm laser (VFL-P-Series, MPB Communications Inc., Pointe-Claire, QC, Canada) for Atto594 excitation.
4. 100× phase-contrast objective lens (NA = 1.35, Olympus).
5. EM-CCD camera (Cascade 512B, Photometrics, Tucson, AZ, USA).
6. Computer installed with control software (Metamorph, Molecular devices, Sunnyvale, CA, USA).
7. Motorized 3D translational stage (MS2000, Applied Scientific Instrumentation, Eugene, OR, USA).
8. Mechanical shutters (LS2 and VMM-D3, Uniblitz, Rochester, NY, USA).
9. Dichroic mirror wheel (IX2-RFACA, Olympus).
10. Filter sets (Di02-R514-25x36 and FF01-542/27-25 for Venus, and FF593-Di03-25x36 and FF01-593/LP-25 for Atto594, Semrock, Rochester, NY, USA).
11. Vibration-free table (M-RS4000, Newport, Bozeman, MT, USA).
12. Lens and mirrors with holders.

3 Methods

3.1 Construction of Microfluidic Device

Standard photolithography and soft-lithography protocols are used to create master molds and microfluidic devices. Carry out all procedures in a clean room and **steps 1** and **2** in a fume hood.

1. Spin-coat UV-curable photoresist with 25 μm thickness on a silicon wafer. Bake the wafer on a hot plate at 65 °C for 3 min and then 95 °C for 6 min.
2. Expose UV-light to the wafer through the photomask and bake at 65 °C for 1 min and 95 °C for 6 min. Immerse the wafer in developer solution until the pattern is developed. Rinse the wafer with isopropyl alcohol and dry it with pressurized nitrogen. Use this as the master mold of the microfluidic device.
3. Place the mold wafer in a plastic petri dish. Mix 40 g base and 4 g curing agent of PDMS and pour the mixture over the mold. Degas until all bubbles are removed. Place the mold in an oven at 60 °C for 60 min.

4. Cut the cured PDMS sheet with a knife and peel off from the mold. Punch ϕ0.75 mm holes with a punching tool through the PDMS sheet at the inlet and outlet positions. Remove any residual PDMS fragments on the PDMS surface using Scotch tape.
5. Expose the PDMS sheet and a coverslip to oxygen plasma in a plasma cleaner. Place the sheet and coverslip in contact to bond the surfaces irreversibly and then in an oven at 60 °C for 15 min.
6. Just before using, inject 2–5 μL of 0.1 % poly-L-lysine solution into each channel to immobilize any cells on the channels using a 20 μL multichannel pipette (Fig. 2) (*see* **Note 4**).

3.2 Automated Imaging System Setup

1. Stably fix an inverted microscope on a vibration-free table. Install an EM-CCD camera, a phase-contrast high-NA objective lens, a motorized 3D translational stage, and an automatic dichroic mirror wheel on the microscope.

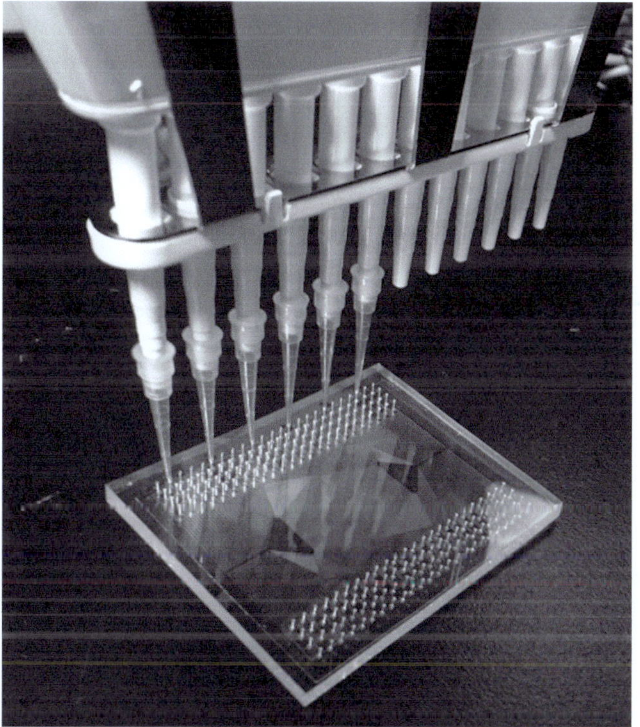

Fig. 2 Injection of the solution into the microfluidic chip. A multichannel pipette is used to inject the solution into the channels. The spacing between four channel inlets is designed to match the spacing between the pipette tips (=9 mm). The elasticity of PDMS works sufficiently to seal space between the ϕ0.75 mm inlet and disposable plastic tips for the solution injection

2. Construct a wide-field illumination using a 514 nm and 580 nm laser with a combination of lens and mirrors, and introduce it into the microscope.
3. Place mechanical shutters in the light paths of the laser illumination and the transmitted light illumination within the microscope.
4. Connect the camera, stage, mirror wheel and mechanical shutters to a computer installed with control software. Set up the software to control these devices synchronically.
5. Create a program to perform a sequence of imaging procedures, such as camera acquisition and stage movement, to image all channels in the microfluidic chip (for step-by-step procedures, refer to Subheading 3.4).

3.3 Sample Preparations for High-Throughput Analysis

1. Add 0.5 mL of supplemented LB medium to each well of a deep 2 mL 96-well plate using an electronic multi-dispensing pipettor. Inoculate a frozen stock of the library strain to the LB using a 10 μL multichannel pipette. Seal the plate with an indented mat lid. Shake overnight at 30 °C.
2. Pipette 1 mL of supplemented M9 medium into each well from **step 1**. Transfer 2.5 μL of each LB sample to the M9 medium plate using a 10 μL multichannel pipette. Seal with a mat lid. Shake at 30 °C for 11–12 h.
3. Spin the plate briefly (<1 min) to remove culture from the lid. Open the lid and remove 200 μL of each sample with a 200 μL multichannel pipette and transfer to a 96-well transparent flat culture dish. Record OD_{600} with a plate reader. Use samples with $OD_{600} = 0.1–0.4$ for the analysis (*see* **Note 5**).
4. Spin the original deep well plate at 3400 × *g* for 10 min. Pour off any liquid. Add 800 μL of 0.85 % NaCl to each well using a 1 mL multichannel pipette.
5. Spin the plate at 3400 × *g* for 10 min. Remove any liquid. Add 50 μL of 0.85 % NaCl to each well using a 200 μL multichannel pipette.
6. Inject 5–10 μL of cell samples to the microfluidic channels using a 20 μL multichannel pipette. Incubate for 15 min.
7. Seal the outlets of the microfluidic channels with Scotch tape. Set the volume of a 20 μL multichannel pipette to 20 μL, but pull only 5–10 μL worth of 0.85 % NaCl into the tips. Insert all the tips into the inlets of the channels and push the plunger of the pipette all the way to create strong pressure inside the channels (*see* **Note 6**).
8. Remove the seal. Inject any remaining 0.85 % NaCl in the tips into the channels to wash out floating cells (*see* **Note 7**).

3.4 High-Throughput Microscope Experiment

1. Place the microfluidic device on the microscope stage (*see* **Note 8**).
2. Run a program that carries out the following steps automatically for measurement.
3. Obtain *XYZ* coordinates in the stage at the following three different points: (1) the beginning point of the 1st channel, (2) near the end point of the 1st channel, and (3) near the beginning point of the last channel. From these values, calculate the tilt angle of the microfluidic device against the stage's coordinate system and obtain proper *XYZ* coordinates for scanning (Fig. 3).

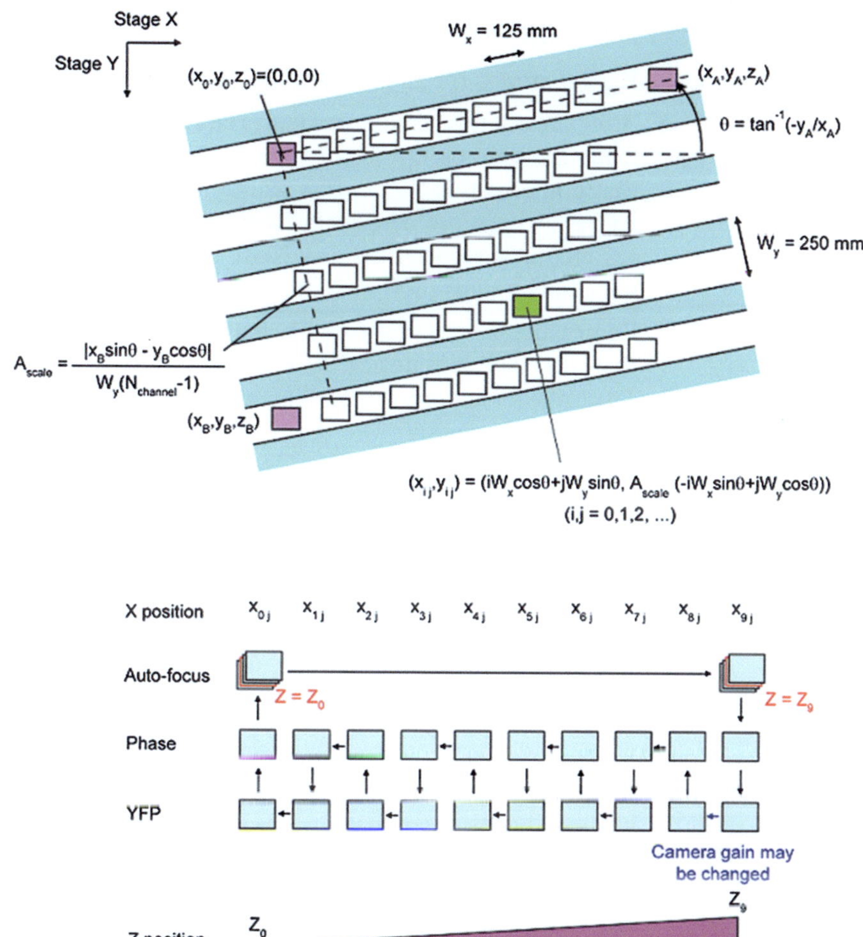

Fig. 3 Scanning method. To scan along the microfluidic channels by an automated stage, the tilt and orientation of the microfluidic chip against the stage *XYZ* coordinates are calibrated. The *XY* positions for scanning are calculated from the *XY* coordinates at (1) the beginning point of the 1st channel, (2) near the end point of the 1st channel, and (3) near the beginning point of the last channel (indicated by *pink squares* in the *upper figure*). The Z positions for scanning are obtained from auto-focusing at the beginning and end points of the channels and calculating the Z position profile assuming a linear landscape (*see* the *lower figure*). Phase-contrast and fluorescent images are taken in turns at the given *XYZ* positions. Adapted from ref. [6]

4. Start the scanning measurement by acquiring ten images at different positions in each channel in turn. In each channel, perform auto-focusing at the beginning and end point of the channels by taking multiple phase-contrast images at different Z positions to find the Z coordinate that provides the highest contrast. Calculate the Z position profile along the channel assuming a linear landscape. Subsequently acquire phase-contrast and fluorescent images using the filter sets for Venus observation sequentially at the given *XYZ* positions along the channel. If a near-saturated pixel is observed in the image, acquire the remaining images at a lower camera gain and subsequently calculate real pixel values with a predetermined scale factor (*see* **Note 9**). Save the images by names with tags that include the channel number, image number, phase-contrast/fluorescent and camera gain.
5. Repeat **step 4** for all channels (*see* **Note 10**).
6. After the scanning measurement, remove the microfluidic chip from the microscope stage. Place another microfluidic chip that contains fluorescein dye solution, 0.85 % NaCl solution, and a background cell strain lacking fluorophores into different channels respectively (*see* **Note 11**). Image the chips with the camera gains used in **step 4**.
7. Remeasure later samples with fewer than 500 cells or those that include many unhealthy cells or lysed cells (*see* **Note 12**). To confirm data reproducibility, measure all samples on at least two different days.

3.5 Large-Scale Image Analysis

Automated image analysis is done through the following steps. Programs can be coded in LabVIEW software (National instruments).

1. Subtract all the library and fluorescein solution images with the background NaCl image. Flatten the images by dividing the background-subtracted library images by the background-subtracted fluorescein solution image (Fig. 4).
2. Process all phase-contrast images through a closed filter and a sharpening convolution filter, and threshold them to create binary images. Segregate each binary image into particles. Filter the particles by the area and minor and major caliper lengths in order to exclude overlapped cells, unhealthy cells, cell debris, and other unexpected objects from the analysis (Fig. 5a).
3. For each particle, obtain the integrated fluorescence intensity within the entire particle area. Normalize the intensities by the cell volume: $(2/3) \times$(particle area)$\times$(particle minor caliper length). Divide the normalized intensities by a value from single molecule fluorescence, which can be obtained by imaging

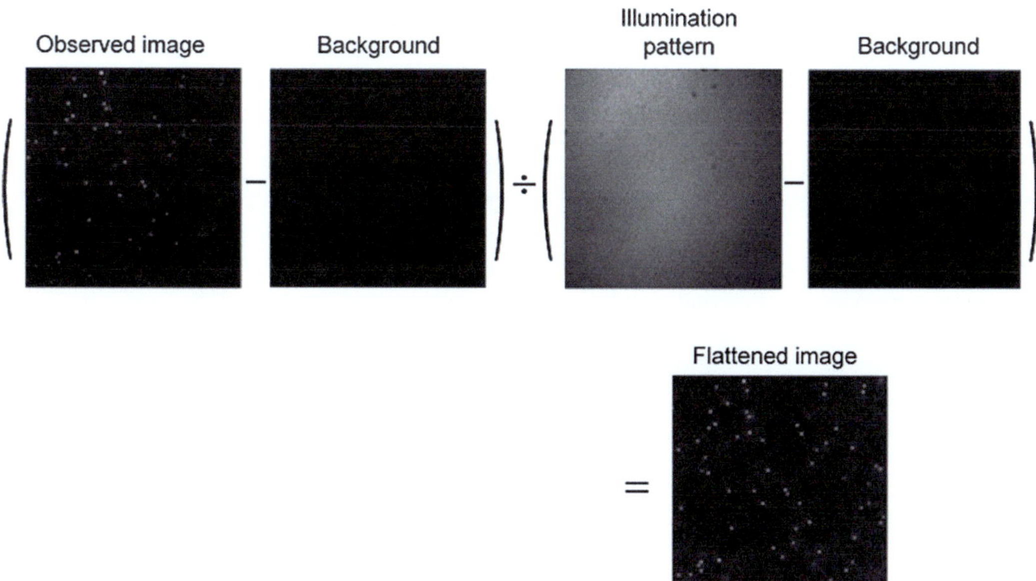

Fig. 4 Compensation of laser illumination pattern. To compensate for the heterogeneity of the laser distribution in the image field, we obtained the distribution pattern by imaging a fluorescein dye solution. *Flattened images* are obtained by dividing the background-subtracted observed images by the background-subtracted laser intensity distribution. Adapted from ref. [6]

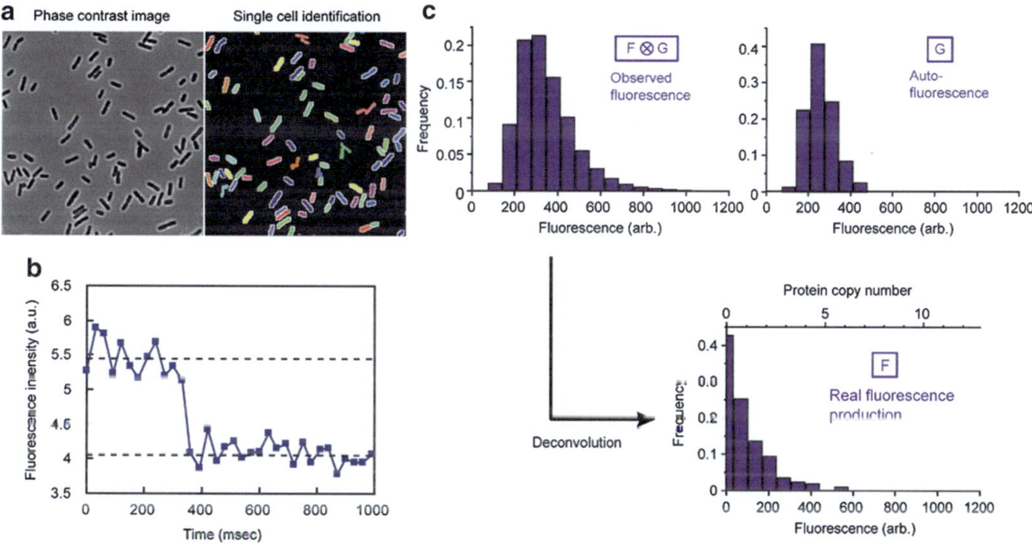

Fig. 5 Image analysis. (**a**) Single cell identification. The phase-contrast images are reduced to binary images through image filters to identify distinct particles that correspond to cells (the *different colored objects* in the *right image*). The particles are then filtered by particle size, area, and shape to identify single cell boundaries (*highlighted in white*). The fluorescence count is integrated for the entire area of each cell. (**b**) Fluorescence time trace of a single membrane-bound Venus molecule that shows abrupt photobleaching. (**c**) Autofluorescence deconvolution. The contribution of cellular autofluorescence background is removed by deconvolution. The net protein copy number distribution (*F*) is calculated by deconvoluting the measured fluorescence histogram ($F \otimes G$) with the cell autofluorescence histogram (*G*). Adapted from ref. [6]

cells that express membrane-localized Venus proteins at low copy numbers (around 1 copy per cell), such as the SX4 strain, which expresses membrane-bound Tsr-Venus [7], and by analyzing the intensities of fluorescent spots that display one-step photobleaching (Fig. 5b).

4. As necessary, calculate localization parameters in each cell. To characterize the membrane/cytoplasm localization, calculate the ratio of fluorescence detected on the edge compared to that inside the cell (E/I). To characterize the punctate localization, apply a top-hat filter and a threshold, and count the number of fluorescent spots.
5. To subtract the influence of cellular autofluorescence, deconvolute the autofluorescence distribution obtained from cells lacking fluorophores from the resulting fluorescence distribution to provide the true protein copy number distributions (*see* **Note 13**, Fig. 5c).
6. As necessary, calculate the mean (μ), standard deviation (σ), skewness, kurtosis, noise (σ^2/μ^2), and Fano factor (σ^2/μ) from the deconvoluted histogram. Provide the errors of these parameters by using a bootstrap resampling method with 1000 samples.
7. As necessary, fit the distributions with physical models, such as a gamma distribution model [11] (Fig. 6). To do this, calculate the error value of each bin by a bootstrap and obtain chi square values and p-values for the fit of each model.

3.6 mRNA Expression Analysis

To analyze mRNA expression levels, substitute the steps between **steps 4–8** in Subheading 3.3 with the following steps.

1. Mix 800 μL of 2× fixation solution rapidly. Incubate at RT for 15 min.
2. Spin the plate at 3400 ×g for 10 min. Wash twice with 800 μL of RNase-free PBS solution (i.e., resuspend in 800 μL of RNase-free PBS, spin at 3400 ×g for 10 min, and remove the supernatant).
3. Resuspend in 800 μL of 70 % ethanol. Incubate at RT for 1 h to permeabilize the cell membrane.
4. Spin at 3400 ×g for 10 min. Wash with 800 μL of wash buffer. Resuspend in 20 μL of wash buffer.
5. Mix with 50 μL of hybridization buffer and 2.5 μL of 30 nM FISH probes dissolved in RNase-free water and 0.2 mg/mL BSA. Incubate at 30 °C for 9 h.
6. Wash twice with 800 μL of wash buffer. Resuspend in 800 μL of wash buffer. Incubate at 30 °C for 1 h. Wash with 800 μL of wash buffer. Wash with 800 μL of RNase-free PBS. Resuspend in 10 μL PBS.

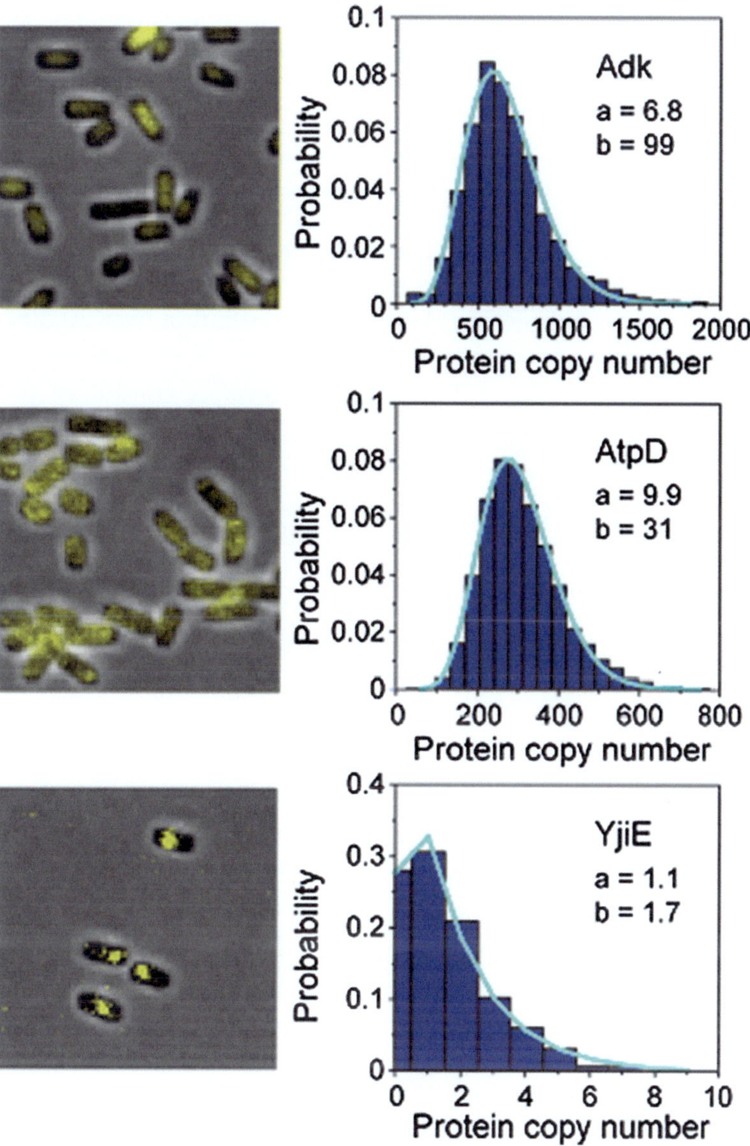

Fig. 6 Example images of three library strains (Adk, AtpD, and YjiE) and corresponding single-cell-protein level histograms. Fluorescence images overlaid on phase-contrast images of three library strains are shown on the *left*. The cytoplasmic protein Adk (*top*) is uniformly distributed within the cell. The membrane protein AtpD (*middle*) is distributed on the periphery of the cell. The predicted DNA binding protein YjiE (*bottom*) is distributed intracellularly where single protein molecules display spots as they are localized. Respective protein copy number distributions are shown on the *right*, where distributions are determined by deconvolution. The protein copy number per average cell volume, or the concentration, is determined as described in the *main text*. The distributions are fit to gamma distributions with parameters *a* and *b*

7. Do **steps 6–8** in Subheading 3.3. Afterwards, wash extensively with RNase-free PBS. After the final wash, fill each well with 140 mM 2-mercaptoethanol in PBS.

4 Notes

1. If there is no equipment for photo-plotting, use commercial services (for example, CAD/Art Services, Inc., Bandon, OR, USA). High resolution over 20,000 dpi is necessary to create the mask.
2. The library currently consists of 1023 strains. In each strain, the coding sequence of Venus is inserted into the C-terminus of a particular gene in the chromosome.
3. You may use a custom oligo service. The oligonucleotide sequence is the reverse complement of the region on the *venus* mRNA that has the least frequency of secondary structures.
4. To generate strong pressure for injection, set the pipette volume to 20 μL, but pull only 2–5 μL of solution into the tips. Insert all the tips to the inlets of the channels and push the plunger of the pipette all the way.
5. Samples out of this range are not used for the analysis, but are reanalyzed another day.
6. Make sure that the solution in the tips is not injected in this process, because of the sealing with Scotch tape. This procedure creates strong pressure inside the channels and causes stable binding of the cells to the channel surface.
7. Make sure that a droplet of solution comes out of the outlet of the channel.
8. For stable measurement, prepare a jig to set the microfluidic device on the stage.
9. The compensation is done by multiplying a scale factor after subtracting offsets. The scale factor is predetermined from the relationship between the camera gain and obtained pixel values when measuring fluorescent samples. This compensation system can virtually increase the dynamic range of the fluorescence measurement.
10. Typically, ten sets of images are taken within 20 s, resulting in the observation of 500–20,000 cells.
11. The fluorescein image is used to compensate for the heterogeneity of the laser illumination in the image field. The NaCl image is used to subtract the dark noise count due to the properties of the EMCCD and the autofluorescence background generated from the channel and immersion oil.

12. Remeasure the data of strains that have bimodal distributions and discard if the measurements are not reproducible. Bimodal distributions are mainly observed in samples that have a mixture of healthy and unhealthy cells.
13. The bin number of the histogram can be set as the maximum bin number where all values of bins in the deconvoluted histogram are positive. We expect that the contribution of the Venus fluorescence distribution to be negligible at the final result, because the standard deviation of the single molecule fluorescence intensity is small (65 %).

Acknowledgement

I thank P. J. Choi, H. Chen, G. W. Li, J. Hearn and S. X. Xie for their key contributions to the protocol described. I thank M. Ohno and K. Nishimura for discussions, and P. Karagiannis for critically reading the manuscript. I acknowledge support from the Grant-in-Aid for Young Scientists, Japan Society for the Promotion of Science (JSPS), the Grant-in-Aid for Scientific Research on Innovative Areas of JSPS, the Mochida Memorial Foundation for Medical and Pharmaceutical Research, the Takeda Science Foundation, the Uehara Memorial Foundation, and the Marubun Research Promotion Foundation.

References

1. Tang F, Barbacioru C, Wang Y, Nordman E, Lee C, Xu N, Wang X, Bodeau J, Tuch BB, Siddiqui A, Lao K, Surani MA (2009) mRNA-Seq whole-transcriptome analysis of a single cell. Nat Methods 6:377–382
2. Tietjen I, Rihel JM, Cao Y, Koentges G, Zakhary L, Dulac C (2003) Single-cell transcriptional analysis of neuronal progeniors. Neuron 38:161–175
3. Irish JM, Hovland R, Krutzik PO, Perez OD, Bruserud Ø, Gjertsen BT, Nolan GP (2004) Single cell profiling of potentiated phospho-protein networks in cancer cells. Cell 118:217–228
4. Newman JR, Ghaemmaghami S, Ihmels J, Breslow DK, Noble M, DeRisi JL, Weissman JS (2006) Single-cell proteomic analysis of S. cerevisiae reveals the architecture of biological noise. Nature 441:840–846
5. Levsky JM, Shenoy SM, Pezo RC, Singer RH (2002) Single-cell gene expression profiling. Science 297:836–840
6. Taniguchi Y, Choi PJ, Li GW, Chen H, Babu M, Hearn J, Emili A, Xie XS (2010) Quantifying *E. coli* proteome and transcriptome with single-molecule sensitivity in single cells. Science 329:533–538
7. Yu J, Xiao J, Ren X, Lao K, Xie XS (2006) Probing gene expression in live cells, one protein molecule at a time. Science 311: 1600–1603
8. Nagai T, Ibata K, Park ES, Kubota M, Mikoshiba K, Miyawaki A (2002) A variant of yellow fluorescent protein with fast and efficient maturation for cell-biological applications. Nat Biotechnol 20:87–90
9. Femino AM, Fay FS, Fogarty K, Singer RH (1998) Visualization of single RNA transcripts in situ. Science 280:585–590
10. Raj A, van den Boogard P, Rifkin SA, van Oudenaaden A, Tyagi S (2008) Imaging individual mRNA molecules using multiple singly labeled probes. Nat Methods 5: 877–879
11. Friedman N, Cai L, Xie XS (2006) Linking stochastic dynamics to population distribution: an analytical framework of gene expression. Phys Rev Lett 97:168302

Chapter 6

Microfluidic Flow Cytometry for Single-Cell Protein Analysis

Meiye Wu and Anup K. Singh

Abstract

Flow-cytometric (FC) detection of proteins in single cells is a rapid, quantitative method for single-cell protein analysis. Recent advancements in microfluidic technologies have leveraged miniaturization and automation to adapt flow cytometry for analyzing single cell protein profiles both for cell surface and intracellular proteins. Here, we describe the method for microfluidic FC, along with instructions to build a microfluidic platform capable of automated cell culture, cell surface receptor immunostaining, intracellular phosphoprotein and intracellular cytokine immunostaining, and analysis using micro-flow cytometry. As a demonstration of our platform and protocol, we detail the profiling of TLR4 receptor activation, ERK1/2 phosphorylation, and TNFα production in LPS stimulated macrophages using the microfluidic platform.

Key words Immunostaining, Phosphoprotein, Cell surface receptors, Cytokine, Single-cell resolution, Microfluidic, Flow cytometry

1 Introduction

With the exception of certain RNAs, proteins carry out all cellular processes—from providing structure, to transporting material, to functioning as enzymes. It has become abundantly clear in recent years that the behavior of single cells differ from that of a heterogeneous mixture of cell populations and hence, measuring proteins at the resolution of single cells will greatly enhance our ability to understand molecular mechanisms underlying many cellular processes [1]. The most established and user-friendly method for single cell protein analysis is flow cytometry (FC). The use of monoclonal antibodies (mAb) that are designed to recognize proteins, sugars, nucleic acids, and even post-translational modifications in combination with flow cytometry is now a classic single cell protein profiling method. Much of what we know regarding immune cell subsets and hematopoietic differentiation has been gathered using flow cytometry of antibody stained cells [2]. The availability of mAbs that bind tyro-

Anup K. Singh and Aarthi Chandrasekaran (eds.), *Single Cell Protein Analysis: Methods and Protocols*, Methods in Molecular Biology, vol. 1346, DOI 10.1007/978-1-4939-2987-0_6, © Springer Science+Business Media New York 2015

sine or serine phosphorylated versions of proteins has made phosphorylation the best understood dynamic post-translational modification (PTM) [3–5]. Entire phosphorylation cascades can be profiled at single cell resolution to monitor disease progression or response to therapy [6]. Moreover, these proteins can be quantified using mAbs by generating a standard curve using varying concentrations of the same mAb probe bound to beads coated with IgG [6].

While conventional FC has been extremely useful in providing single cell protein analysis, there are shortcomings to the method. Typically, sample preparation for conventional FC requires large numbers of cells (~10^6/sample), with correspondingly large fluorescent labeled antibody requirement, which can be prohibitively expensive when profiling many proteins. The sample preparation process also involves numerous centrifugation and aspiration steps, where sample loss of up to 50 % is the norm. These limitations make conventional FC unsuitable for studying rare samples such as rare primary cell types and stem cells. Recent advancements in microfluidic engineering has enabled miniaturized, chip based sample preparation and micro-flow cytometry (μFC), where small number of single cells (~100–1000) can be cultured, prepared, and analyzed in integrated microfluidic systems to provide multiparameter protein analysis at single cell resolution [7–9]. In this chapter, we provide materials list and instructions to build the integrated microfluidic platform, and provide demonstration of microfluidic multiparameter μFC sample preparation and analysis by monitoring TLR4 receptor activation, ERK phosphorylation, and TNFα production in macrophages stimulated with lipopolysaccharide (LPS).

2 Materials

2.1 Reagents for On-Chip Cell Culture and Stimulation (See Note 1)

1. RAW 264.7 mouse macrophage cell line (ATCC, Manassas, VA).
2. DMEM.
3. FBS.
4. 100× penicillin–streptomycin.
5. RAW growth media: 450 ml of DMEM, 50 ml of FBS, 5 ml of 100× penicillin–streptomycin.
6. Vented cap non-treated tissue culture flask, T75.
7. PBS based enzyme-free cell dissociation solution (Gibco, Grand Island, NY).
8. 0.02 N acetic acid, sterile filter using Steriflip.
9. Rat tail collagen type I, in 0.02 N acetic acid.
10. PBS based Enzyme Free Cell Dissociation Solution.
11. Deionized (DI) water, sterile filtered.
12. 0.22 μm Steriflip filter.

13. 10 % bleach in sterile filtered DI water.
14. Lipopolysaccharide (LPS), HPLC purified, lyophilized. Resuspend lyophilized LPS in sterile DI water at 1 mg/ml.

2.2 Reagents for Immunostaining of Cell Surface TLR4 Receptor, Intracellular Phospho-ERK, and Intracellular TNFα Cytokine

1. Make 1.6 % paraformaldehyde solution: 1 ml 16 % paraformaldehyde, 9 ml PBS.
2. Triton X-100.
3. CHAPS detergent.
4. Permeabilization buffer: 0.1 % Triton X-100, 0.05 % CHAPS, in PBS.
5. PBS, pH 7.4, sterile filtered.
6. PE labeled anti-TLR4-MD2 antibody (BioLegend, San Diego, CA). Make 1:15 dilution of PE conjugated anti-TLR4-MD2 antibody: 5 μl of antibody, 70 μl of PBS.
7. Alexa-488 labeled phospho-ERK1/2 antibody (Cell Signaling, Danvers, MA). Make 1:20 dilution of Alexa-488 conjugated anti-phospho-ERK antibody: 5 μl of antibody, 95 μl of PBS.
8. Cy5 labeled anti-mouse TNFα antibody (eBiosciences, San Diego, CA). Dilute Cy5 labeled TNFα antibody at 1:10: 5 μl antibody, 45 μl PBS.
9. Golgi-Plug reagent with Brefeldin A (BD Biosciences, San Jose, CA).
10. Golgi-Plug/LPS solution: 1 μl of 1 mg/ml LPS, 1 μl of Golgi-Plug reagent, 998 μl of growth medium to make 1 μg/ml LPS and 1× Golgi-Plug.

2.3 Reagents for Cell Detachment

1. Purified elastase, ≥8 units/mg protein (Worthington Biochemical Corporation, Lakewood, NJ).
2. 0.1 M Tris buffer.
3. PBS with Ca^{2+} and Mg^{2+}, pH 7.4, sterile filtered.
4. PBS without Ca^{2+} and Mg^{2+}, pH 7.4, sterile filtered.
5. Elastase stock solution: make 1 mg/ml stock solution of elastase in 0.1 M Tris buffer, pH 8.0. Divide into 100 μl aliquots and store at −20 °C.
6. Working elastase solution: immediately before cell detachment, thaw one aliquot on ice, and add 900 μl of PBS with Ca^{2+} and Mg^{2+}, and warm up to 42 °C.
7. 100 mM EDTA solution: 1 ml of 0.5 M EDTA, 4 ml PBS without Ca^{2+} and Mg^{2+}.

2.4 Components for Microfluidic Chip Setup

Performing microfluidic sample preparation is now possible in traditional molecular biology laboratories with little prior engineering expertise. Commercially available ready-to-use microfluidic chips,

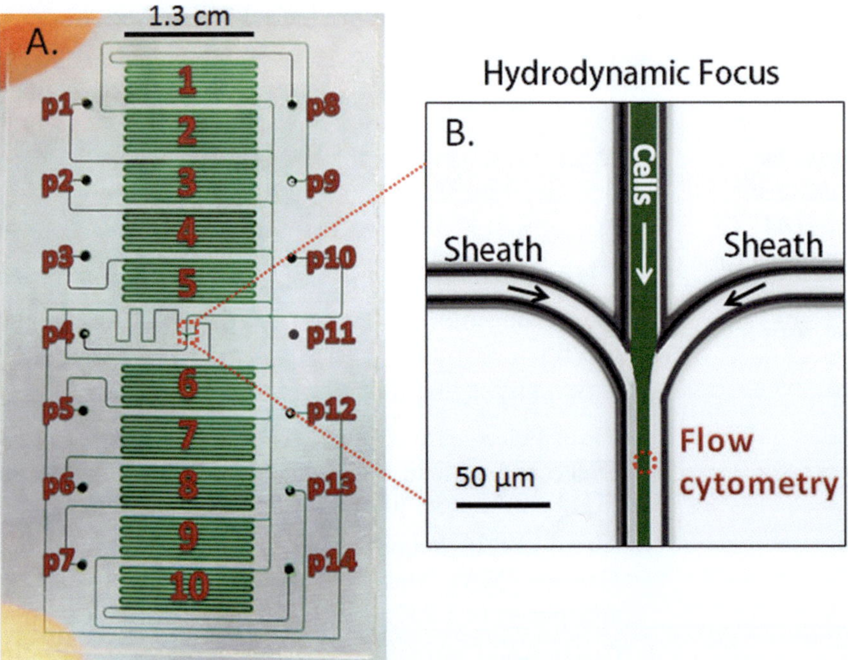

Fig. 1 (**a**) 10-chamber microfluidic chip, each chamber, numbered 1 through 10 can hold up to 2000 cells and is fluidically isolatable to enable ten different experimental conditions in one experiment. An array of 14 ports, labeled p1 through p14 serve as inlets for fluid connections to the chip's microchannels. The center of the chip shown in (**b**) is where cells can be hydrodynamically focused into single file for on-chip laser induced fluorescence and flow cytometry

fittings, valves, and pumps can be purchased and configured for microfluidic cell preparation and signal acquisition. Versions of the following components are commercially available, and customization can be achieved through the vendors.

1. Microfluidic chip. A custom-fused silica microfluidic chip with 14 ports labeled 1 through 14 is shown in Fig. 1a. The design was made using AutoCAD (Autodesk Inc. San Rafael, CA), photomasks were generated at PhotoSciences (Torrence, CA), and fused silica chips were fabricated by Caliper Life Sciences (Hopkinton, MA) (*see* **Note 2**). The cell holding channel dimensions are 200 μm wide by 30 μm deep, total length of ~12 cm. The dimensions of the sheath channels should be 30 μm wide and 30 μm deep. An array of 14 holes 500 μm in diameter, seven on each side, provides for fluid inlet to the cell holding chambers and the sheath channels. The center of the chip is shown magnified in Fig. 1b, where cells can be detached after immunostaining, and hydrodynamically focused for on-chip micro-flow cytometry. The fused silica chips can be reused after cleaning with 15 min flush of all chambers with sterile filtered 10 % bleach, followed by 5 min rinse with DI water.

2. Custom plastic chip manifold (Delrin) (*see* **Note 3**).
3. 1/32 in o.d., 125 μm i.d. PEEK tubing (Upchurch Scientific, Oak Harbor, WA).
4. Fittings for PEEK tubing (Upchurch Scientific, Oak Harbor, WA).
5. 14 custom-built Shut-off electronic valves, one for each port on the chip (commercial version available from LabSmith, Livermore, CA).
6. Airtight pressurizable reagent reservoir.
7. Electronic pressure controllers (Parker Hannifin, Cleveland, OH).
8. Premixed 5 % CO_2, 95 % air canister.
9. Thermoelectric hot plate.
10. Proportional integral controller for hot plate.

2.5 Micro-flow Cytometer Components (See Note 4)

1. 20-mW diode pumped solid state laser at 488-nm (85-BCF-020-112; CVI Melles Griot, Carlsbad, CA) in an epi-fluorescence configuration for excitation.
2. Long pass dichroic mirror (LPD01-488S; Semrock, Rochester, NY).
3. Aspheric lens (5722-H-B; New Focus, Santa Clara, CA).
4. Photomultiplier tube (PMT) based detector (H5784-20; Hamamatsu, Bridgewater, NJ).
5. Custom-made sculpted tip silica optical fiber (1000 μm core, 2000 μm sculpted spherical tip; Polymicro Technologies, Phoenix, AZ).
6. Eight-channel Hamamatsu linear multi-anode (LMA) PMT coupled with filter optics (H9797TM; Hamamatsu, Bridgewater, NJ).
7. Three dichroic mirrors (DMT560, DMT650, and DMT740, Hamamatsu, Bridgewater, NJ).
8. Four bandpass filters (BPF534_30, BPF585_40, and BPF692_40, and BPF785_62, Semrock, Rochester, NY).

3 Methods

3.1 Microfluidic Chip Setup, Priming, Cleaning, and Collagen Coating

1. To prime the chip, preload sterile filtered DI water into each channel by placing a droplet of ~100 μl DI water onto an inlet port and letting capillary action drive water into the microfluidic channel. Vacuum can be applied to an opposite port to suck water into the microchannel as an alternative. The goal is to fill as much of the microchannels with water as possible before assembly of the chip into the manifold. If a few bubbles remain in the microchannel, they can be driven out by positive pressure once the chip is assembled into the valve setup.

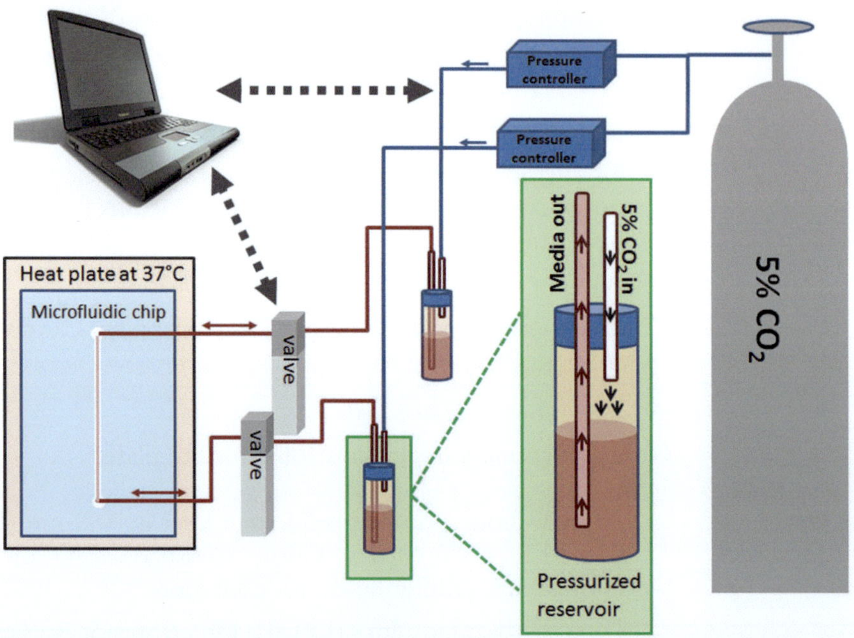

Fig. 2 Simplified schematic of a rudimentary microfluidic cell assay system. Pressure controllers (*blue boxes*) are connected to air source such as house nitrogen (*not shown*), or custom-premixed air canister, and are linked via tubing to preloaded, airtight reagent/sample reservoirs. The pressure controllers generate regulated pressure that is delivered into the reservoirs (*blue arrows*). The mechanism by which fluid moves out of the reservoir is shown in the magnified reservoir diagram in the *green box*. By applying air into the sample reservoir through the short tubing, the positive pressure generated by the air drives fluid out through the longer tubing that is submerged in the preloaded reagent/sample. Fluid then moves through tubing and reach programmable valves that are connected in-line between the reservoirs and the microfluidic chip. A hot plate can be attached to the chip manifold so that the chip can be heated to 37 °C for cell culture. The user can control which inlet on the chip the fluid can access by controlling the opening and closing of the in-line valves. A computer controls the pressure controllers, microvalves, and temperature controller. The user can script operational programs to automate sample preparation sequences such as washing and staining of cells in the microfluidic chip

2. To set up the chip (*see* **Note 5**), visually align the custom manifold to the chip and each port is connected to one end of the PEEK tubing via Upchurch fittings (*see* Fig. 2 for chip schematic). The electronic shut-off valves are used in-line with the tubing to control the movement of fluid into each chip port. Submerge the other end of the PEEK tubing in an airtight reservoir and pressurized using premixed 5 % CO_2 pumped through electronic pressure controller (*see* **Note 6**). The velocity at which the fluid flow into the chip is controlled by the pressure controller, higher pressure generates faster fluid flow velocity. Once assembled, load 1 ml DI water into reservoir at p10 (Figure 1a), and flush each chamber and sheath channels with DI water to drive out all air bubbles. To load sample/reagent into the chip, positive pressure generated by flowing

5 % CO_2 into the pressurized reservoir drives the media or reagent from the reservoir into the chip, and the pressure can be adjusted using the pressure controller between 0.2 psi and 15 psi. The hot plate with temperature control is integrated into the chip manifold in order to heat the chip to desired temperature. The temperature is set to 37 °C for the duration of cell culture, and then to room temperature for the fixation of cells and all subsequent steps in sample preparation, and finally to 42 C for enzymatic cell detachment.

3. To address a specific cell chamber, load reagent in reservoir at p10, open valve to p10 and open valve to outlet port for the targeted chamber, and shut off all other valves. Turn on pressure controller to apply house nitrogen to pressurized reservoir at p10, and reagent in the reservoir will move from the reservoir into the target chamber. For example, to target chamber 1, first shut off all 14 valves, apply pressure to p10, open valves to p10 and p8, the only fluid flow will be from reservoir at p10 through chamber 1, exiting p8.
4. To clean the chip, flush all channels with 10 % sterile filtered bleach in DI water for 10 min at 10 psi at room temperature, followed by 10 min flush with DI water at 10 psi.
5. Immediately before use, dilute Rat Tail Collagen to 50 μg/ml in 0.02 N acetic acid (*see* **Note** 7).
6. Flow 50 μg/ml collagen at 10 psi for 2 min into each cell holding chamber, then turn down the pressure to 0.5 psi and flow very slowly for 1 h.
7. Wash all chambers for 10 min at 5 psi using sterile filtered PBS. Chip is now ready for cell loading and on-chip cell culture.
8. Use the collagen-coated chip within 1 day.

3.2 Building the Micro-flow Cytometer

A simplified schematic of the micro-flow cytometer is shown in Fig. 3. To build the micro-flow cytometer, use a 20-mW diode pumped solid state laser at 488-nm in an epi-fluorescence configuration.

1. Place the laser beneath the microchip, and align the laser beam to reflect and focus upon the detection region at the center of the chip by using a long pass dichroic mirror and an aspheric lens.
2. Install the optical fiber with PMT and align on top of chip, position it slightly behind the path of the laser beam to collect forward scatter from passing cells.
3. Install aspheric lens, dichroic mirror and PMT modules beneath the chip so that scatter and fluorescence signal from passing cells will first be focused by the lens and subsequently passed via the dichroic mirror for detection in the PMT module.

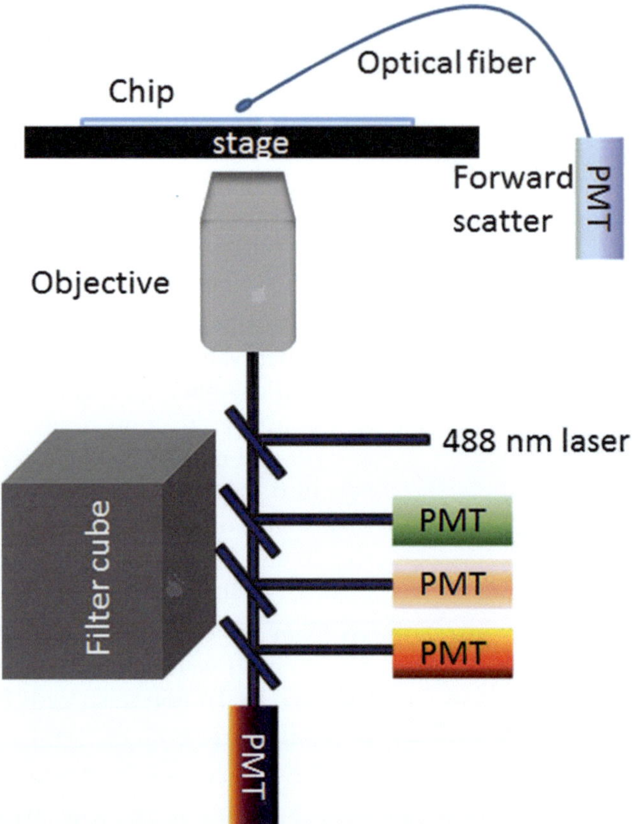

Fig. 3 Schematic of micro-flow cytometer. The optical fiber, and PMTs are aligned vertically to the area on the chip where cells are hydrodynamically focused into a single file line. As the cells pass through the laser beam, PMTs record scatter and laser induced fluorescence signals from the cells

4. For multiplexed fluorescence detection, use an eight-channel Hamamatsu linear multi-anode (LMA) PMT coupled with filter optics. For a four-color detection setup, use four channels of the LMA PMT, and select the filtering for green, yellow, red, and far-red fluorescence detection.
5. For green fluorescence detection, cascade a portion of the aggregate florescence signal upon the first LMA PMT channel via a long-pass dichroic mirror and filter through to the detection region of that channel using a band-pass filter (BPF534_30).
6. For yellow fluorescence detection, cascade a portion of the remaining florescence signal upon the second LMA PMT channel using another long-pass dichroic mirror (DMT650) and filter through to the respective channel's detection region using a second band-pass filter (BPF585_40).
7. For red fluorescence detection, use the third channel of the LMA PMT via a third long-pass dichroic (DMT740) and filter using a third band-pass filter (BPF692_40).

8. For far-red fluorescence detection, cascade the remaining florescence signal onto the fourth LMA PMT channel using a mirror, and filtered onto that channels detection region using a fourth band-pass filter (BPF785_62).

3.3 RAW Cell Culture and Stimulation on Microfluidic Chip (See Note 8)

1. Maintain RAW264.7 cells in growth medium in a humidified cell culture incubator maintained at 5% CO_2 and 37 °C. Seed two million RAW cells into untreated plastic T75 vented flask for RAW cell culture.
2. To detach RAW 264.7 cells, aspirate media using a serological pipet, and place media in a sterile 50 ml conical tube.
3. Wash monolayer of cells with 10 ml PBS, discard PBS.
4. Apply 5 ml of Cell Dissociation Solution that has been equilibrated to room temperature.
5. Incubate at room temperature for 10 min.
6. Close the cap of the flask tightly, tilt the flask to 45° with the cap on top, and repeatedly tap the bottom of the flask to dislodge the cells. Try not to get the vent cap wet while tapping the flask.
7. Transfer the dislodged cells to the conical tube containing media from **step 2**.
8. Tightly close the tube and centrifuge at 300 × *g* for 5 min to pellet the cells.
9. Aspirate the supernatant and resuspend the pellet in sterile PBS at 10^7/ml for loading into the collagen-coated microfluidic chip.
10. Load cells into reservoir at p10, set temperature to 37 °C, connect pressure controller to 5 % CO_2 canister.
11. Flow cells into each chamber at 5 psi for 2 min, then stop the flow for 15 min to let cells settle and attach to the microchannel surface.
12. Replace cells with growth medium at reservoir p10 and flow growth media into each cell holding chamber at 2.5 psi for 5 min to exchange the PBS with growth medium.
13. For maintaining cells on the chip, flow medium for 5 min at 2.5 psi into each chamber, then stop flow for 1 h, and repeat sequence of 5 min flow and 1 h stop flow for the duration of the culture experiment.
14. Load Golgi-Plug/LPS solution (*see* **Note 9**) into reservoir and flow into microchannels containing cells to stimulate TLR4 pathway activation in RAW cells. Flow at 5 psi for 2 min. Stop flow.

3.4 Cell Fixation and Cell Surface TLR4 Receptor Staining

1. Place 1 ml of 1.6 % paraformaldehyde into reservoir in port 10.
2. Assign one cell holding chamber to each time point in a LPS stimulation time-course (0 s, 30 s, 1 m, 5 m, 30 m, 2 h, 4 h), flow PFA at 5 psi through designated chambers at appropriate time-points to fix cells. Flow PFA for 2 min at 5 psi, stop flow and incubate for 8 min at room temperature (*see* **Note 10**).
3. Replace 1.6 % PFA reservoir tube with reservoir tube containing 1 ml of PBS, flow PBS into chambers containing PFA and displace all PFA by flowing at 5 psi for 5 min.
4. Flow PE-conjugated anti-TLR4-MD2 antibody into desired chamber at 5 psi for 2 min, stop flow and incubate at room temperature for 20 min.
5. Wash unbound TLR4-MD2 antibody from cells by flowing PBS into chambers at 2 psi for 10 min at room temperature.

3.5 Cell Permeabilization, and Intracellular Phospho-ERK and Cytokine Immunostaining

1. After immunostaining with TLR4-MD2 antibody, permeabilize cells by flowing permeabilization buffer into each cell holding chambers at 5 psi for 2 min, stop flow, and incubate at room temperature for 10 min.
2. Wash permeabilized cells by flowing PBS into each cell holding chamber at 2.5 psi for 10 min.
3. Flow Alexa 488-conjugated anti-phospho-ERK antibody into each cell holding chamber at 5 psi for 2 min, stop flow, and incubate at room temperature for 30 min.
4. Wash off any unbound antibody by flowing PBS into each cell holding chamber at 2.5 psi for 10 min.

3.6 Intracellular Cytokine Immunostaining

1. Flow Cy5 labeled anti-TNFα antibody into microchannel containing fixed and permeabilized RAW cells at 5 psi for 2 min, stop flow.
2. Incubate for 15 min at room temperature.
3. Wash all chambers with PBS at 2.5 psi for 5 min.

3.7 Cell Detachment

After on-chip sample preparation and image acquisition, cells can be detached from the microchannel surface by a combination of proteolytic cleavage and shear force.

1. Flow warmed (42 °C) elastase solution into a desired cell holding chamber with cells to be detached by opening the appropriate exit port. For example, to detach cells in chamber 1, flow elastase from p8 to p4 for 3 min at 5 psi, stop flow and incubate at 42 °C for 5 min. During incubation, place collection tube containing 100 μl of 100 mM EDTA in Ca^{2+}/Mg^{2+}-free PBS at p4. After incubation, flow elastase solution at 15 psi from p8 to p4, the combination of shear force and elastase cleavage will detach cells in chamber 1 and deposit them into collection tube

at p4. The EDTA in the collection tube will stop the elastase activity from destroying the fluorescent signal on the detached cells. Detach cells from each chamber as described, changing collection tube at p4 for cells detached from each chamber.

2. To analyze samples using a μFC, follow the protocol in Subheading 3.8. Alternatively, to use a commercial flow cytometer, follow Subheading 3.9, immediately after cells are detached.

3.8 Cell Analysis Using Micro-flow Cytometer

The micro-flow cytometry is performed at the center of the chip (Fig. 1b), where the sheath fluid focuses the sample stream containing the detached cells into the path of the laser beam coming from underneath the chip.

1. Load PBS without Mg^{2+} and Ca^{2+} at p12 to function as sheath fluid.
2. When detaching cells with elastase solution as described in Subheading 3.7, flow PBS from p12 to p4 at 5 psi to hydrodynamically focus the detached cells for on-chip flow cytometry. Adjust fluid pressure coming from p12 between each sample to ensure the focusing is consistent across all chambers (*see* **Note 11**).
3. Collect the laser induced forward scatter and fluorescence signals from the passing cells with the optical fiber and PMT located above the chip (Fig. 3).
4. Collect the scatter and fluorescence signals for all channels (488-nm scatter, green, yellow, red and far-red) by the data acquisition module located beneath the chip (Fig. 3).
5. Generate scatter and fluorescence histograms for each passing cell, use LabView to script an application (peak finder) to discriminate the peak from the scatter and fluorescence histograms to generate a numerical value representing each cell.
6. Export the peak values generated by the "peak finder" program into Microsoft Excel or another equivalent software, plot the values in each sample set to generate histograms for each experimental condition, and overlay the histograms to show changes in the fluorescence between each experimental condition (Fig. 4).

3.9 Flow-Cytometric Analysis Using Commercial Flow Cytometer

1. Pre-label new collection tubes with sample identification.
2. Aliquot 100 μl of 100 mM EDTA solution into each collection tube, and place on ice for at least 10 min prior to cell detachment.
3. Detach cells in each holding chamber as indicated in Subheading 3.7, and collect each chamber in its own collection tube. Place detached cells on ice immediately.
4. Bring total volume up to 200 μl with ice cold 100 mM EDTA solution, immediately run in a commercial flow cytometer and analyze the data using appropriate analysis software.

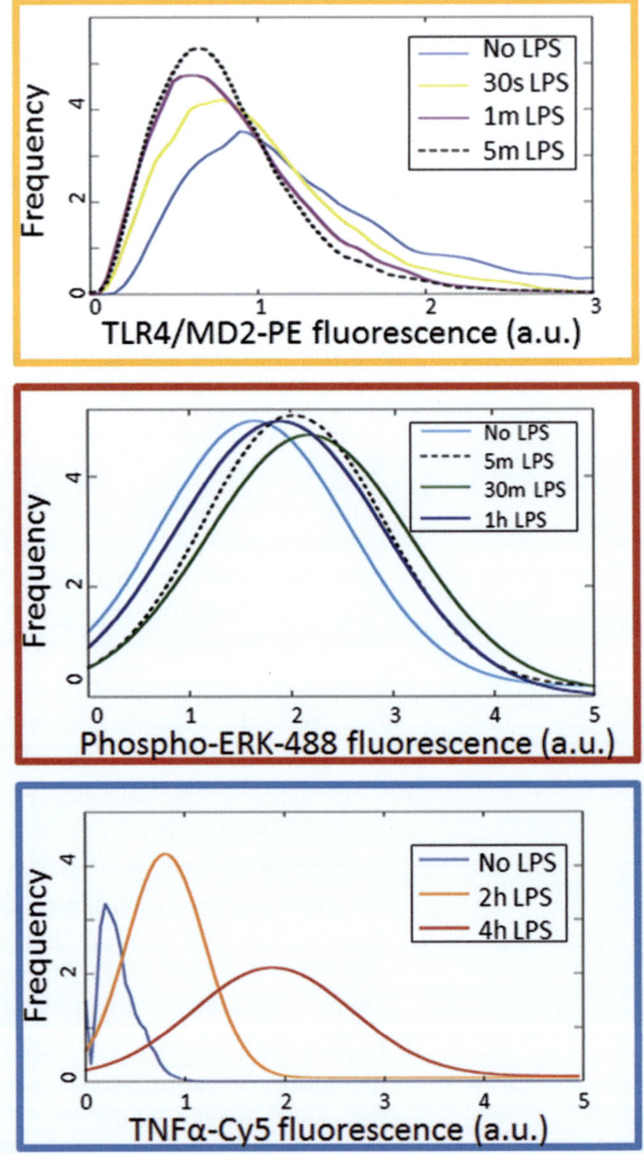

Fig. 4 Flow-cytometric overlays of TLR4 receptor activation, ERK phosphorylation, and TNFα production as measured using the microfluidic platform

4 Notes

1. Sterile techniques should be strictly followed when making cell culture media and culturing mammalian cells. Always clean incubator shelves by autoclaving, and replace the water reservoir in the incubator every week. Cell passaging should always be done in a biosafety cabinet and never open any tube containing cells outside of the biosafety cabinet. Always use

sterile filtered media and PBS, the sterile filter pore size should be equal or smaller than 0.22 μm for mammalian cell culture reagents.

2. We used fused silica chips fabricated by Caliper, now Perkin-Elmer. We have not tested chips fabricated by any other vendor, hence cannot attest to their effectiveness in cellular assays. We recommend using fused silica or glass microfluidic chips, we have not tested other types of chips such as those fabricated from PDMS.
3. We have custom-fabricated microvalves, reagent reservoirs, chip manifold, and computer control unit in our microfabrication facility. Commercial versions of the microvalves, reservoirs, and chip connectors can be purchased from Fluigent (Paris, France) or LabSmith (Livermore, CA, USA), among other microfluidic component vendors. We, however, have not performed experiments with any commercial microfluidic parts.
4. While setting up the chip and fluid control station can be done with minimal engineering knowledge, building a custom micro-flow cytometer requires technical expertise in microfluidics, mechanical and optical engineering. If such engineering capabilities are not available, samples can be detached from the chip and analyzed using a commercial flow cytometer. If using conventional flow cytometer for analysis, skip steps in Subheading 3.8, and follow Subheading 3.9.
5. We have built a 14-valve setup with temperature control to accommodate our 10-chamber microfluidic chip (shown in Fig. 1a). The number of chip inlets, valves, reservoirs, and pressure controllers can be customized for different chip designs.
6. For cell culture in the device, use premixed 5 % CO_2 and 95 % air instead of house nitrogen to drive fluid into the chip to provide the necessary CO_2 level in the platform to maintain proper pH of the media, both in the reservoir as well as inside the chip. Maintaining CO_2 levels at ~5 % is necessary for health of the cells inside the chip beyond 30 min. The hotplate and the premixed 5 % CO_2 make the microfluidic platform a suitable long term cell culture device.
7. BD Collagen I, rat tail, is insoluble at neutral pH. Make 50 ml of 50 μg/ml Rat Tail Collagen in 0.02 N Acetic acid and sterile filter using Steriflip. Divide the collagen solution into 1 ml single-use aliquots and store at 4 °C.
8. All manipulations involving cells should be done in a biosafety cabinet. Use only sterile tubes as reservoirs, and load all cells/reagents into the reservoirs in the biosafety cabinet. Once assembled, the microfluidic chip and manifold should be completely airtight, and can be transported outside of the biosafety cabinet.

9. Golgi-plug has to be administered while cells are alive to entrap nascent cytokine inside the cell. Predetermine which cell holding chamber will be used for cytokine measurement, and treat those chambers with Golgi-plug prior to fixation and permeabilization. Do not treat cells with Golgi-Plug or any other Golgi release inhibitor for longer than 8 h. Prolonged Golgi-Plug treatment is toxic to the cells. We kept Golgi-Plug treatment to less than 4 h in our experiments.
10. With a 10-chamber microfluidic chip, up to ten different assay conditions can be performed in one experiment. Only seven chambers are used in the LPS time-course experiment discussed in this chapter, the rest of the chambers can be empty or used as technical replicates.
11. The pressure settings for creating consistent hydrodynamic focusing between chambers can be determined empirically by using 10 % glycerol solution in PBS in the sheath channel. Load the glycerol solution into p12, and flow towards p4 at 5 psi. Then flow PBS one chamber at a time into p4 at 15 psi and visually determine how much pressure to apply at p12 so that the pinched PBS stream remains in the center of the channel, and at a consistent width between each chamber.

Acknowledgements

The authors would like to thank Matthew Piccini, Ron Renzi, JimVan De Vreugde, and Jim He for integration and automation of microfluidic platform. This work was supported by the following grants: R01 DE020891, funded by the NIDCR (NIH); The MISL Grand Challenge Laboratory Directed Research and Development program at Sandia National Laboratories; P50GM085273 (The New Mexico Spatiotemporal Modeling Center) funded by the NIGMS(NIH). Sandia is a multiprogram laboratory operated by Sandia Corporation, a Lockheed Martin Company, for the US Department of Energy under contract DE-AC04-94AL85000.

References

1. Wu M, Singh AK (2012) Single-cell protein analysis. Curr Opin Biotechnol 23(1):83–88
2. Tario JD Jr et al (2011) Tracking immune cell proliferation and cytotoxic potential using flow cytometry. Methods Mol Biol 699:119–164
3. Gibbs KD Jr et al (2011) Single-cell phospho-specific flow cytometric analysis demonstrates biochemical and functional heterogeneity in human hematopoietic stem and progenitor compartments. Blood 117(16):4226–4233
4. Krutzik PO et al (2011) Phospho flow cytometry methods for the analysis of kinase signaling in cell lines and primary human blood samples. Methods Mol Biol 699:179–202

5. Wu S et al (2010) Development and application of 'phosphoflow' as a tool for immunomonitoring. Expert Rev Vaccines 9(6):631–643
6. Chang RL, Yeh CH, Albitar M (2010) Quantification of intracellular proteins and monitoring therapy using flow cytometry. Curr Drug Targets 11(8):994–999
7. Liu P et al (2011) Microfluidic fluorescence in situ hybridization and flow cytometry (muFlowFISH). Lab Chip 11(16):2673–2679
8. Srivastava N et al (2009) Fully integrated microfluidic platform enabling automated phosphoprofiling of macrophage response. Anal Chem 81(9):3261–3269
9. Wu M et al (2012) Microfluidically-unified cell culture, sample preparation, imaging and flow cytometry for measurement of cell signaling pathways with single cell resolution. Lab Chip 12(16):2823–2831

Chapter 7

Microfluidic Image Cytometry for Single-Cell Phenotyping of Human Pluripotent Stem Cells

Yasumasa Mashimo and Ken-ichiro Kamei

Abstract

A microfluidic human pluripotent stem cell (hPSC) array has been developed for robust and reproducible hPSC culture methods to assess chemically defined serum- and feeder-free culture conditions. This microfluidic platform, combined with image cytometry, enables the systematic analysis of multiple simultaneously detected marker expression in individual cells, for screening of various chemically defined media across hPSC lines, and the study of phenotypic responses.

Key words Microfluidic image cytometry, Human pluripotent stem cells, Microfluidic hPSC array, Chemically defined media, Matrigel, Immunofluorescence, Basic fibroblast growth factor

1 Introduction

High-content screening (HCS) measures biological activity in individual cells or whole organisms following treatment with thousands of agents, such as drug candidates or siRNA, in a high-throughput fashion. This is done by integrating automated and quantitative fluorescent microscopy with multiparametric analyses. HCS has been used as an early drug-discovery platform for defining the functions of genes, proteins and other biomolecules in in vitro cellular models [1, 2]. Furthermore, by integrating HCS and microfluidic technology, it is possible to generate a novel kind of platform called microfluidic image cytometry (MIC) (Fig. 1), which enables cell-based experiments on a sub-microliter scale in the laboratory using readily available materials and equipment [3, 4]. Such a platform would allow integration and automation of cell-sample preparation, cell culture, fluorescent labeling, image acquisition, image processing and analysis. Combined with information management and knowledge mining, the MIC platform will have a significant impact in basic biomedical research and in drug discovery. The MIC platform offers the advantages of reduced

Anup K. Singh and Aarthi Chandrasekaran (eds.), *Single Cell Protein Analysis: Methods and Protocols*, Methods in Molecular Biology, vol. 1346, DOI 10.1007/978-1-4939-2987-0_7, © Springer Science+Business Media New York 2015

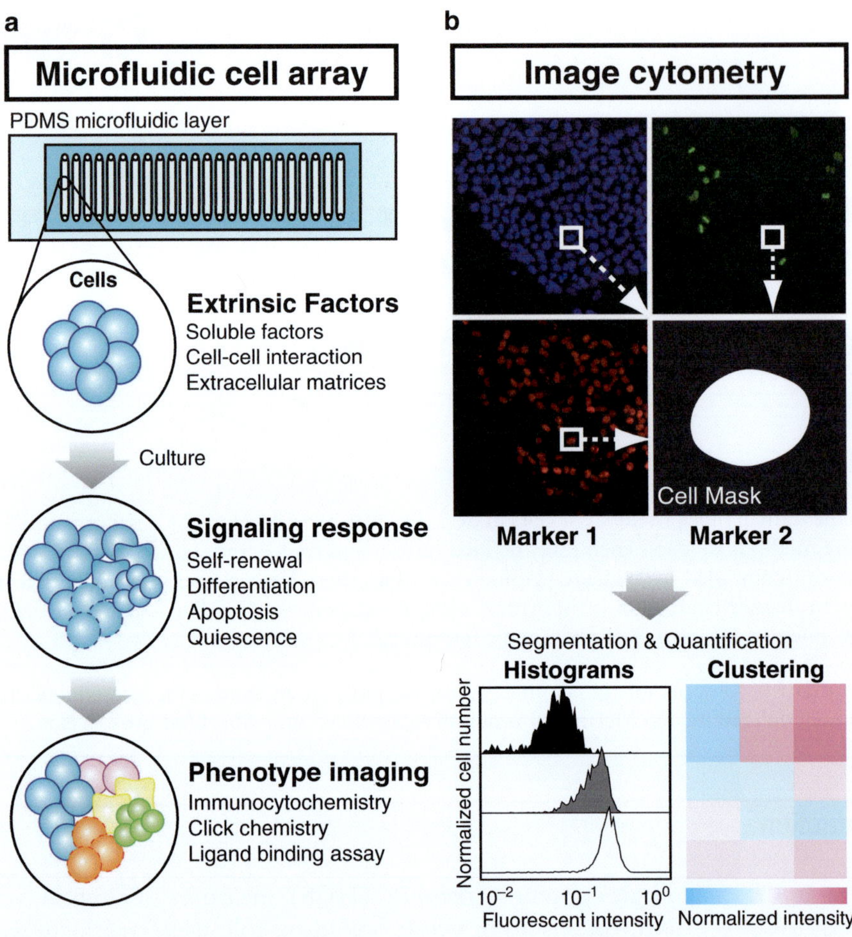

Fig. 1 (**a**) Schematic illustration of a microfluidic hPSC array for hPSC culture and phenotype assay. (**b**) Image cytometry was conducted, followed by segmentation and quantification analysis. Reproduced from ref. [4] with permission from The Royal Society of Chemistry

sample volumes (1/100 to 1/1000) and reagent consumption, and improved reproducibility and sensitivity, resulting in allowing easier handling of larger number of samples than normal HCS.

In this chapter, we present a protocol that applies MIC technology to evaluate the culture conditions for human pluripotent stem cells (hPSCs), i.e., embryonic stem cells (hESCs) and induced pluripotent stem cells (hiPSCs). hPSCs have distinctive characteristics, the capability for unlimited self-renewal and differentiation into most cell types in the body, and have great potential as a source of cells or tissues for drug development and regenerative medicine. To achieve this goal, precise control of hPSC fate (e.g., self-renewal, differentiation, apoptosis, and quiescence) is essential. Until now, many studies have uncovered extrinsic factors that can influence the stability of the state of hPSCs, and contribute to the determination of their fate [5]. These extrinsic factors are the critical components of the stem cell microenvironments ("niches") of the human body, which are generally divided into three categories: (1) soluble factors (i.e., growth factors and bioactive gaseous molecules), (2) insoluble factors (i.e., extracellular matrix proteins and three-dimensional topological cues), and (3) cell-cell interactions (i.e., tight junctions, GAP junctions, and adherens junctions). Furthermore, cell lines respond to the extrinsic factors differently because of the differences between their genetic and epigenetic backgrounds, and these results in variations in the phenotypes. However, few studies have analyzed how the state of an hPSC line responds to a variety of extrinsic factors. Such studies are necessary to determine the optimal microenvironmental components that will guide hPSCs to a desired fate, thus providing a source of a specific cell type. Thus, there is a clear need to develop an MIC platform for performing such complicated analyses, and which will allow the enormous number of interactions between the microenvironment and the hPSC line to be investigated (Fig. 2).

Here, we introduce a procedure to carry out MIC measurements, (1) to fabricate a microfluidic hPSC array, (2) to perform hPSC culture, and (3) to quantitatively analyze cellular phenotypes

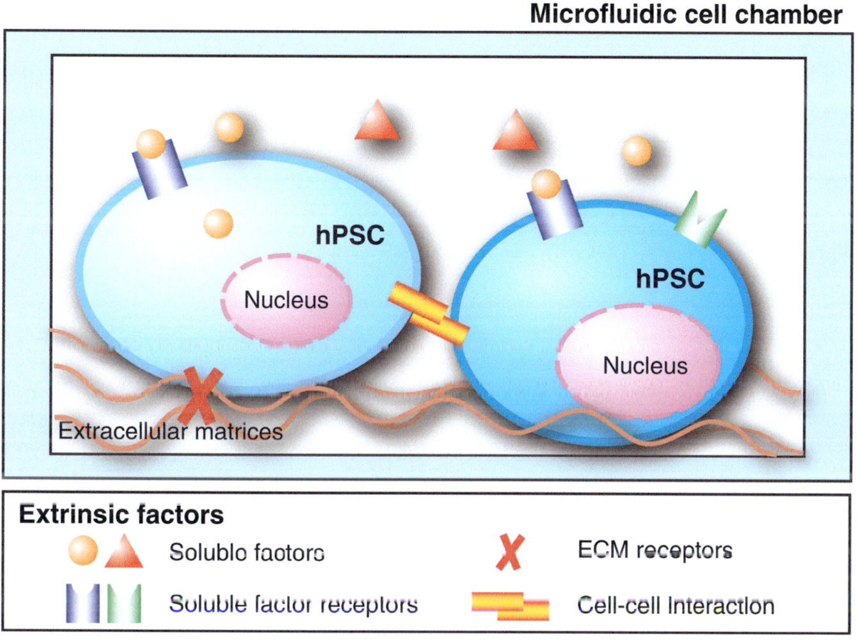

Fig. 2 Extrinsic factors, such as soluble growth factors, cell-cell interactions, and the ECM, play an important role in controlling stem cell fate in each microenvironment. Reproduced from ref. [4] with permission from The Royal Society of Chemistry

on a chip, using small amount of samples, such as cells (300–3000 cells), media and reagents (4 μL) for each microfluidic cell culture chamber. First, the method optimized the concentration of fundamental extrinsic factors (an extracellular matrix (Matrigel)) and a supplemental soluble factor (basic fibroblast growth factor) to maintain pluripotency of hPSCs. Second, we performed a side-by-side comparison of the hPSC phenotypic responses across the available hPSC lines and chemically defined media (CDM) by measuring the expression of pluripotent marker proteins in individual cells. These analyses demonstrate that the microfluidic hPSC array allows examining individual hPSC phenotypes under various conditions, and shows advantages of multiparametric single-cell measurements. These unique features of a microfluidic hPSC array are beneficial for stem cell biology and its application in regenerative medicine and drug screening.

2 Materials

2.1 Fabrication of the Microfluidic Device

1. Polydimethylsiloxane (PDMS) precursors (Sylgard 184, Dow Corning).
2. Glass slide (thickness 1.0 mm, length 76 mm, width 26 mm; Matsunami).
3. UM-113 solder paste mixer/softener (PDMS Mixer, Blaze Technology).
4. Spin coater MS-A150 (Mikasa).
5. Corona discharge generator CFA-500 (Shinko Electric and Instrumentation).

2.2 Human Pluripotent Stem Cell (hPSC) Culture

1. OCT4-enhanced green fluorescent protein (EGFP) knock-in HSF1 cell line (HSF-OCT4-EGFP) [3].
2. hESC lines (HSF1 and H1).
3. hiPSC lines (iPSA1 and iPSB2) [4].
4. Mitomycin-C-treated mouse embryonic fibroblasts (MEFs).
5. hPSC medium [DMEM/F-12 medium supplemented with 20 % Knockout Serum Replacement (KSR), 1 % nonessential amino acids (NEAA), 0.1 mM β-mercaptoethanol, and 10 ng/ml basic fibroblast growth factor (bFGF)].
6. Freezing medium [2 M dimethyl sulfoxide (DMSO), 1 M acetamide, 3 M propylene glycol (1,2-propanediol) in hPSC medium].
7. Dulbecco's Modified Eagle Medium: Nutrient Mixture F-12 media (DMEM/F12).
8. hESC-certified BD Matrigel.

9. Collagenase type IV.
10. StemPro hESC fully defined, serum- and feeder-free medium (SFM) [6].
11. mTeSR-1 medium [7, 8].
12. N2B27 medium [9].

2.3 Cell Manipulation in a Microfluidic Device

1. An electronic pipette.

2.4 On-Chip Immunocytochemistry (ICC)

1. 4 % Paraformaldehyde (Electron Microscope Science).
2. 0.5 % Triton X-100 (Fluka) in PBS.
3. Blocking solution [5 % normal goat serum (Vector Laboratory), 5 % normal donkey serum (Jackson ImmunoResearch Laboratories), 3 % bovine serum albumin (Fraction V, Sigma) and 0.1 % *N*-dodecyl-β-D-maltoside (Pierce)].
4. 0.5 % Tween 20 in PBS (PBS-T).
5. Human specific antibodies for OCT4 (mouse monoclonal IgG; Santa Cruz Biotechnology).
6. Human specific antibodies for SSEA4 (mouse monoclonal IgG; Santa Cruz Biotechnology).
7. Human specific antibodies for TRA-1-60 (mouse monoclonal IgM; Santa Cruz Biotechnology).
8. Human specific antibodies for NANOG (rabbit polyclonal IgG).
9. FITC-conjugated donkey anti-mouse IgG (H+L; Jackson ImmunoResearch).
10. Rhodamine Red-X (RRX)-conjugated goat anti-mouse IgM (Jackson ImmunoResearch).
11. 4′,6-diamidino-2-phenylindole (DAPI).

2.5 Image Acquisition

1. Nikon ECLIPSE Ti inverted fluorescence microscope (Nikon).
2. NIS Elements ver. 4.0 (Nikon) for microscope control and image acquisition.
3. ORCA-R2 charge-coupled device (CCD) camera (Hamamatsu).
4. Intensilight mercury lamp (Nikon).
5. Ti-S-ER motorized stage with encoders (Nikon).
6. Objective lenses (Plan Fluor 10×/0.30 Ph1 DL and Plan Fluor ELWD 20×/0.45 Ph; Nikon).
7. Filters for four fluorescent channels [W1 (UV excitation, DAPI, Nikon), W2 (Blue excitation, GFPHQ, Nikon), W3 (Green excitation, TRITC, Nikon), W4 (Orange excitation, CY5, Nikon)].

2.6 Image Processing and Statistical Analysis

1. CellProfiler (ver. 2.1.0, open software, Broad Institute, http://www.cellprofiler.org/) [10].
2. *R* (ver. 3.1.0, open software, R Core Team, http://www.r-project.org/).
3. Cluster 3.0 (ver. 1.52, open software, University of Tokyo, human genome center, http://bonsai.hgc.jp/~mdehoon/software/cluster/index.html).
4. Java TreeView (developed by Alok Saldanha, http://jtreeview.sourceforge.net/).

3 Methods

Soft lithography is a promising method for the production of miniaturized microfluidic hPSC arrays consisting of a PDMS upper layer mounted on a glass slide. The PDMS precursors are mixed in the specified ratio, and the resulting solution is poured into a mold that has been produced by soft lithography. The PDMS layer and the glass slide must be cleaned in nitrogen gas, and are then bonded together by activating the surfaces with a corona discharge. The surface of the glass can be modified with BD Matrigel, a standard substrate for maintaining hPSCs, which facilitates the culture of hESCs. The use of the array is relatively simple. The hPSC suspension is applied to the array using a handheld pipette, and the cells will attach to the glass surface. Since the volume of a microfluidic array is very small, the medium will be depleted more quickly in a microfluidic array than under standard cell culture conditions. Therefore, it is important to change the medium on a regular basis, ideally once a day. After the predetermined incubation time, the microfluidic array can be immunohistochemically processed and stained for marker proteins. Finally, the array can be read using a high content screening system, and image analysis software can be used to produce single-cell measurements. The resulting measurements can then be summarized using hierarchical clustering analysis, which identifies the similarities among the large number of samples.

3.1 Design and Fabrication of the Microfluidic hPSC Array

The microfluidic hPSC array consists of 24 cell-culture chambers, each with dimensions of 1000 μm (l) × 700 μm (w) × 100 μm (h), with a total volume of 700 nl, and is fabricated using the soft lithography method. Firstly, a mold of the culture chamber is produced, which is used to create a PDMS-based microfluidic component, and this component is directly attached to a glass slide (Fig. 3).

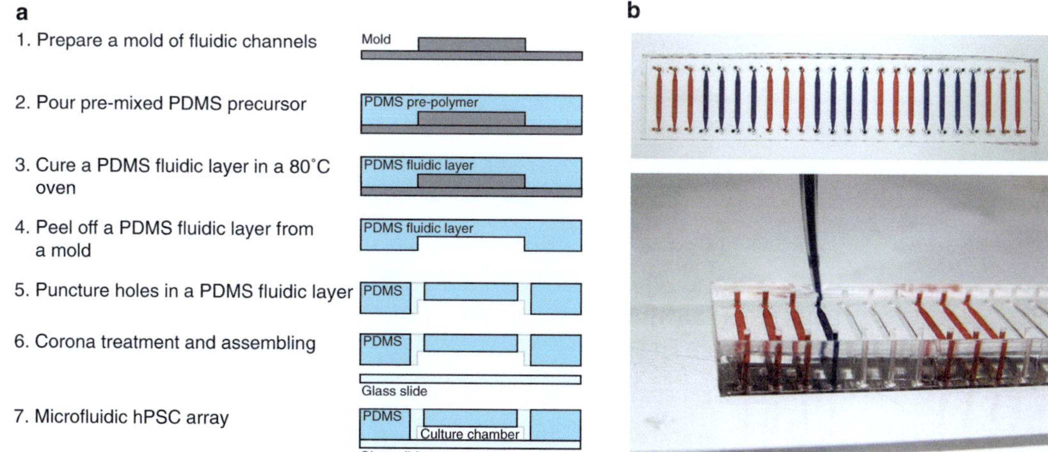

Fig. 3 A microfluidic hPSC array. (**a**) Schematic of the fabrication process of a microfluidic hPSC array using a soft lithography method. (**b**) A digitally controllable manual Matrix™ pipette with 12.5 μl pipette tips, aligned with microfluidic inlets allows precise cell, fluid and reagent delivery. Reproduced from ref. [4] with permission from The Royal Society of Chemistry

1. Spin-coat a 100-μm-thick negative photoresist (SU 8 2100) onto a silicon wafer.
2. Expose a coated negative photoresist with UV light.
3. Solidify a UV-exposed photoresist with SU-8 developer to obtain a mold with rectangular-profiled patterns for the culture chambers.
4. Pour a well-mixed PDMS pre-polymer (Sylgard 184, A:B = 10:1 weight ratio) onto a negative silicon wafer mold with the microchannel patterns.
5. After vacuum degassing and pre-curing at 80 °C for 2 h, peel off a microfluidic component from the mold.
6. Perforate the microfluidic component at the ends of the molded microchannels using a hole puncher that produces holes with a diameter that matches the size of the pipette tip.
7. Trim and clean the microfluidic component.
8. To assemble the microfluidic component and a glass slide, treat their surface with corona discharge.
9. Then, assemble the microfluidic component on a glass slide.
10. Bake in an oven at 80 °C for 16 h.

3.2 Human Pluripotent Stem Cell Culture

hPSCs are cultured using the knockout serum replacement (KSR) that is widely used with bFGF to support feeder-based hPSC culture. The maintenance of stemness in all hPSC lines is confirmed by measuring the expression of pluripotent markers, embryoid body formation assay, teratoma formation assay, and karyotyping.

1. Warm up hPSC culture medium at room temperature (RT) prior to rapid-thawing procedure.
2. Take the vials containing hPSCs in 200 μl of freezing medium from a liquid nitrogen tank.
3. Add 800 μl of hPSC culture medium to the vial under sterile conditions, and pipette (2–3 times).
4. Transfer cells into a sterile 15-ml conical tube containing 9 ml of hPSC culture medium at RT.
5. Centrifuge cells at 200×*g* for 5 min at 4 °C, and aspirate the supernatant.
6. Resuspend cells in 4 ml of hPSC culture medium.
7. Aspirate MEF culture medium from a 35-mm cell culture dish with mitomycin-C-treated MEF feeder cells (*see* **Note 1**).
8. Transfer hPSC suspension to the culture dish with MEFs.
9. Put the dish back into a 37 °C, 5 % CO_2 incubator.
10. Change hPSC culture medium daily until hPSCs reached to approx. 80 % confluence.
11. For passage hPSCs, aspirate the medium from the dish and washed with PBS.
12. Treat cells with 1 ml of collagenase type IV solution (1 mg/ml) in an incubator at 37 °C (*see* **Note 2**).
13. Add hPSC culture medium to harvest detached hPSCs from the dish by pipetting.
14. Transfer hPSCs to a sterile 15-ml conical tube.
15. Wait for 5 min to settle down cells on the tube bottom by gravity at RT.
16. Remove the supernatant and resuspend hPSCs in hPSC culture medium.
17. Pipette the cell suspension to make hPSC aggregates into smaller pieces. Avoid making too small cell aggregates, which might cause massive cell death.
18. Plate hPSCs into a fresh 35-mm cell culture dish.

3.3 Optimization of the Extracellular Matrices and Supplements in a Microfluidic hPSC Array

The culture conditions in the microfluidic hPSC array are optimized by evaluating the concentration of an extracellular matrix (ECM), Matrigel (a standard ECM for maintaining hPSCs), with a supplement, bFGF. Here, the OCT4-enhanced green fluorescent protein (EGFP) knock-in hESC line (HSF1-OCT4-EGFP) can be used to monitor the pluripotency and morphology of the colony during the culturing period.

1. For preparing MEF conditioned medium (CM), culture MEFs in hPSC culture medium and harvest the conditioned medium. Until use, store CM at −20 °C.

2. Prior to use, sterilize a microfluidic hPSC array by exposure to UV light for 15 min.
3. Introduce Matrigel at different concentrations (20, 50 and 100 μg/ml) in DMEM/F12 medium into the cell culture chambers of a microfluidic hPSC array (*see* **Note 3**).
4. Incubate at 4 °C overnight in dark.
5. Rinse the cell culture chambers with DMEM/F12 medium twice at 37 °C.
6. Harvest HSF1-OCT4-EGFP cells on a feeder layer with either mechanically or enzymatically with collagenase type IV, transfer cells with a diameter of 100 μm and of roughly 200 cells into a sterile 15-ml conical tube, and then separate from MEFs by gravity.
7. Remove the supernatant, are then resuspend in CM.
8. Introduce cells into a microfluidic hPSC array, gently. Each cell culture chamber contains four to ten small hPSC colonies.
9. Place the microfluidic hPSC array with cells in a sterilized box with vent holes.
10. Pour 4 ml of sterilized water into the box to prevent dry out.
11. Place the box into an incubator with 5 % CO_2 at 37 °C (*see* **Note 4**).
12. Three hours after cell loading, introduce CM supplemented bFGF at various concentrations (10, 20, and 100 ng/ml) with into the cell culture chambers to optimize hPSC culture conditions in a microfluidic hPSC array.
13. Change medium every 12 h.
14. Monitor the morphology of hPSC colonies and EGFP expression in cells during 5-day culture to determine the optimal hPSC culture condition (Fig. 4) (*see* **Note 5**).

3.4 Cell Culture on a Microfluidic hPSC Array

Small screening and evaluation of the chemically defined medium (CDM) can be performed to demonstrate the MIC protocol for single-cell profiling of hPSCs. Three different chemically defined media are screened: StemPro hESC SFM, mTeSR-1, and N2B27. The process is as follows:

1. Sterilize a microfluidic hPSC array by UV exposure for 15 min.
2. Dilute hESC-certified Matrigel with DMEM/F12 medium, then introduce into the cell culture chambers of a microfluidic hPSC array, and incubated at 4 °C overnight in dark.
3. Prior to introducing cells, rinse the cell culture chambers with DMEM/F12 medium at 37 °C.
4. Harvest HSF1-OCT4-EGFP cells on a feeder layer with either mechanically or enzymatically with collagenase type IV, transfer

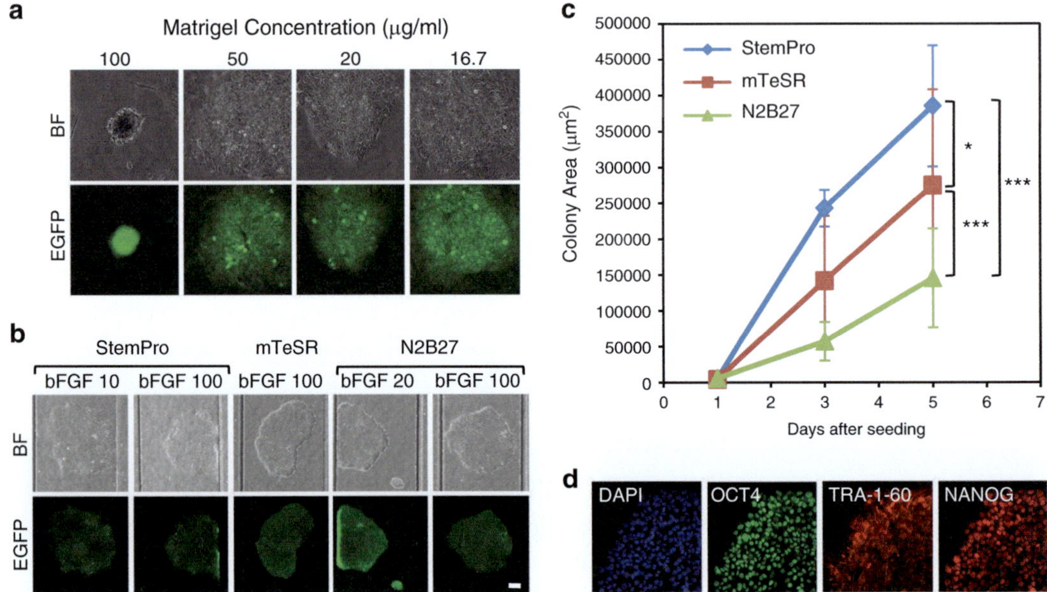

Fig. 4 The establishment of serum- and feeder-free chemically defined hPSC culture conditions for CDM screening. (**a**) Optimization of the Matrigel coating conditions. HSF1-OCT4-EGFP cells were cultured on Matrigel of various concentrations (16.7, 20, 50, or 100 μg/ml). (**b**) The morphology of the colonies of HSF1-OCT4-EGFP cells in different CDM (StemPro, mTeSR, and N2B27) with different bFGF concentrations (10, 20 and 100 ng/ml). HSF1-OCT4-EGFP cells were cultured on glass slides coated with 20 μg/ml Matrigel. (**c**) Quantitative comparison of the growth curves of HSF1-OCT4-EGFP cells cultured in three CDM. *Each dot* represents mean ± S.D. (**p* < 0.05, ****p* < 0.001). (**d**) Fluorescent images of HSF1 cells cultured in StemPro, with nuclei immunostained (DAPI) and pluripotent markers (OCT4, TRA-1-60, NANOG). Reproduced from ref. [4] with permission from The Royal Society of Chemistry

cells with a diameter of 100 μm and of roughly 200 cells into a sterile 15-ml conical tube, and then separate from MEFs by gravity.

5. Remove the supernatant, are then resuspend in hPSC culture medium.
6. Introduce cells into a microfluidic hPSC array, gently. Each cell culture chamber contains four to ten small hPSC colonies.
7. Place the microfluidic hPSC array with cells in an incubator with 5 % CO_2 at 37 °C.
8. Three hours later, introduce the medium of interests (StemPro hESC SFM, mTeSR-1, and N2B27) into the cell culture chambers (Fig. 4). The microfluidic hPSC culture was conducted under static conditions.
9. Place the microfluidic hPSC array with cells in a sterilized box with vent holes.
10. Pour 4 ml of sterilized water into the box to prevent dry out.

11. Place the box into an incubator with 5 % CO_2 at 37 °C.
12. Change medium every 12 h.
13. Monitor the morphology of hPSC colonies and EGFP expression in cells during 5-day culture to determine the optimal hPSC culture condition.

3.5 Immunocytochemistry (ICC) on a Microfluidic hPSC Array

1. To fix hPSC colonies, introduce 2 μl of 4 % paraformaldehyde solution into the cell culture chambers of a microfluidic hPSC array, and then incubate at RT for 15 min.
2. Rinse cells with 4 μl PBS, and then permeabilize with 4 μl 0.5 % Triton X-100 in PBS for 30 min.
3. To prepare a blocking solution, mix 5 % normal goat serum, 5 % normal donkey serum, 3 % bovine serum albumin (Fraction V) and 0.1 % *N*-dodecyl-β-D-maltoside in PBS.
4. Introduce 2 μl of the blocking solution into the cell culture chambers, and then incubate at RT for 1 h.
5. Rinse cells with 4 μl of PBS containing 0.5 % Tween 20 (PBS-T).
6. Incubate hPSC colonies with primary antibody solution contained with 2 μl human-specific antibodies for OCT4 (2 μg/ml mouse monoclonal IgG), SSEA4 (2 μg/ml mouse monoclonal IgG), TRA-1-60 (2 μg/ml mouse monoclonal IgM) or NANOG (2 μg/ml rabbit polyclonal IgG) in a blocking solution at 4 °C for 24 h.
7. Rinse the cell culture chambers with 4 μl blocking solution twice at RT for 15 min.
8. Introduce 2 μl of the respective secondary antibodies [FITC-conjugated donkey anti-mouse IgG (H+L; 3.75 μg/ml), Rhodamine Red-X (RRX)-conjugated goat anti-mouse IgM (0.375 μg/ml), or DyLight649-conjugated donkey anti rabbit IgG (3.75 μg/ml)] into the cell culture chambers.
9. Incubate a microfluidic hPSC array at RT for 1 h.
10. After incubation, rinse cells twice washed with 4 μl PBS-T at RT for 15 min (*see* **Note 6**).
11. Introduce 4 μl of DAPI solution (300 nM) into the cell culture chambers, and keep at RT for 30 min to stain cell nuclei, followed by a rinse in PBS.

3.6 Image Acquisition and Processing

1. Mount the microfluidic hPSC array with stained cells at a defined position on a Nikon ECLIPSE Ti inverted fluorescence microscope.
2. Take images of immunostained cells with the setting listed below.

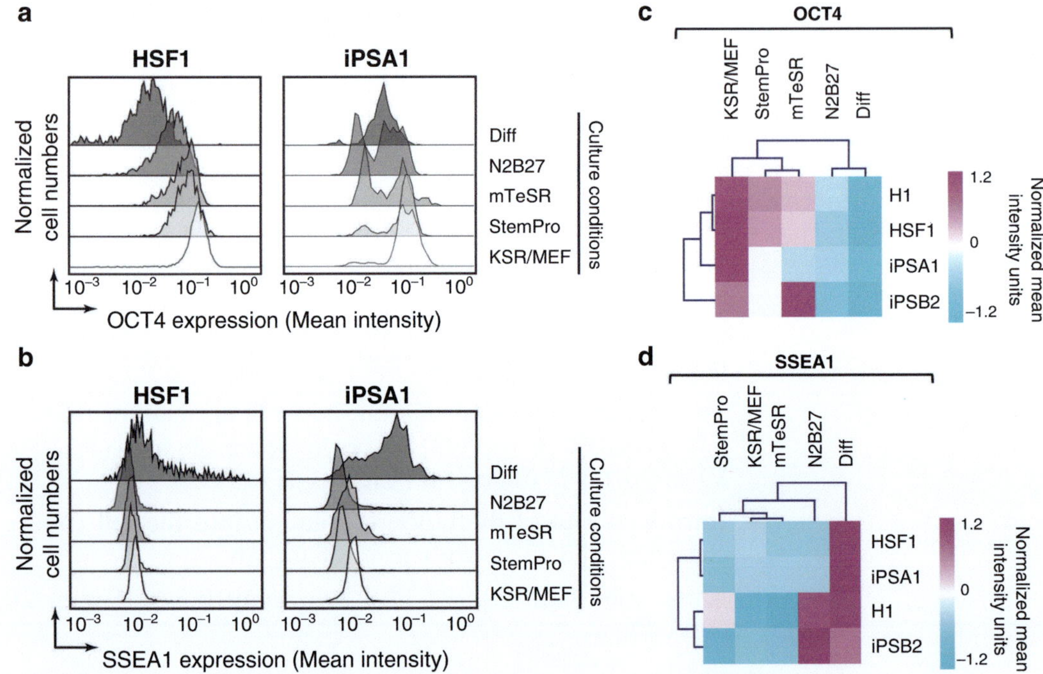

Fig. 5 Evaluation of pluripotent/differentiation marker expression in a microfluidic hPSC array. (**a**, **b**) Histograms of the immunofluorescence from OCT4 and SSEA1 expression in individual HSF1 cells, iPSA1 cells cultured in KSR/mEF, StemPro, mTeSR, and N2B27, and in different states of differentiation. (**c**, **d**) Heat maps based on the quantified expression levels of OCT4 and SSEA1. The protein expression level was normalized among samples of H1, HSF1, iPSA1, and iPSB2 cultured in KSR/mEF, StemPro, mTeSR, N2B27 and with different states of differentiation. These data were analyzed by Euclidean distance hierarchical clustering to categorize similar groups together. *Each row* represents the expression level of the marker in each cell line. Reproduced from ref. [4] with permission from The Royal Society of Chemistry

   (a) Exposure times: 40 msec for W1 (DAPI), 0.5 s for W2 (OCT4), and 0.1 s for W3 (SSEA1).
   (b) Objective lens: 20×.

3. Load a set of fluorescent images into CellProfiler using the "LoadImages" module.
4. Subtract background signals from all images using the "SubtractBackground" module.
5. Use DAPI images to identify nuclei as "primary objects" using the "IdentifyPrimAutomatic" module. Use "Otsu global" thresholding method in the "IdentifyPrimAutomatic" module in the range of 0.8–1.5 of the "threshold correction factor".
6. Use the "IdentifySecondary" module to identify the area (object) of the fluorescence derived from each marker (OCT4 and SSEA1).
7. Measure fluorescence intensities in individual cells with the "MeasureObjectIntensity" module.

8. Export the obtained results to Excel spreadsheet files with the "ExportToExcel" module.
9. Use R to plot a histogram with the quantified fluorescence intensity values of each marker in individual cells (Fig. 5a and b).
10. Use the mean value of the fluorescence intensities of an individual cell is used to generate unsupervised hierarchical clusters and heat maps, by using the average-linkage method based on the Pearson correlation in Cluster and Java TreeView (Fig. 5a and b).

4 Notes

1. The use of mice to obtain MEF cells and its protocol was approved in strict accordance with the recommendations in the guidelines for the care and use of animal models for the Committee on the Ethics of Animal Experiments. Fetal mice were obtained from pregnant E12.5 ICR mice. Then, heads and visceral organs were removed. The remaining tissues were cut and minced, and trypsinized at 37 °C for 30 min to harvest MEFs. MEFs were thawed and cultured in a tissue culture dish in DMEM supplemented with 10 % (v/v) FBS, 100 U/ml penicillin–streptomycin, and 1 % (v/v) NEAA in a humidified incubator at 37 °C with 5 % CO_2. Prior to use MEFs, they were treated with 10 μg/ml of mitomycin C (Wako) for 2 h for inactivation and seeded at 1.8×10^5 cells in a 35-mm culture dish.
2. An electronic pipette is preferable for loading precise volumes of hPSC cell suspension and reagents, with a constant ejection speed in all experiments involving microfluidic devices.
3. hPSC passage is carried out with collagenase IV, because it digests collagen gentler than trypsin to prevent undesired cell death.
4. To prevent drying of the medium in a microfluidic hPSC array, the array needs to be placed in a highly humid environment. Compared with conventional culture settings, the array uses only small amount of culture medium (in μl to nl range). Furthermore PDMS is very porous and cause quick medium evaporation, which would cause cell stress due to medium evaporation.
5. HSF1-OCT4-EGFP colonies maintained their growth for a maximum of 7 days on 20 μg/ml Matrigel. When the bFGF concentration was in the range of 10–100 ng/ml, the growth of undifferentiated hPSCs was sustained in the three chemically defined conditions.

6. These washing steps are important to avoid a nonspecific background signal in each staining and obtain reproducible data. At this step it is possible to confirm the levels of the background signals by using the fluorescence microscope. If the background fluorescence of individual chambers on the array is higher than the average, the washing step can be repeated to reduce the background signal.

Acknowledgements

This work was supported by the Eli and Edythe Broad Center of Regenerative Medicine and Stem Cell Research at the Institute of Molecular Medicine at University of California, Los Angeles. Funding was also provided by the Terumo Life Science Foundation. The WPI-iCeMS is supported by the World Premier International Research Centre Initiative (WPI), the Ministry of Education, Culture, Sports, Science and Technology (MEXT), Japan.

References

1. Giuliano KA et al (1997) High-content screening: a new approach to easing key bottlenecks in the drug discovery process. J Biomol Screen 2(4):249–259
2. Abraham VC, Taylor DL, Haskins JR (2004) High content screening applied to large-scale cell biology. Trends Biotechnol 22(1):15–22
3. Kamei K et al (2009) An integrated microfluidic culture device for quantitative analysis of human embryonic stem cells. Lab Chip 9(4):555–563
4. Kamei K et al (2010) Microfluidic image cytometry for quantitative single-cell profiling of human pluripotent stem cells in chemically defined conditions. Lab Chip 10(9):1113–1119
5. Morrison SJ, Spradling AC (2008) Stem cells and niches: mechanisms that promote stem cell maintenance throughout life. Cell 132(4): 598–611
6. Wang L et al (2007) Self-renewal of human embryonic stem cells requires insulin-like growth factor-1 receptor and ERBB2 receptor signaling. Blood 110(12):4111–4119
7. Ludwig TE et al (2006) Feeder-independent culture of human embryonic stem cells. Nat Methods 3(8):637–646
8. Ludwig TE et al (2006) Derivation of human embryonic stem cells in defined conditions. Nat Biotechnol 24(2):185–187
9. Yao S et al (2006) Long-term self-renewal and directed differentiation of human embryonic stem cells in chemically defined conditions. Proc Natl Acad Sci U S A 103(18): 6907–6912
10. Carpenter AE et al (2006) Cell Profiler: image analysis software for identifying and quantifying cell phenotypes. Genome Biol 7:R100

Chapter 8

Characterizing Phenotypes and Signaling Networks of Single Human Cells by Mass Cytometry

Nalin Leelatian, Kirsten E. Diggins, and Jonathan M. Irish

Abstract

Single cell mass cytometry is revolutionizing our ability to quantitatively characterize cellular biomarkers and signaling networks. Mass cytometry experiments routinely measure 25–35 features of each cell in primary human tissue samples. The relative ease with which a novice user can generate a large amount of high quality data and the novelty of the approach have created a need for example protocols, analysis strategies, and datasets. In this chapter, we present detailed protocols for two mass cytometry experiments designed as training tools. The first protocol describes detection of 26 features on the surface of human peripheral blood mononuclear cells. In the second protocol, a mass cytometry signaling network profile measures 25 node states comprised of five key signaling effectors (AKT, ERK1/2, STAT1, STAT5, and p38) quantified under five conditions (Basal, FLT3L, SCF, IL-3, and IFNγ). This chapter compares manual and unsupervised data analysis approaches, including bivariate plots, heatmaps, histogram overlays, SPADE, and viSNE. Data files in this chapter have been shared online using Cytobank (http://www.cytobank.org/irishlab/).

Key words Single cell biology, Mass cytometry (CyTOF), Human, Immunophenotyping, Signaling network profile, Phospho-specific flow cytometry (phospho-flow)

1 Introduction

Computational tools and instrumentation advances have introduced a new era of single cell systems biology research where it is straightforward to comprehensively characterize all cell types, known and unknown, in primary tissues [1]. Traditional aggregate analysis techniques, such as Western blotting, characterize cellular features with the assumption that the total signal reflects the sum of a homogeneous underlying population of cells. However, even among clonally derived cell lines, biologically meaningful cell-to-cell differences in protein expression and phosphorylation are the rule, not the exception [2]. The presence of multiple cell subpopulations with distinct expression signatures is common in primary samples taken directly from the body, such as healthy tissue and

Anup K. Singh and Aarthi Chandrasekaran (eds.), *Single Cell Protein Analysis: Methods and Protocols*, Methods in Molecular Biology, vol. 1346, DOI 10.1007/978-1-4939-2987-0_8, © Springer Science+Business Media New York 2015

tumor specimens [3, 4]. In the case of blood cancers, cancer cell subsets defined by abnormal signaling are associated with patient clinical outcomes, including overall survival [5, 6]. Single cell tools such as flow cytometry enable high content single cell measurements of cellular identity and functional response [7, 8]. Overlap in the emission spectra of conventional fluorescent probes can create experiment design and data analysis challenges that are particularly a drawback in quantitative single cell comparisons [9]. Moreover, autofluorescence of some tissue types can overlap with fluorophore emission spectra and complicate quantitative analysis.

Mass cytometry employs metal isotope reporters (mass tags) that are not normally found in biological specimens [10]. Typically, mass tags are coupled to an antibody or other target-specific probe so that their abundance in a cell corresponds to the abundance of a target of interest. Intercalator reagents containing iridium or rhodium are routinely used to mark single cells [11]. In addition to making cells detectable as an event to the instrument, comparison of event length and uptake of a cell marking intercalator helps distinguish single cells, cell doublets, and other particles (Fig. 1). Intercalator reagents aim to provide a consistent, strong signal that is minimally variable with experimental conditions. Regular marking of events with an invariant signal helps to identify cell events when parsing mass cytometry data into flow cytometry standard (FCS) format, gating single cells (Fig. 1), and analyzing cells with computational tools (Figs. 2 and 3).

While mass spectrometry avoids fluorescence associated problems, there are aspects of the technology that can be valuable to monitor and test. Mass cytometry issues include (1) impure isotopic mass tags, (2) spillover between closely spaced spectral channels when signal is very abundant (+1 and −1 spillover), (3) variable oxide formation (primarily +16 spillover), and (4) other less common confounding signals not originating from the cells (e.g., barium in buffers, gadolinium contrast agent from patient magnetic resonance imaging). This chapter will not specifically address these aspects of the technique except to say that they can be minimized by following best practices for instrument use, reagent quality control, and experiment design [2, 9].

A key advantage of mass cytometry is the multiplexed detection of many features of each cell. Typical experiments measure approximately 35 features of every cell [10, 12–14], with 42 being state of the art [12]. The theoretical limit on the instrument has not been approached and is likely between 100 and 200 features per cell using the current technology. Mass cytometry therefore, allows single-cell deep profiling of cell identity, phenotype, response, and functional outcome. Relative to microscopy, mass cytometry is high content and high throughput at the single cell level: a typical experiment quantifies 35 features on each of ≥ 100,000 cells from a sample in ~15–20 min. Mass cytometry has

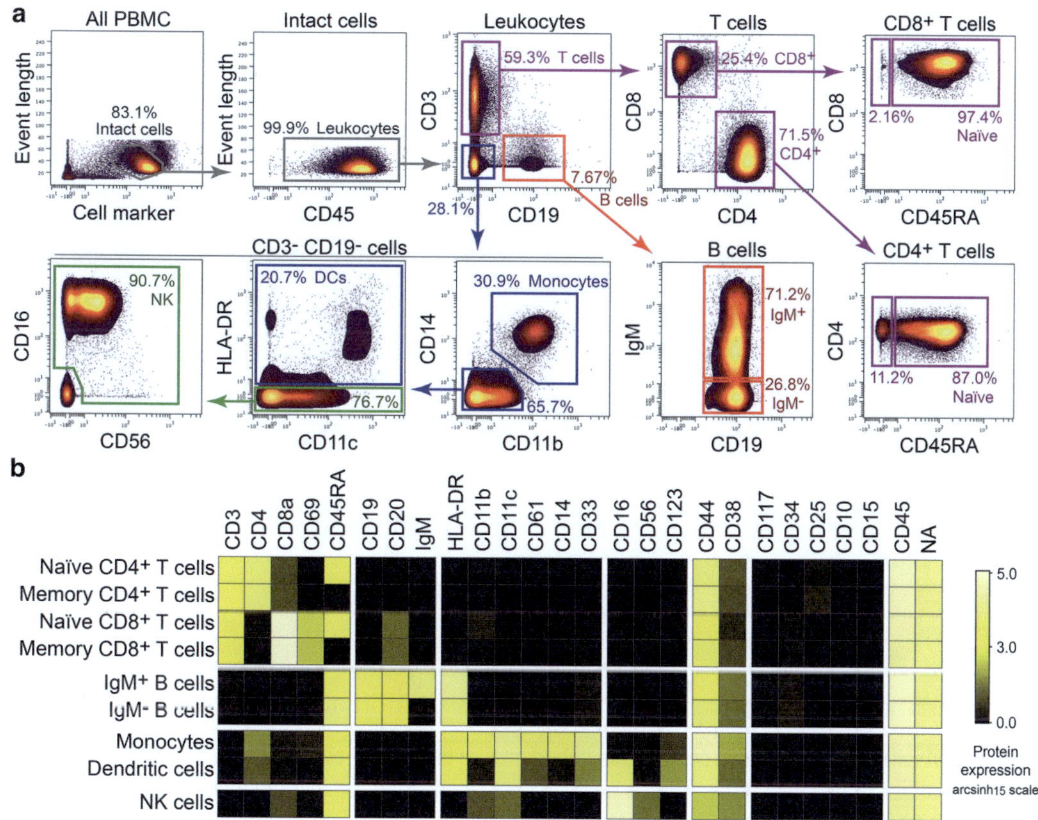

Fig. 1 Phenotyping human PBMC subsets with traditional bivariate gating & heatmap analysis. (**a**) Bivariate plots compare features measured on healthy human PBMCs by mass cytometry. Heat corresponds to proportional cell abundance in a plot region for the population indicated above the plot. Intact cells (*grey gate*) were defined using event length and an iridium based cell marker (Ir-191 intercalator, NA). Among the intact cells, leukocytes were defined as CD45+ events. CD3 and CD19 were then used to identify T cells and B cells, respectively. Subsets of T cells and B cells were identified using additional markers (CD4, CD8, CD45RA, and IgM). The non-T non-B cells ($CD45^+$ $CD3^-$ $CD19^-$) were gated as monocytes, dendritic cells, and natural killer (NK) cell using CD14, CD11b, CD11c, HLA-DR, CD16, and CD56. (**b**) A heatmap compares expression of 27 measured features on the same cells populations shown in (**a**) using a log-like $arcsinh_{15}$ scale. The heat corresponds to the $arcsinh_{15}$ fold difference in median expression for a given marker compared to the table minimum

numerous applications for characterizing the cellular heterogeneity of healthy and diseased tissues and for tracking changes in populations over time in primary tissue samples [2].

Here we present protocols for two mass cytometry experiments: (1) quantifying cell surface biomarkers expressed on healthy human peripheral blood mononuclear cells (PBMCs) and (2) quantifying intracellular signaling network responses in Kasumi 1 cells using phospho-flow [15, 16]. Data from experiments provided in this chapter are available online (http://www.cytobank.org/irishlab). In addition, computational tools are an integral part of analyzing multidimensional datasets. In this chapter we provide examples of multidimensional data visualization. As data analysis

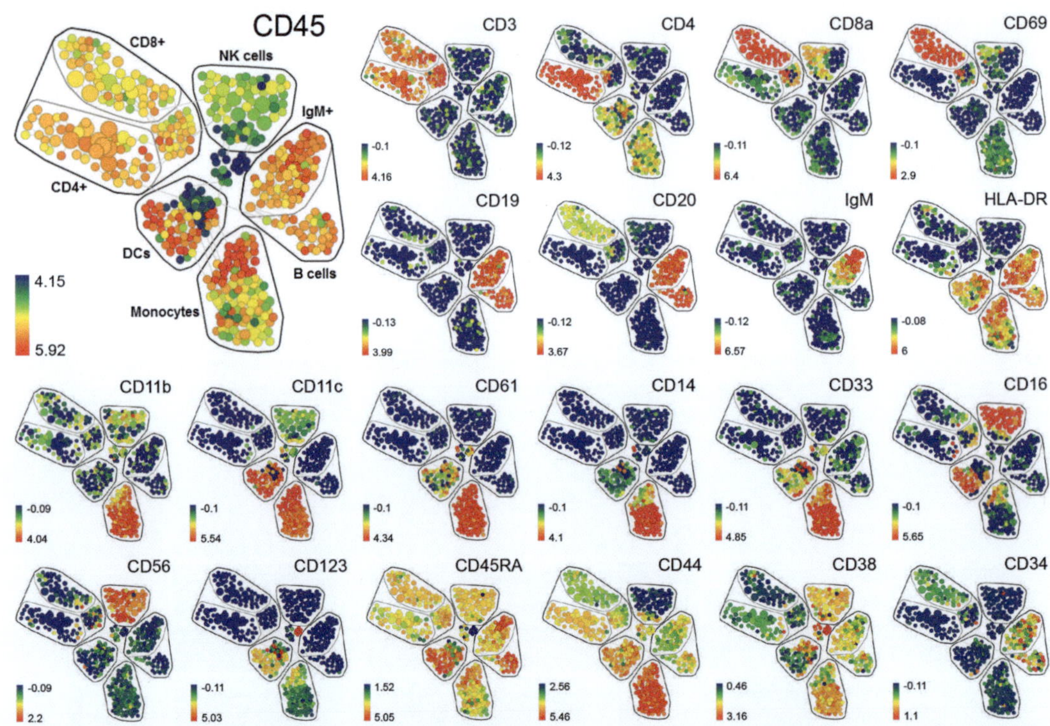

Fig. 2 SPADE clusters PBMCs into populations based on similar marker expression. SPADE plots show clustered populations of healthy human PBMCs (*circles*) connected using a minimum spanning tree. *Each circle* represents a population of cells with a similar phenotype for the 21 markers shown. The size of each circle is proportional to the number of cells in that population. The *heat color for each circle* corresponds to the median expression of the indicated marker on the cells within that circle. Heat corresponds to the arcsinh_{15} fold difference in median expression for a given marker. Note the scale min and max differ for each marker. *Black outlines* termed "bubbles" and associated population labels derive from manual interpretation of cellular identity based on marker expression

can be daunting in 25-dimensional datasets, this chapter compares analysis of the human PBMC cell surface immunophenotyping dataset by three methods: (1) traditional bivariate gating, heatmaps, and histogram overlays [17], (2) Spanning-Tree Progression Analysis of Density-Normalized Events (SPADE [18]), and (3) visualization of t-Stochastic Neighbor Embedding (viSNE [19]).

2 Materials

1. Ficoll-Paque solution.
2. 15 mL and 50 mL conical tubes.
3. Cell culture medium: RPMI 1640 containing 10 % fetal bovine serum (FBS), 100 U/mL penicillin, and 100 μg/mL

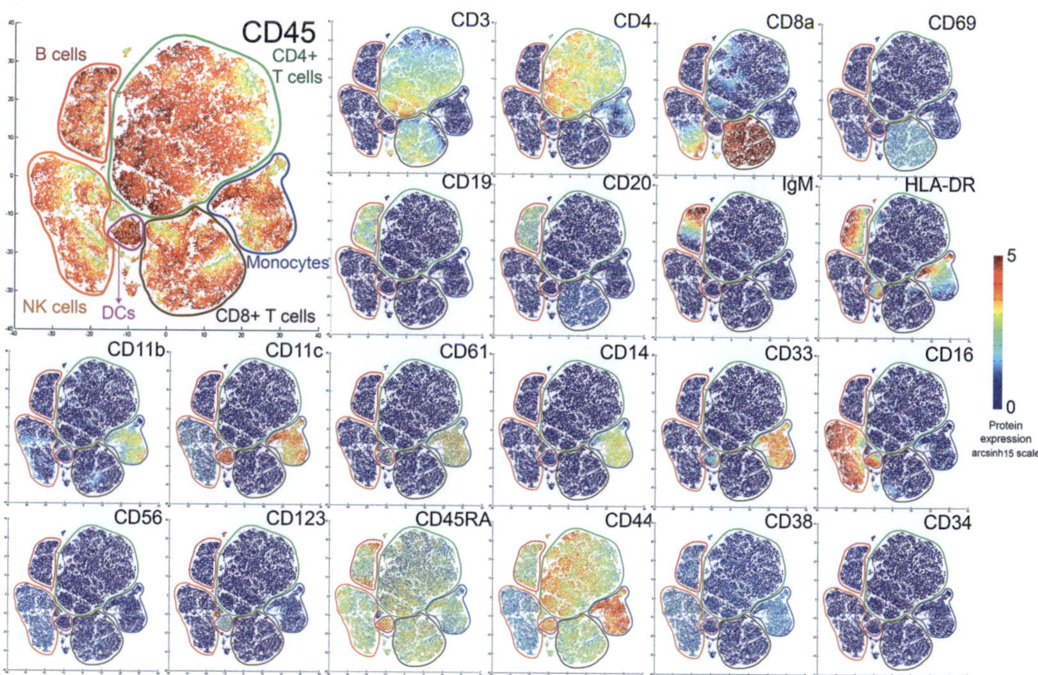

Fig. 3 viSNE arranges cells in a 2D map representing phenotypic similarity. viSNE maps show healthy human PBMCs arranged according to phenotypic similarity for the 21 displayed markers measured by mass cytometry. The axes are unitless dimensions that reflect phenotypic differences. The distance between any two cells on the map corresponds to how similar or different the cells are from each other in high-dimensional space. Heat corresponds to the arcsinh_{15} fold difference in median expression for a given marker. This allows for a global single-cell view of every parameter in every cell. Cell populations can then be identified through various techniques, including automated clustering or manual analysis of well-characterized markers

streptomycin, stored at 4 °C. Heat by immersing in 37 °C water bath for 15–20 min.

4. Freezing medium: 12 % DMSO and 88 % FBS, keep cold on ice.
5. Cryopreservation tubes, 1.8 mL.
6. 12×75 mm round-bottom polystyrene cytometry tubes.
7. Water bath set at 37 °C.
8. Cell culture incubator set at 37 °C with 5 % CO_2.
9. Absolute methanol stored at −20 °C or lower.
10. 1× phosphate buffered saline (PBS).
11. Staining medium: 1 % bovine serum albumin (BSA) in phosphate buffered saline (PBS).
12. Deionized water.

13. Intercalator: 500 μM iridium. Prepare a 50× working solution (12.5 μM) by diluting with PBS.
14. Cytometry tubes with 35-μm cell strainer caps.
15. Trypan blue, prepared as recommended by manufacturer.
16. Hemocytometer.
17. 16 % paraformaldehyde (PFA) aqueous solution.
18. Antibodies: refer to Tables 1 and 2.
19. Stimuli: refer to Table 3.
20. 15–30 mL of human peripheral blood.
21. Kasumi-1 cells (ATCC; CRL-2724).
22. Mass cytometer (CyTOF) (Fluidigm).
23. Data in this manuscript were analyzed using Cytobank (http://www.cytobank.org).

3 Methods

3.1 Immunophenotyping of Surface Markers of Cell Identity in Human PBMCs

Human PBMCs are heterogeneous populations of cells that can be defined and categorized into distinct cell types by different surface marker expression levels. CyTOF allows simultaneous measurement of all the key surface markers of cell identity to enable deep profiling of cell subsets in human PBMCs. The protocol outlined below describes the method for live-cell surface staining of human PBMCs to characterize expression of 25 different cell surface markers.

3.1.1 Preparation and Freezing of PBMCs

1. Collect 15–30 mL of human peripheral blood from each donor in sterile tubes with heparin or EDTA anticoagulant.
2. Transfer the blood to a 50 mL conical tube and add an equal amount of 1× PBS into the same tube and pipet to mix.
3. Add Ficoll-Paque solution into two new 50 mL conical tubes at 12.5 mL per tube (*see* **Note 1**).
4. Slowly overlay 20 mL of diluted blood sample onto the Ficoll-Paque solution in each tube (*see* **Note 2**).
5. Slowly add PBS into each tube to increase the volume in each tube to 50 mL.
6. Centrifuge at 400 × *g* for 30 min at room temperature. Set the deceleration on the centrifuge to the lowest speed to avoid mixing the PBMCs with the rest of the solution.
7. Check for a white ring of PBMCs (buffy coat) right above the red blood cell pellet. Gently aspirate and discard the majority of the plasma above the ring. Avoid disturbing the buffy coat.

Table 1
Antibody panel for cell identity

#	Target	Tag	Clone	μL[a]	Catalog #[b]
1	CD19	142Nd	HIB19	1	3142001B
2	CD117	143Nd	104D2	1	3143001B
3	CD11b	144Nd	ICRF44	1	3144001B
4	CD4	145Nd	RPA-T4	1	3145001B
5	CD8a	146Nd	RPA-T8	1	3162001B
6	CD20	147Sm	2H7	1	3147001B
7	CD34	148Nd	581	1	3148001B
8	CD61	150Nd	VI-PL2	1	3150001B
9	CD123	151Eu	6H6	1	3151001B
10	CD45RA	153Eu	HI100	1	3153001B
11	CD45	154Sm	HI30	1	3154001B
12	CD10	156Gd	HI10a	1	3156001B
13	CD33	158Gd	WM53	1	3158001B
14	CD11c	159Tb	Bu15	1	3159001B
15	CD14	160Gd	M5E2	1	3160001B
16	CD69	162Dy	FN50	1	3162001B
17	CD15	164Dy	W6D3	1	3164001B
18	CD16	165Ho	3G8	1	3165001B
19	CD44	166Er	BJ18	1	3166001B
20	CD38	167Er	HIT2	1	3167001B
21	CD25	169Tm	2A3	1	3169003B
22	CD3	170Er	UCHT1	1	3170001B
23	IgM	172Yb	MHM-88	1	3172004B
24	HLA DR	174Yb	L243	1	3174001B
25	CD56	176Yb	CMSSB	1	3176003B

[a]Microliter of antibody per 100 μL total volume
[b]Source was Fluidigm for all antibodies

8. Gently pipette the buffy coat using a wide bore pipet tip and transfer the buffy coat into a new 15 mL conical tube. Combine buffy coat from all the tubes of the same blood donor.
9. Add PBS to the 15 mL conical tube to increase the volume in each tube to 15 mL.

Table 2
Antibody panel for cell signaling

#	Target	Tag	Clone	μL[a]	Catalog #[b]
1	p-STAT5	150Nd	47	1	3150005A
2	p-AKT	152Sm	D9E	1	3152005A
3	p-STAT1	153Eu	58D6	1	3153003A
4	p-p38	156Gd	D3F9	1	3156002A
5	p-ERK1/2	167Er	D13.14.4E	1	3167005A

[a]Microliter of antibody per 100 μL total volume
[b]Source was Fluidigm for all antibodies

Table 3
Stimulation conditions

Stimulus	Time (min)	50× Dose[a]
FLT3L	15	10 μg/mL
SCF	15	25 μg/mL
IL-3	15	10 μg/mL
IFNγ	15	10 μg/mL

[a]Add 20 μL of 50× to 1 mL cells to yield 1×

10. Centrifuge at 300 × *g* for 10 min at room temperature. Set the deceleration on the centrifuge to high brake speed.
11. Aspirate to discard the supernatant.
12. Resuspend cell pellet in 2 mL of PBS. Count cells with trypan blue (*see* **Note 3**).
13. Add 10 mL of PBS to each tube and centrifuge at 300 × *g* for 10 min at room temperature again. Repeat the wash as needed until the supernatant in relatively clear.
14. Aspirate to discard the supernatant and avoid disturbing the cell pellet.
15. Resuspend the cells in freezing medium to yield a concentration of 10–15 × 10^6 cells/mL.
16. Aliquot cell suspension into cryopreservation tubes at 1 mL per tube.
17. Freeze the cells slowly at the rate of −1 °C/min in −80 °C freezer (*see* **Note 4**). Transfer the cryopreservation tubes into liquid nitrogen the next day for long-term storage.

3.1.2 Live-Cell Surface Staining of PBMCs

1. Remove a cryopreservation tube with PBMCs from liquid nitrogen. Immediately immerse the cryopreservation tube in 37 °C water bath. Remove the cryopreservation tube form the water bath once the cell suspension is completely thawed. This step should not take longer than 1–2 min.
2. Add 10 mL of warm cell culture medium into a 15 mL conical tube. Transfer thawed PBMCs form the cryopreservation tube into the same 15 mL conical tube. Pipet to mix cells with the cell culture medium.
3. Pellet the PBMCs at $300 \times g$ for 5 min at room temperature. Aspirate to discard the supernatant.
4. Add 5 mL of warm cell culture medium and pipet to mix the PBMCs with the medium. Count the cells using trypan blue (*see* **Note 3**) and adjust the volume to yield a concentration of $1–2 \times 10^6$ live cells/mL. Pipet the cells to get a single cell suspension.
5. Label two cytometry tubes as "Intercalator only" and "Surface panel."
6. Aliquot the PBMC suspension into cytometry tubes at 1 mL per tube. Put the tubes with cell suspension into a 37 °C incubator with 5 % CO_2. Rest cells for 15 min before staining.
7. Take the cytometry tubes from the incubator. Pellet the cells at $300 \times g$ for 5 min at room temperature.
8. Check for cell pellets and discard supernatant by decanting (*see* **Note 5**).
9. For the tube labeled "Intercalator only," vigorously vortex the cell pellets in void volume and add 1 mL of ice-cold methanol (stored at −20 °C or lower) to permeabilize the cells. Pipet or briefly vortex to mix (*see* **Note 6**) and store the tube at −20 °C for at least 10 min (*see* **Note 7**).
10. For the tube labeled "Surface panel," resuspend the cell pellets with 75 μL of staining medium (*see* **Note 8**).
11. Transfer 75 μL of cell suspension from the tube labeled "Surface panel" to a new cytometry tube. Add antibodies from Surface Marker Panel (Table 1) to the tube to a final volume of 100 μL. Pipet or briefly vortex the cells with the antibodies to mix. Incubate the cells with the antibodies at room temperature for 30 min.
12. Add 1 mL of 1× PBS and pipet or briefly vortex to mix. Pellet the cells at $300 \times g$ for 5 min at room temperature.
13. Check for cell pellets and discard supernatant by decanting (*see* **Note 5**). Repeat **steps 12** and **13** once.
14. Vigorously vortex the cell pellet in void volume and add 1 mL of ice-cold methanol (stored at −20 °C or lower) to permeabilize the cells, pipet or briefly vortex to mix (*see* **Note 6**) and store the tube at −20 °C with the tube labeled "Intercalator only" for at least 10 min.

3.1.3 Cell Rehydration and DNA Intercalation

1. Take the tubes from −20 °C.
2. Add 1 mL of staining media to each tube, pipet or briefly vortex to mix. Pellet the cells at 800 × *g* for 5 min at room temperature.
3. Check for cell pellets and discard supernatant by decanting (*see* **Note 5**). Repeat **steps 2** and **3** once.
4. Add 200 μL of 1× PBS to each cytometry tube, pipet or briefly vortex to mix.
5. Add 4 μL of 50× Intercalator to each tube. Pipet or briefly vortex to mix the cells. Incubate cells with intercalator for 30 min at room temperature.
6. Add 1 mL of 1× PBS to each tube, pipet or briefly vortex to mix. Pellet the cells at 800 × *g* for 5 min at room temperature.
7. Check for cell pellets and discard supernatant by decanting (*see* **Note 5**). Repeat **steps 6** and **7** once.
8. Add 1 mL of deionized water to each tube, pipet or briefly vortex to mix.
9. Strain the cells through cytometry tubes with 35-μm cell strainer cap. Cells are now ready for CyTOF.

3.2 Phospho-flow Signaling in Kasumi-1 Human AML Cell Line

A phospho-flow signaling experiment can characterize the cellular heterogeneity of intracellular signaling responses to various stimuli among cells within a given cell population. Here, we are using Kasumi-1, a human AML cell line, to demonstrate the concept. This experiment includes addition of various stimuli (Fms-related tyrosine kinase 3 ligand (FLT3L), stem cell factor (SCF), interleukin-3 (IL-3), and interferon-γ (IFNγ)) to the cells to stimulate signaling, followed by fixation and immunostaining with antibodies specific for phosphorylated proteins (Fig. 4).

3.2.1 Cell Culture

1. Culture Kasumi-1 cells according to the manufacturer's recommendation.
2. The day before the experiment, passage cells and resuspend the cells at 0.5–1.0 × 10^6 cells/mL in the recommended cell culture media. Keep the cells in a 37 °C incubator with 5 % CO_2.

3.2.2 Stimulation of Cells

1. Label 6 cytometry tubes as "Intercalator only," "Unstimulated," "FLT3L," "SCF," "IL-3," and "IFNγ."
2. Take Kasumi-1 cells from the incubator and aliquot 1 mL of cell suspension into each cytometry tube.
3. Prepare the stimuli while the cells are resting in the incubator. Make a 50× solutions of each stimulus by diluting them in PBS (Table 3).

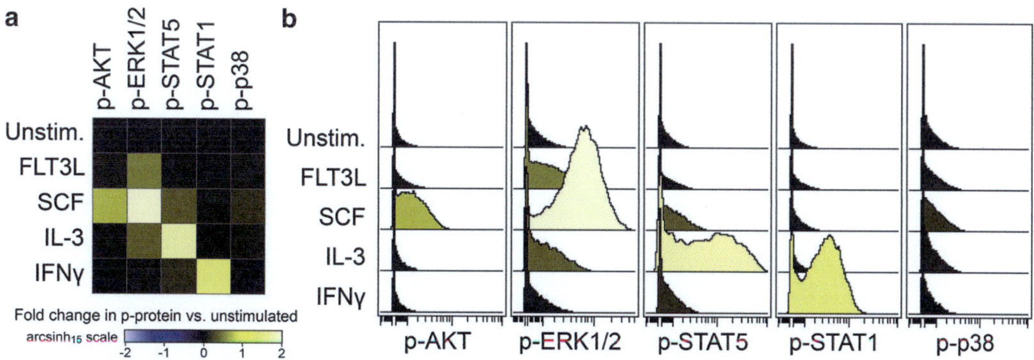

Fig. 4 Mass cytometry phospho-flow analysis of AML cell signaling responses. (**a**) A heatmap compares phospho-protein abundance in Kasumi-1 AML cells following 15 min of stimulation by FLT3L, SCF, IL-3, or IFNγ. *Each row* in the heatmap or histogram overlay corresponds to a stimulation condition and each column corresponds to a phospho-protein (p-AKT, p-ERK1/2, p-STAT5, p-STAT1, or p-p38). Heat corresponds to the arcsinh_{15} fold difference in median expression for a given marker compared to the unstimulated condition (Unstim.). (**b**) The same data as in (**a**) are shown in histogram overlay format. Histogram overlays illustrate the distribution of marker expression within a population and highlight heterogeneity (e.g., p-STAT5 response to IL-3)

4. Add 20 μL of the corresponding 50× stimuli to each cytometry tubes. For tubes labeled "Intercalator only" and "Unstimulated," do not add any stimulus. Place all the tubes back in the incubator and incubate for 15 min (*see* **Notes 9** and **10**).

3.2.3 Fixation and Permeabilization of Cells

1. At the end of the stimulation timepoint, add 100 μL of 16 % PFA to each cytometry tube and vortex the tube to mix (*see* **Note 11**). Allow cells to fix for at least 5 min at room temperature.
2. Wash the cells by adding 1 mL of PBS to each tube and pipet or vortex to mix. Pellet the cell suspension at 800 × *g* for 5 min at room temperature. Check for cell pellets and discard supernatant by decanting (*see* **Note 5**).
3. Vigorously vortex the cell pellets in void volume to resuspend the cells.
4. Add 1 mL of ice-cold methanol to each tube (stored at −20 °C or lower) to permeabilize the cells. Pipet or briefly vortex to mix (*see* **Note 6**) and store the tube at −20 °C for at least 10 min.

3.2.4 Intracellular Phospho-protein Staining and DNA Intercalation of Cells

1. Take the tubes from −20 °C.
2. Add 1 mL of staining media to each tube, pipet or briefly vortex to mix. Pellet the cells at 800 × *g* for 5 min at room temperature.
3. Check for cell pellets and discard supernatant by decanting (*see* **Note 5**). Repeat **steps 2** and **3** once.

4. Add 200 μL of 1× PBS to the tube labeled "Intercalator only." Pipet or briefly vortex to mix and set the tube aside.
5. For the remaining tubes, resuspend the cell pellets with 95 μL of staining media per tube (*see* **Note 8**).
6. Transfer 95 μL of cell suspension to new cytometry tubes. Add intracellular antibodies from the signaling panel (Table 2) to each tube. Pipet or briefly vortex the cells with the antibodies to mix. Incubate the cells with the antibodies at room temperature for 30 min.
7. Add 1 mL of 1× PBS to each tube and pipet or briefly vortex to mix. Pellet the cells at 800 × *g* for 5 min at room temperature.
8. Check for cell pellets and discard supernatant by decanting (*see* **Note 5**). Repeat **steps 7** and **8** once.
9. Add 200 μL of 1× PBS to all the tubes (except the tube labeled "Intercalator only") and pipet or briefly vortex to mix.
10. Add 4 μL of 50× intercalator to all the tubes including the tube labeled "Intercalator only." Pipet or briefly vortex to mix. Incubate the cells with intercalator for 30 min at room temperature.
11. Add 1 mL of 1× PBS to each tube, pipet or briefly vortex to mix. Pellet the cells at 800 × *g* for 5 min at room temperature.
12. Check for cell pellets and discard supernatant by decanting (*see* **Note 5**). Repeat **steps 11** and **12** once.
13. Add 1 mL of deionized water to each tube, pipet or briefly vortex to mix.
14. Strain the cells from each tube through a cytometry tube with 35-μm cell strainer cap. Cells are now ready for CyTOF.

3.3 High-Dimensional Data Analysis

CyTOF simultaneously measures 30+ features per cell, providing a global view of biological events in individual cells. However, such high dimensional data is difficult to analyze effectively with conventional bivariate data visualization and serial manual gating alone. Computational tools help users visualize relationships in high dimensional space and separate rare cells into populations based on distinct, multi-feature phenotypes. As data analysis can be daunting in 25-dimensional datasets, this chapter compares analysis of the human PBMC cell surface immunophenotyping dataset by three methods: (1) traditional bivariate gating (*see* **Note 12**), heatmaps, and histogram overlays (using Cytobank [17], Fig. 1, *see* **Note 13**) Spanning-Tree Progression Analysis of Density-Normalized Events (SPADE [18], Fig. 2, *see* **Note 14**) visualization of t-Stochastic Neighbor Embedding (viSNE [19], Fig. 3, *see* **Note 15**).

4 Notes

1. Avoid spilling Ficoll-Paque solution on the side of the 50 mL conical tubes.
2. Gently overlay blood and PBS mixture on top of the Ficoll-Paque solution. This can be done by adjusting the release speed of the serological pipet aid to minimum. Do not mix the Ficoll-Paque solution and the diluted blood sample since this would minimize PBMCs isolation.
3. To count live cells, mix 10 μL of trypan blue working solution with 10 μL cell suspension and count cells under a microscope using a hemocytometer.
4. To maximally preserve cell viability, cells should be frozen down at a slow rate close to −1 °C/min. This can generally be achieved by placing cryopreservation vials in plastic containers that are immersed in isopropanol prior to transfer to −80 °C. Commercially available products can also be used. Transfer viably cryopreserved cells from −80 °C to liquid nitrogen within 48 h.
5. To avoid losing the cell pellet, decant only once and try to discard as much supernatant as possible. Avoid re-decanting after inverting the cytometry tubes to an upright position since this can loosen the cell pellet or detach it from the bottom of the tube.
6. Cell pellets may clump when methanol is added if they are not resuspended well. To prevent cell clumping, vigorously vortex the pellets in void volume prior to adding ice-cold methanol.
7. Following permeabilization ice cold methanol for 10 min, samples can be maintained long term at −20 °C or −80 °C for months without additional apparent changes to target epitopes.
8. The goal of this step is to resuspend each cell pellet to yield the *total* volume indicated in the protocol and to avoid losing cells in the following steps. Therefore, less staining media than what is indicated in the protocol might be needed depending on the void volume in the tube.
9. Stimuli should be added as soon as the cells are taken out of the incubator to avoid cooling the cells below 37 °C.
10. In general, 15-min stimulations work well for many signaling responses, including the cytokine stimulations used here. However, if this protocol should be adapted for other stimulations, it is crucial to find optimal timepoints and concentrations for each signaling response.
11. Adding 100 μL of 16 % PFA to each tube would yield a final concentration of approximately 1.6 % PFA. Cells must be in single-cell suspension prior to adding PFA to avoid fixing cells in clumps.

12. Bivariate analysis: employs pairwise comparisons of markers to select, or gate, cell populations. Identification of cell subsets has traditionally relied on sequential bivariate gating. Here, we use the PBMC surface immunophenotyping data to demonstrate the concept (Fig. 1) before using the high dimensional population identification tools SPADE and viSNE. In this example, established cell surface markers were used for sequential manual gating of well-characterized human blood mononuclear cell populations, including B cells, T cells, monocytes, dendritic cells, and natural killer (NK) cells.
13. Heatmaps and histogram overlays: are ways to visualize a statistic, such as median expression or fold change, across a large number of populations or experimental variables. Unlike bivariate analysis and viSNE, these tools do not provide a single-cell view of the data. Instead, these tools summarize a key population feature using color. Histogram overlays illustrate the distribution of a single marker for a population and shade the graphic based on the statistic, whereas heatmaps generally just show the color. To demonstrate these data visualization tools, we have used the Kasumi-1 phospho-flow dataset (Fig. 4). Kasumi-1 phospho-flow experiment data are available online (www.cytobank.org/irishlab).
14. SPADE: With high content single cell experiments measuring more than 25 features per cell, sequential bivariate gating can overlook populations of cells with low or unexpected patterns of marker expression. When applied to flow data, SPADE clusters cells based on phenotypic similarities in high-dimensional space. SPADE is a powerful complement to manual analysis approaches. After the initial clustering step, SPADE creates a minimum-spanning tree that illustrates the relationships between cell populations in two dimensions. Each node of the tree is a cluster of cells, and nodes can be colored to reflect the expression intensity of each marker [13]. Here, we subjected the human PBMC immunophenotyping dataset to SPADE analysis to create a minimum-spanning tree (Fig. 2). Known cell subsets in human PBMCs were identified as "bubbles" (Fig. 2) through manual review of biomarker expression on different populations on the tree.
15. viSNE: is a computational tool that projects cells onto a two dimensional map such that the distances between cells in 2D reflect the distance between them in high-dimensional space [14]. Thus, cells that are close together on a viSNE map are phenotypically similar for the markers used to create the map. Users then identify and characterize populations of cells based on the groups formed in the viSNE map. We performed a viSNE analysis of the human PBMCs immunophenotyping dataset (Fig. 3). As with SPADE, expression intensity of each parameter can be visualized as heat intensity on the viSNE

map, enabling identification of known cell subsets in human PBMCs. Importantly, viSNE allowed a global single-cell view of the data. viSNE enabled identification of rare cell subsets and cell populations without supervision or prior knowledge of the expected population distribution.

Acknowledgements

The authors thank P.B. Ferrell for use of Kasumi-1 mass cytometry data. This work was supported by the NIH/NCI R00 CA143231, NIH/NCI R25 CA136440 (K.E.D.), the Vanderbilt International Scholars Program (N.L.), and Vanderbilt-Ingram Cancer Center (VICC NIH/NCI P50 CA68485) pilot grants including a Young Ambassador award.

Conflict of interest disclosure: J.M.I. declares a competing financial interest (cofounder and board member of Cytobank Inc.).

References

1. Irish JM (2014) Beyond the age of cellular discovery. Nat Immunol 15:1095–1097
2. Irish JM, Doxie DB (2014) High-dimensional single-cell cancer biology. Curr Top Microbiol Immunol 377:1–21
3. Marusyk A, Almendro V, Polyak K (2012) Intra-tumour heterogeneity: a looking glass for cancer? Nat Rev Cancer 12(5):323–334
4. Kleppe M, Levine RL (2014) Tumor heterogeneity confounds and illuminates: assessing the implications. Nat Med 20(4):342–344
5. Irish JM et al (2004) Single cell profiling of potentiated phospho-protein networks in cancer cells. Cell 118(2):217–228
6. Irish JM et al (2010) B-cell signaling networks reveal a negative prognostic human lymphoma cell subset that emerges during tumor progression. Proc Natl Acad Sci U S A 107(29): 12747–12754
7. Krutzik PO et al (2004) Analysis of protein phosphorylation and cellular signaling events by flow cytometry: techniques and clinical applications. Clin Immunol 110(3):206–221
8. Irish JM, Kotecha N, Nolan GP (2006) Mapping normal and cancer cell signalling networks: towards single-cell proteomics. Nat Rev Cancer 6(2):146–155
9. Bendall SC et al (2012) A deep profiler's guide to cytometry. Trends Immunol 33(7): 323–332
10. Bendall SC et al (2011) Single-cell mass cytometry of differential immune and drug responses across a human hematopoietic continuum. Science 332(6030):687–696
11. Ornatsky O et al (2010) Highly multiparametric analysis by mass cytometry. J Immunol Methods 361(1-2):1–20
12. Bendall SC et al (2014) Single-cell trajectory detection uncovers progression and regulatory coordination in human B cell development. Cell 157(3):714–725
13. Newell EW et al (2012) Cytometry by time-of-flight shows combinatorial cytokine expression and virus-specific cell niches within a continuum of CD8+ T cell phenotypes. Immunity 36(1):142–152
14. Becher B et al (2014) High-dimensional analysis of the murine myeloid cell system. Nat Immunol 15(12):1181–1189
15. Krutzik PO, Nolan GP (2003) Intracellular phospho-protein staining techniques for flow cytometry: monitoring single cell signaling events. Cytometry A 55(2):61–70
16. Bodenmiller B et al (2012) Multiplexed mass cytometry profiling of cellular states perturbed by small-molecule regulators. Nat Biotechnol 30(9):858–867
17. Kotecha N, Krutzik PO, Irish JM (2010) Web-based analysis and publication of flow cytometry experiments. Curr Protoc Cytom Chapter 10:Unit10.17
18. Qiu P et al (2011) Extracting a cellular hierarchy from high-dimensional cytometry data with SPADE. Nat Biotechnol 29(10):886–891
19. el Amir AD et al (2013) viSNE enables visualization of high dimensional single-cell data and reveals phenotypic heterogeneity of leukemia. Nat Biotechnol 31(6):545–552

Chapter 9

Multiplexed Peptide-MHC Tetramer Staining with Mass Cytometry

Mei Ling Leong and Evan W. Newell

Abstract

Mass cytometry is flow cytometry based on single cell mass spectrometry with decreased crosstalk between channels and an ability to probe >40 parameters per cell, making it well suited for multiplexed assays. Peptide major histocompatibility (MHC) tetramer staining allows direct detection of antigen specific cells and is also amenable to multiplexing/combinatorial approaches. Here we describe methods for multiplexed pMHC-tetramer staining using mass cytometry.

Key words Mass cytometry, Peptide-MHC tetramer, Antigen-specific T cell, Combinatorial tetramer staining

1 Introduction

T cell mediated immune responses are initiated though interactions between T cell receptor and cognate peptide antigen presented in the context of major histocompatibility complex (MHC) expressed by antigen-presenting cells [1]. Antigen-specific T cells can be identified indirectly with functional assays using antigen to stimulate specific cells and elicit a response such as cytokine production or proliferation. Alternatively, antigen-specific cells can be directly identified through the use of recombinantly expressed peptide-MHC proteins linked together as tetramers (or multimers) [2]. Direct identification of antigen-specific cells using pMHC-tetramers has the advantage of allowing for simultaneous analysis of phenotypes without prior perturbation [3] and is amenable for use in conjunction with high dimensional cellular analysis methods such as mass cytometry [4, 5]. Such in-depth and high dimensional analysis of cellular phenotype and function is particularly critical for T cells because of the large number of ways that they can be delineated [6]. Mass cytometry is particularly well suited for this purpose for its ability to assess >40 different parameters per cell with reduced cross-talk between channels [7, 8] compared to traditional flow cytometry.

Anup K. Singh and Aarthi Chandrasekaran (eds.), *Single Cell Protein Analysis: Methods and Protocols*, Methods in Molecular Biology, vol. 1346, DOI 10.1007/978-1-4939-2987-0_9, © Springer Science+Business Media New York 2015

Minimal crosstalk between channels also makes this approach very amenable to multiplexing approaches, not limited to the one described here [9].

Influenced by a wide range of factors, the identities of peptide epitopes targeted by T cells during immune responses can be difficult to predict [10]. Thus, methods for simultaneously assessing a broader range of T cell epitopes are needed and this need is being met through the use of multiplexed (or combinatorial) peptide-MHC tetramer/multimer staining approaches [11–13]. These methods work by tagging each peptide-MHC tetramer/multimer reagent (consisting of MHC loaded with a single peptide sequence) with a unique combination of multiple fluorophores or heavy-metal labels. Thus, the number of unique combinations and the number of T cell antigen specificities that can be probed in the same sample is determined by number of unique combinations that can be put together. For instance, up to 15 different antigen-specificities can be probed with only four different tags if all possible combinations of the four tags are used 2^N-1 [11]. Additional stringency can be achieved if each antigen-specificity is tagged with the same number of different tags (e.g., if each species of peptide-MHC multimer is labeled with two out of a possible five different tags, then up to ten different antigen specificities can be probed simultaneously) [12, 13]. Here we describe the technical details of an approach that uses five different heavy metal labeled-streptavidin conjugates to probe ten different T cell antigen specificities by coding each with a unique combination of two metal isotopes. We demonstrate this approach using cryopreserved peripheral blood mononuclear cells (PBMCs). For this protocol, the cells are thawed, cultured overnight and then subjected to the staining procedure while still alive. In addition to peptide-MHC tetramers, the cells are simultaneously stained with various probes that allow for identification and characterization of viable single cells. As part of this analysis, 34 additional heavy metal-conjugated antibody probes are used, which allows for in-depth phenotypic comparison of antigen-specific cells.

2 Materials

2.1 Streptavidin Conjugation Reagents and Components for Streptavidin Conjugation

1. MAXPAR antibody conjugation kits with DN3 polymer (DVS Sciences).
2. Centrifugal filter 3 kD MWCO, 500 μL capacity, flat-bottom.
3. Centrifugal filter 30 KD MWCO, 500 μl capacity, flat-bottom.
4. Metal chloride stock (If not available form DVS): Dissolve monoisotopic lanthanide chloride stocks in L-buffer (from DVS) to a concentration of 100 mM.
5. Centrifugal filter 0.1 μm pore size, 500 μL capacity.
6. Cysteine-streptavidin protein [13–15].

2.2 Peptide MHC Tetramerization Reagents and Components

1. Centrifugal filter 0.1 μm pore size, 500 μL capacity.
2. 96-well round bottom tissue-culture-treated plate.
3. 96-well PCR plate.
4. UV cross-linker 365 nm (UVP, USA) or equivalent UV source.
5. Biotinylated HLA-A*11:01 MHC protein loaded with UV-cleavable peptide [16] expressed and purified as previously described [2, 17] (Sequence: RVFAJSFIK, where J is the 3-amino-3-(2-nitro)phenyl-propionic acid-linker).
6. Peptide Stock: Dissolve the peptides in DMSO to a stock concentration of 10 mM. The ten peptides and their amino acid sequence are as follows. Peptides synthesized and tested for purity >70 %.

   EBV-EBNA3Bv1 (AVFDRKSDAK).

   EBV-EBNA3Bv2 (IVTDFSVIK).

   HIV-nef (AVDLSHFLK).

   DV-NS3 (GTSGSPIVNR).

   HBV-P (LVVDFSQFSR).

   HBV-C (STLPETTVVRR).

   Flu-NP (SVQRNLPFER).

   CMV-tegument (ATVQGQNLK).

   EBV-LMP2 (SSCSSCPLSK).

   Flu-PB2 (SFSFGGFTFK).
7. Metal conjugated streptavidin prepared in Subheading 3.1.

2.3 Mass Cytometry Cell Staining Reagents and Components

1. Human blood.
2. Complete RPMI: RPMI 1640 supplemented with 10 % heat-inactivated fetal bovine serum (HI-FBS), 55 μM 2-mercapto-ethanol, 100 units/ml penicillin, 100 μg/ml streptomycin, 292 μg/ml L-glutamine, 10 mM HEPES.
3. Phosphate buffered saline, pH 7.4 (PBS). Store at 4 °C.
4. Dasatinib stock: 50 mM dasatinib in DMSO. Store aliquots at −20 °C and avoid repeated freeze/thaw cycles.
5. Washing buffer: 4 % fetal bovine serum (FBS), 2 mM ethylene-diaminetetraacetic acid (EDTA), 0.05 % sodium azide in PBS. Filter through a 0.2 μm bottle filter unit. Store at 4 °C.
6. Cisplatin stock: 100 mM cisplatin in DMSO. Store aliquots at −80 °C and avoid repeated freeze/thaw cycles.
7. Tetramer cocktail: This makes 50 μL of tetramer cocktail, sufficient for staining one well of cells. Prepare 50 μL of tetramer cocktail for each well of cells to be stained. Combine the

tetramers that were prepared in Subheading 3.2 in a 50 kD centrifugal concentrator, accordingly (*see* **Note 1**):

125 μL washing buffer.

10 μM D-biotin.

5 μL EBV-EBNA3Bv1 (AVFDRKSDAK).

5 μL EBV-EBNA3Bv2 (IVTDFSVIK).

5 μL HIV-nef (AVDLSHFLK).

5 μL DV-NS3 (GTSGSPIVNR).

5 μL HBV-P (LVVDFSQFSR).

5 μL HBV-C (STLPETTVVRR).

5 μL Flu-NP (SVQRNLPFER).

5 μL CMV-tegument (ATVQGQNLK).

5 μL EBV-LMP2 (SSCSSCPLSK).

5 μL Flu-PB2 (SFSFGGFTFK).

500 nM dasatinib.

Concentrate in a 50 kD MWCO centrifugal filter by centrifuging at 3250 × *g* at 4 °C until the volume in the filter is less than 50 μL. While concentrating, resuspend the cocktail every 5 min by pipetting (*see* **Note 2**). Transfer the cocktail to a 0.1 μm centrifugal filter and top up the volume to 50 μL with washing buffer. Centrifuge at 14,000 × *g* for 5 min at 4 °C.

8. Primary surface antibody cocktail: Prepare 50 μL of primary surface antibody cocktail for each well of cells to be stained. Make this cocktail up in washing buffer, accordingly:

   1:10 dilution FITC-anti-CLA (BioLegend clone: HECA-452).

   1:10 dilution APC-anti-PD-1 (eBioscience clone: eBioJ105).

   2:25 dilution PE-anti-GDTCR (Invitrogen clone: 5A6.E9).

   500 nM dasatinib (*see* **Note 3**). Filter through a 0.1 μm centrifugal filter (*see* **Note 4**).

9. Secondary surface antibody cocktail: The secondary surface antibody cocktail is a mixture of the metal conjugated antibodies purchased in ready to use pre-conjugated form or prepared using MAXPAR conjugation kits used as suggested. Prepare 70 μL of the cocktail for each well of cells to be stained. Make this cocktail up in washing buffer, accordingly:

   1:160 dilution anti-CD14 Qdot 800 (Invitrogen clone:TüK4).

   1 μg/ml Ln-115 conjugated anti-CD57 (BioLegend clone: HCD57).

   7 μg/ml La-139 conjugated anti-CD45 (BioLegend clone: HI30).

10 μg/ml Pr-141 conjugated anti-FitC (BioLegend clone: FIT-22).

8 μg/ml Nd-142 conjugated anti-HLA-DR (BioLegend clone: L243).

5 μg/ml Nd-143 conjugated anti-CD38 (BioLegend clone: HIT2).

10 μg/ml Nd-144 conjugated anti-CD16 (BioLegend clone: 3G8).

5 μg/ml Nd-145 conjugated anti-CD62L (BioLegend clone: DREG-56).

3 μg/ml Nd-146 conjugated anti-CD8 (BioLegend clone: SK1).

6 μg/ml Sm-147 conjugated anti-CD45RO (BioLegend clone: UCHL1).

7 μg/ml Nd-148 conjugated anti-CD85j (R&D Systems clone: 292303).

10 μg/ml Eu-151 conjugated anti-CD27 (eBioscience clone: LG.7F9).

7 μg/ml Sm-152 conjugated anti-CD5 (BioLegend clone: UCHT2).

6 μg/ml Eu-153 conjugated anti-CDCD49a (BioLegend clone: TS2/7).

4 μg/ml Eu-154 conjugated anti-CD3 (BioxCell clone: UCHT1).

4 μg/ml Gd-156 conjugated anti-CD19 (BioLegend clone: HIB19).

6 μg/ml Gd-158 conjugated anti-CD56 (BD clone: NCAM16.2).

6 μg/ml Gd-160 conjugated anti-CD28 (BioLegend clone: CD28.2).

5 μg/ml Dy-162 conjugated anti-CD4 (BioLegend clone: SK3).

7 μg/ml Dy-164 conjugated anti-CD95 (BioLegend clone: DX2).

5 μg/ml Ho-165 conjugated anti-TCR Vα7.2 (BioLegend clone: 3C10).

7 μg/ml Er-166 conjugated anti-CXCR5 (BD clone: RF8B2).

5 μg/ml Er-167 conjugated anti-CD103 (eBioscience clone: B-Ly7).

9 μg/ml Er-168 conjugated anti-CCR7 (R&D Systems clone: 150503).

10 μg/ml Er-170 conjugated anti-CD161 (BioLegend clone: HP-3G10).

5 μg/ml Yb-171 conjugated anti-CCR6 (BioLegend clone: G034E3).

10 μg/ml Yb-172 conjugated anti-APC (BioLegend clone: APC003).

5 μg/ml Yb-174 conjugated anti-CD45RA (BD clone: HI100).

5 μg/ml Lu-175 conjugated anti-CD69 (BioLegend clone: FN50).

7 μg/ml Yb-176 conjugated anti-PE (BioLegend clone: PE001).

500 nM dasatinib (*see* **Note 3**). Filter through a 0.1 μm centrifugal filter (*see* **Note 4**).

10. 16 % paraformaldehyde (PFA). Store as single-use aliquots at −80 °C.
11. Permeabilization buffer: 1× permeabilization wash buffer (BioLegend) in water. Store at 4 °C.
12. Intracellular antibody cocktail: The intracellular antibody cocktail is a mixture of the metal conjugated antibodies purchased in ready to use pre-conjugated form or prepared using MAXPAR conjugation kits as suggested. Prepare 70 μL of the cocktail for each well of cells to be stained. Make this cocktail up in washing buffer, accordingly:

3 μg/ml Sm-149 conjugated anti-Granulysin 3 μg/ml (R&D catalog # AF3138).

2 μg/ml Sm-150 conjugated anti-GranzymeB 2 μg/ml (Abcam clone: CLB-GB11).

2 μg/ml Gd-155 conjugated anti-CD152 (BD clone: BN13).

5 μg/ml Yb-173 conjugated anti-Perforin (Abcam clone: B-D48).

Filter through a 0.1 μm centrifugal filter (*see* **Note 4**).

13. DNA Interchelator-Ir (DVS Sciences Inc).
14. Cell counting chambers.
15. 24-well tissue-culture-treated plate.
16. 96 well non-tissue-culture-treated plate.
17. Nylon mesh cell strainer 35 μm pore size.

3 Methods

3.1 Streptavidin Conjugation to Lanthanide Metal

Carry out all procedures on wet ice and perform all centrifugation steps at 14,000 × *g* with a fixed angle rotor for 5 min at 4 °C, unless otherwise specified (*see* **Note 5**).

1. Equilibrate DN3 polymer to room temperature and centrifuge tube for 10 s to ensure that the reagent is pelleted at the bottom of the tube (*see* **Note 6**).

2. Add 95 μL of L-buffer to the tube and dissolve the polymer by vortexing. Centrifuge the tube for 5 s to collect the liquid at the bottom of the tube.
3. Add 5 μL of lanthanide metal solution to the tube (If using metals that are not purchased from DVS, adjust the volume of metal solution and L-buffer accordingly such that the final concentration of metal is 2.5 mM in a 100 μL volume) and mix by vortexing.
4. Incubate for 1 h in a 37 °C water bath.
5. Transfer the metal-polymer to a 3 kD centrifugal concentrator. Wash out the tube by adding 150 μL of L-buffer to the tube and transferring to the concentrator. The volume in the 3 kD concentrator should be about 250 μL.
6. Centrifuge for 30 min and discard the flow through.
7. Add 250 μL of L buffer, centrifuge 30 min and discard the flow through.
8. Add 200 μL of L-buffer, centrifuge 30 min and discard the flow through.
9. Dilute 50 μg of streptavidin (SAv) to 90 μL with W-buffer and add 10 μL of 1 M HEPES (*see* **Note 7**).
10. Filter through a 0.1 μM centrifugal filter.
11. Transfer the flow through to the 3 kD concentrator containing the metal-polymer by pipetting.
12. Mix by tapping the side of the concentrator.
13. Centrifuge for 1 h 30 min (*see* **Note 8**).
14. Seal the concentrator using a sealing film (e.g., Parafilm) to prevent evaporation during the subsequent steps.
15. Incubate for 2 h in a 37 °C water bath.
16. Remove concentrator from water bath and incubate overnight at 4 °C.
17. Add 450 μL of W-buffer, resuspend by pipetting and transfer to a 0.1 μM centrifugal filter.
18. Centrifuge and transfer the flow through to a new 30 kD centrifugal concentrator by pipetting.
19. Centrifuge and discard the flow-through.
20. Add 450 μL of W-buffer, centrifuge and discard the flow-through.
21. Repeat **step 20** another 3 times.
22. Add 100 μL of W-buffer and resuspend the metal conjugated streptavidin by pipetting. Transfer to a new 1.5 ml microfuge tube.

23. Measure the absorbance at 280 nm using a NanoDrop™ (*see* **Note 9**).
24. Dilute to 200 μg/ml with W-buffer and store at 4 °C (*see* **Note 10**).

3.2 Peptide MHC Class I Tetramer Preparation

The measurements in the protocol given here will make enough tetramer to perform cytometry staining for 11 donor samples. There will be a total of ten tetramer specificities that are combinatorially labeled with two out of a total of five metal conjugated streptavidin. Carry out all procedures on wet ice, unless otherwise stated.

1. Synthesize, biotinylate and purify HLA-A*11:01 MHC complex according to the established protocols [2, 17].
2. Dilute biotinylated HLA-A11:01 MHC complex to 0.1 mg/ml with PBS. Set up one exchange reaction for each of the ten desired peptides. Since the reaction volume is 50 μL, 550 μL of diluted A11:01 will suffice (*see* **Note 11**).
3. Filter through a 0.1 μm centrifugal filter.
4. To each well of a 96-well round bottom plate, add 50 μL of diluted A11:01 and 0.25 μL of the desired peptide (10 mM DMSO stock) (*see* **Note 12**). Mix, by pipetting, immediately on addition of the peptide. Centrifuge the plate at 525 × *g* for 2 min at 4 °C to remove any bubbles.
5. Place plate on wet ice and remove the lid of the plate. UV irradiate at 365 nm for 5 min (*see* **Note 13**).
6. Rotate the plate 90° clockwise and UV irradiate for another 5 min (~120,000 μJ total UV exposure) (*see* **Note 14**).
7. Prepare streptavidin mixtures for combinatorial tetramer staining by mixing two differentially metal-conjugated streptavidin (from Subheading 3.1) in equimolar proportions. For instance, to make a streptavidin 157–159 combination, combine 5 μL of Gd-157 conjugated SAv with 5 μL of Tb-159 conjugated SAv. Make all the ten possible combinations with the five differentially conjugated SAv (5-choose-2) (*see* **Note 15**).
8. Transfer 50 μL of exchange reaction to a PCR plate. Add 1.9 μL of streptavidin mixture, mix by pipetting and incubate for 10 min at room temperature.
9. Add another 1.9 μL of streptavidin mixture, mix by pipetting and incubate for 10 min at room temperature.
10. Repeat **step 9** another two times (*see* **Note 16**).
11. Seal the plate and incubate overnight at 4 °C (*see* **Note 17**).

3.3 Mass Cytometry Cell Staining

The protocol described here will take several days, but there are several stopping points, which will be highlighted in Subheading 4. Carry out all centrifugation steps at 525 ×*g* for 2 min at 4 °C with a swing out rotor, unless otherwise specified.

1. Isolate peripheral blood mononuclear cells (PBMC) from human blood by ficoll gradient and cryopreserve according to established procedures [18, 19].
2. Type the donors for their HLA phenotype status and bank donors who are HLA-A*11:01 positive.
3. Thaw cryopreserved PBMC of HLA-A*11:01 positive donors into 10 ml of complete RPMI in a 15 ml conical tube.
4. Centrifuge at 335 ×*g* for 10 min.
5. Pour off the supernatant and resuspend PBMC in 1.5 ml of media by pipetting.
6. Transfer PBMC to one well of a 24-well TC plate and incubate 37 °C, overnight (*see* **Note 18**).
7. Harvest rested PBMC by transferring from the 24-well plate into 10 ml of PBS in a 15 ml conical tube. Count cells at this step using a counting chamber.
8. Centrifuge at 335 ×*g* for 10 min.
9. Pour off supernatant and resuspend PBMC at a concentration of 60×10^6 PBMC/ml in PBS.
10. Transfer 100 μL of PBMC into each well of a 96-well non-TC round bottom plate (*see* **Note 19**).
11. Add 100 μL of 100 nM dasatinib to a final concentration of 50 nM and mix by pipetting (*see* **Note 20**).
12. Incubate for 30 min at 37 °C (*see* **Note 21**).
13. Fill to 200 μL with washing buffer and centrifuge. Remove the supernatant by flicking the plate over a waste container then dab on a paper towel to remove any residual liquid (*see* **Note 22**).
14. Dilute cisplatin stock to 200 μM in PBS and resuspend PBMC in 100 μL of the diluted cisplatin by pipetting (*see* **Note 23**).
15. Incubate for 10 min on wet ice.
16. Add 100 μL washing buffer, centrifuge and flick plate.
17. Resuspend PBMC in 50 μL of tetramer cocktail by pipetting.
18. Incubate for 1 h at room temperature.
19. Add 50 μL of primary surface antibody cocktail and mix by pipetting.
20. Incubate for 30 min on wet ice.
21. Add 100 μL of washing buffer, centrifuge and flick plate.

22. Add 200 μL of washing buffer, centrifuge and flick plate.
23. Resuspend PBMC in 70 μL of secondary surface antibody cocktail by pipetting.
24. Incubate for 30 min on wet ice.
25. Add 130 μL of washing buffer, centrifuge and flick plate.
26. Add 200 μL of washing buffer, centrifuge and flick plate.
27. Add 200 μL of PBS, centrifuge and flick plate.
28. Dilute PFA stock to 2 % with PBS and resuspend PBMC in 200 μL of 2 % PFA by pipetting (*see* **Note 24**).
29. Seal plate and incubate overnight at 4 °C (*see* **Note 25**).
30. Centrifuge and flick out the 2 % PFA.
31. Add 200 μL of permeabilization buffer and incubate for 10 min at room temperature.
32. Resuspend in 70 μL intracellular antibody cocktail and mix by pipetting.
33. Incubate for 45 min at room temp.
34. Add 130 μL of permeabilization buffer, centrifuge and flick plate.
35. Add 200 μL of washing buffer, centrifuge and flick plate.
36. Dilute PFA stock to 2 % with PBS and add 1:2000 dilution of Ir-interchelator. Resuspend PBMC in 100 μL of this by pipetting (*see* **Note 26**).
37. Incubate for 20 min at room temperature (*see* **Note 27**).
38. Add 100 μL of washing buffer, centrifuge and flick plate.
39. Add 200 μL of washing buffer, centrifuge and flick plate (*see* **Note 28**).
40. Add 200 μL of water, centrifuge and flick plate.
41. Add 200 μL of water, centrifuge and flick plate.
42. Resuspend in 200 μL of water by pipetting and filter through a 35 μm mesh strainer. Count PBMC using a counting chamber.
43. Dilute to 0.5×10^6/ml cells in water for running on CyTOF (*see* **Note 29**).

3.4 Data Analysis and Interpretation

1. Use Fluidigm acquisition software to create .fcs files for subsequent analysis [20].
2. Standard cytometry analysis software packages including FlowJo, Cytobank, and DVS-Cytobank can be used for mass cytometry data. In this case, FlowJo (MAC version 9.7.5) software is used with bioexponential (logicle) transform parameters as previously described [13].

3. It is important to inspect the data to validate that the antibody stains are properly titrated and staining the cells specifically. Knowledge about expected staining patterns is critical for this assessment.
4. To identify antigen specific cells, gating is first performed to isolate live $CD8^+$ T cells (Fig. 1a). Then, manual gating can be use to de-convolve each of the ten T cell antigen specificities (each coded with a unique combination of two out of five possible metals). Here we demonstrated this approach with two HLA-A*11:01+ donors. For donor 1, a population of EBV_EBNA3Bv2-specific cells were identified by co-staining with tetramers loaded with this peptide and labeled with Gd-157 and with tetramers loaded with this same peptide and labeled with Tm-169. Cells that stain with any of the other metal-labeled tetramers are excluded (Fig. 1b). This analysis is repeated for each of the ten specificities probed (Fig. 1c). Frequencies for all ten antigen-specificities are calculated as a percentage of total $CD8^+$ T cells and plotted (Fig. 1d).
5. To compare phenotypic properties of the identified antigen-specific cells, a large number of antibody probes are measured and can be considered. To take all of these into account, high dimensional analysis techniques can be used. In this case linear principal component analysis (PCA) is applied to all $CD8^+$ T cells acquired as previously described [5] (Fig. 2). To annotate the PCA plot, cells are gated based on traditional definitions of naïve, central memory (Tcm), effector memory (Tem) and effector cells (Temra) based on expression of CD45RA and CCR7. In addition, mucosal associated invariant T (MAIT)-like cells can be delineated by high expression of CD161 and are also annotated (Fig. 2a, b). To compare phenotypes of antigen specific cells to the global profiles of $CD8^+$ T cell, contour plots are overlaid on the PCA plots for each antigen-specificity and for each donor (Fig. 2c, *see* **Note 30**).

4 Notes

1. The order that the reagents are added to the filter is important. It is essential that biotin be added before any of the tetramers to ensure that any free biotin sites on the streptavidin are blocked. Mix by pipetting after the addition of each tetramer. Take care not to touch the filter as it may cause the filter membrane to tear.
2. Watch out for precipitation of the tetramers. If lots of precipitation occurs, stop concentrating and consider staining in a volume larger than 50 μL for a longer period.

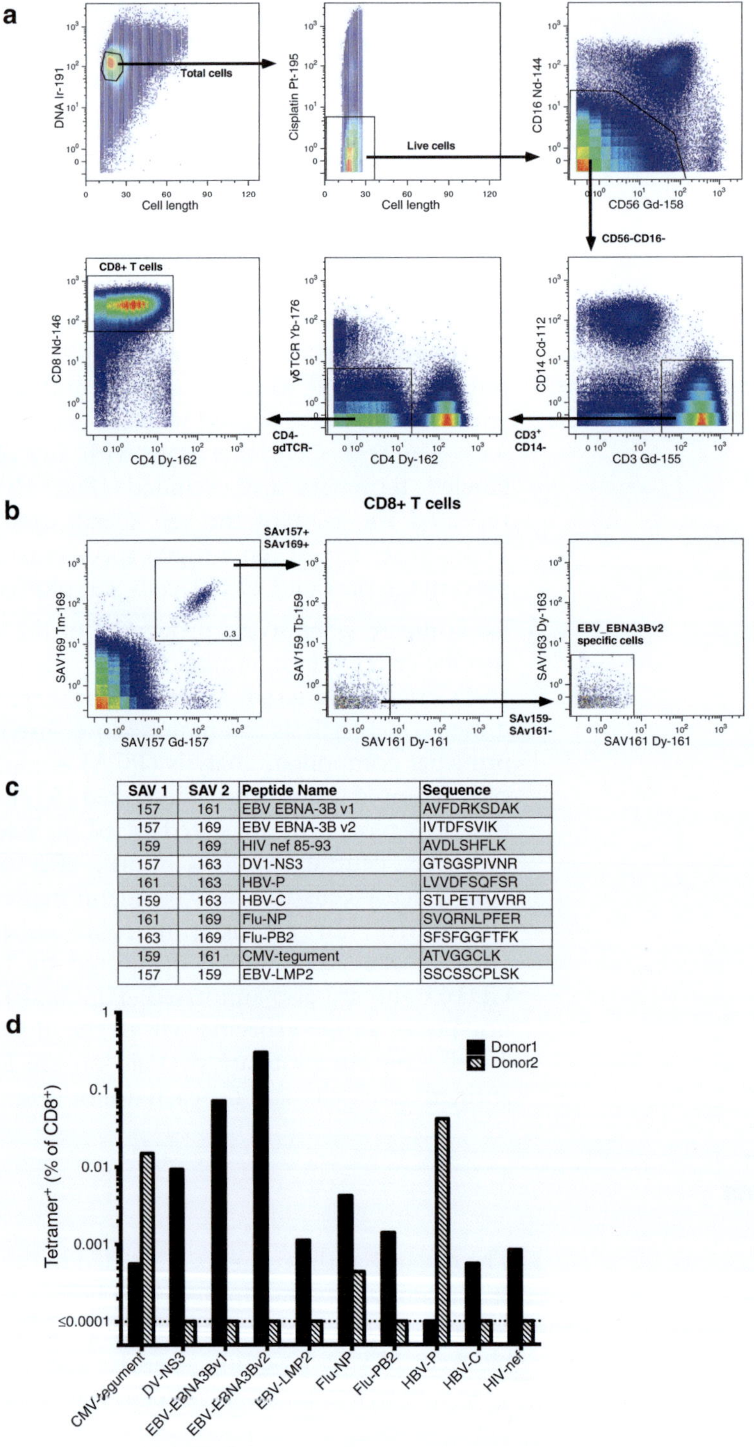

SAV 1	SAV 2	Peptide Name	Sequence
157	161	EBV EBNA-3B v1	AVFDRKSDAK
157	169	EBV EBNA-3B v2	IVTDFSVIK
159	169	HIV nef 85-93	AVDLSHFLK
157	163	DV1-NS3	GTSGSPIVNR
161	163	HBV-P	LVVDFSQFSR
159	163	HBV-C	STLPETTVVRR
161	169	Flu-NP	SVQRNLPFER
163	169	Flu-PB2	SFSFGGFTFK
159	161	CMV-tegument	ATVGGCLK
157	159	EBV-LMP2	SSCSSCPLSK

Fig. 1 Identifying and quantifying antigen-specific T cells using mass cytometry. (**a**) Gating strategy shown for isolating live CD8+ T cell events. First, cells are gated based on DNA content and "Cell length" which is a measure of the number of scans collected for each event. Then live cells are gated based on lack of cisplatin incorporation. Next, NK cells expressing CD16 or high levels of CD56 are excluded before excluding CD14 expressing monocytes. CD3+ T cells are then identified before removal of CD4+ and γδ-TCR+ cells.

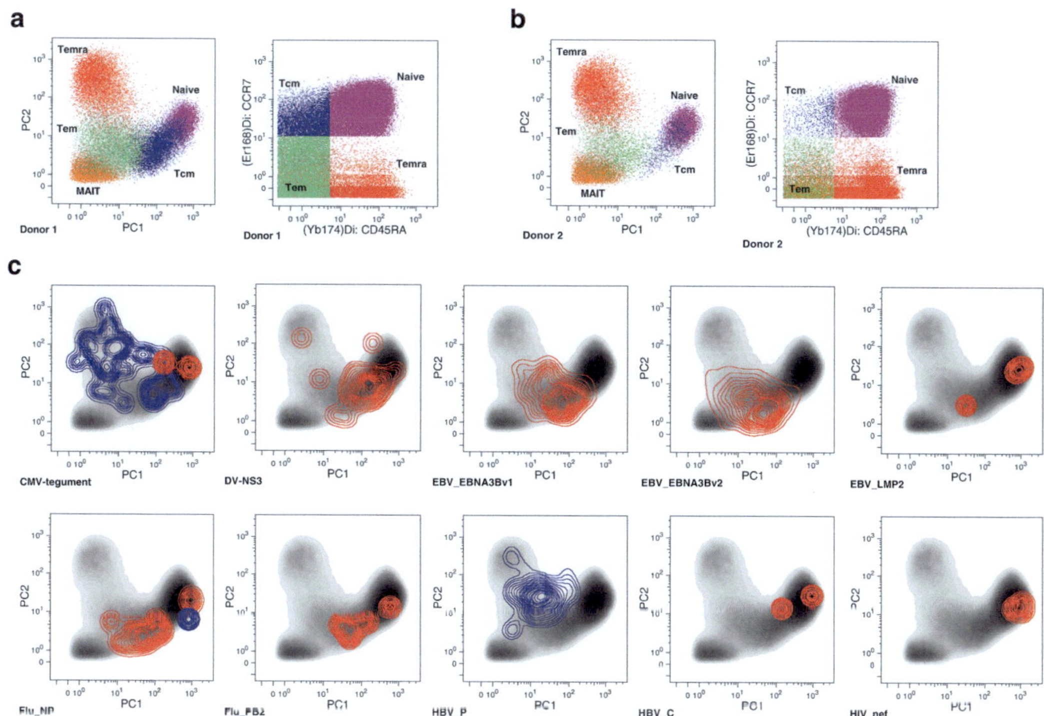

Fig. 2 Phenotypic analysis of antigen-specific cells compared to bulk CD8+ T cells. (**a**, **b**) Coordinates derived from principal component analysis (PCA) are plotted in left panels for donor 1 (**a**) and donor 2 (**b**). Color-coding for these plots are based on CD45RA and CCR7 expression (*right panels*). Orange cells correspond to CD161+ MAIT-like T cells. (**c**) For each antigen-specificity, contour plots of PCA coordinates of antigen specific cells for donors 1 (*red*) and 2 (*blue*) are overlaid on gray-scale density plots of PCA coordinates for each donor's total CD8+ T cells

3. Dasatinib is a reversible protein tyrosine kinase inhibitor that has been shown to improve tetramer staining. Pre-incubating PBMC in 50 nM dasatinib for 30 min has been shown to improve tetramer staining [21].
4. The cocktail must be made fresh and used on the same day.

   The final concentration of each metal-conjugated antibody in the cocktail should be titrated for optimal results. We typically use the antibodies at a range of 1–10 μg/ml.

Fig. 1 (continued) Lastly cells are selected based on high expression of CD8. (**b**) Example tetramer gating for cells specific for one EBV-EBNA-3B-v2 epitope. To identify cells that stain with EBV-EBNA-3B-v2 epitope loaded tetramers (Gd-157 and Tm-169 loaded), cells that are $157^+169^+159^-161^-163^-$ are gated as shown. (**c**) Cells specific for the other nine epitopes using the tabulated codes are gated using a similar strategy (not shown). (**d**) Frequencies expressed as percentages of total CD8+ T cells are plotted for each donor and for each of the ten specificities probed

5. We typically conjugate streptavidin on "bright" channels (atomic masses > 150) as the combinatorial strategy of labeling tetramers means that the cytometry signal strength is divided equally across three channels.
6. Fluidigm's MAXPAR X8 conjugated streptavidin (SAv) has a weaker cytometry staining signal compared to Fluidigm's MAXPAR DN3 conjugated SAv.
7. Streptavidin should be kept in 10 mM tris(2-carboxyethyl) phosphine (TCEP). Once TCEP is removed or diluted out, as it is in this step, the engineered cysteine tails on streptavidin will cross-link. To minimize this, proceed with **steps 10–13** in Subheading 3.1 without delay.
8. There will be a very small amount of liquid left over in the filter unit after centrifugation. This concentration step is necessary for effective conjugation between SAv and polymer.
9. The concentration can be calculated using the derivation of the Beer-Lambert law defined by the equation $A = \varepsilon cl$, where A is the absorbance at 280 nm, ε is the extinction coefficient (0.3 ml/mg cm for streptavidin), c is the concentration in mg/ml, and l is the path length (1 cm; the absorbance on the NanoDrop is automatically corrected for a path length of 1 cm.)
10. Conjugated streptavidin may be stored up to several months. However, the cytometry staining signal will decrease slowly over time.
11. The scale may be adjusted to accommodate the desired number of donor samples. However, it is recommended to perform the peptide exchange reaction within volumes ranging from 50 μL to 110 μL for efficient UV cleavage.
12. The peptide is in huge molar excess to the MHC complex to ensure saturation of the MHC complexes with the desired peptide on cleavage of the UV peptide.
13. If using a different UV source, ensure an equivalent UV exposure of ~120,000 μJ in total. Heat is generated during UV irradiation hence the necessity to keep the wells in contact with wet ice throughout the UV process. If the bottom of the plate becomes warm to touch, consider resting the plate on wet ice for a few minutes before the next round of UV irradiation.
14. The walls of the plate wells may block UV light, hence the plate is rotated in between UV exposure to ensure maximum UV exposure. This is especially important if using a single bulb or tube UV source that will not provide uniform exposure.
15. It is possible to make more of each SAv mixture for future usage and store at 4 °C until required. The mixtures have the same life-span as non-combined SAv, a period of several months.

16. The stepwise addition of SAv maximizes the formation of tetramers as opposed to monomer, dimer, and trimer MHC-SAv complexes. If pressed for time, SAv can be added in three steps instead of four. The final reaction will contain SAv:peptide MHC in the molar ratio of 1:4.
17. An overnight incubation ensures that the exchange reaction proceeds to completion. This means that the desired peptides have entered the binding groove of all free MHC complexes in the solution. Perform the exchange reaction and streptavidin tetramerization 1 day before cytometry staining to achieve best results. Older tetramers may be used for cytometry staining but the signal may be diminished compared to a freshly made batch.
18. Resting the PBMCs overnight allow the recovery of certain cell surface markers, hence the ability to stain and detect these markers by mass cytometry. These markers include, but are not limited to, CD45RA, CCR7, and CD62L.
19. A range of three to six million PBMC may be seeded in each well of the 96-well plate. Top up the volume in each well to 100 μL with PBS.
20. Dasatinib is diluted fresh from a 50 mM frozen stock.
21. Pre-incubating PBMC in 50 nM dasatinib for 30 min has been shown to improve tetramer staining [21].
22. Flick the plate in one hard and abrupt motion to minimize cell loss and remove most of the supernatant. Take care to prevent splashing by lining the waste container with paper towels to absorb any liquid that is flicked out. The number of washes in this protocol may seem excessive, but are essential to ensure that the reagents are washed out before the next one is added.
23. Cisplatin stock must be diluted just prior to use. Cisplatin has a mass of 195 and is used to discriminate live cells from dead cells. Live cells will stain less brightly than dead cells on the 195 mass channel [22].
24. Dilute PFA just prior to use.
25. Stop point: PBMC must be incubated in 2 % PFA for a minimum duration of one night. However, they may be incubated for several days without any adverse effects.
26. Dilute PFA just prior to use. The Ir-interchelator stains DNA with the 191 and 193 isotopes of Iridium.
27. Do not exceed 20 min as the DNA staining may become so bright that it overwhelms the CyTOF detector.
28. Stop point: If not analyzing immediately, retain the washing buffer in this step and pellet the PBMC by centrifuging. Complete **steps 40–43** in Subheading 3.3 on the day that the PBMC are analyzed on the CyTOF. PBMC may be stored for several days in washing buffer.

29. The concentration to dilute the cells to is a balance between the acquisition time and data quality. On the one hand, if the cells are too concentrated, they will not resolve well as single cells and there will be a large number of doublets. Doublets can be gated out when analyzing the data in FlowJo, but there will be a loss of cells. On the other hand, it will take longer to analyze the cells if they are excessively diluted. In saying this, however, it is recommended to dilute the cells to a concentration of less than 0.5×10^6 cells/ml if there are lots of debris, as determined when counting the cells under a light microscope, to improve the data quality.
30. Phenotypic profiles of cells vary according to antigen specificity and rarely overlap with MAIT-like cells. As would be expected, only very low abundance antigen-specific T cells overlay with naïve-like cells.

References

1. Davis MM et al (1998) Ligand recognition by alpha beta T cell receptors. Annu Rev Immunol 16(9099):523–544
2. Altman JD et al (1996) Phenotypic analysis of antigen-specific T lymphocytes. Science 274 (5284):94–96
3. Davis MM, Altman JD, Newell EW (2011) Interrogating the repertoire: broadening the scope of peptide-MHC multimer analysis. Nat Rev Immunol 11(8):551–558
4. Newell EW, Davis MM (2014) Beyond model antigens: high-dimensional methods for the analysis of antigen-specific T cells. Nat Biotechnol 32(2):149–157
5. Newell EW et al (2012) Cytometry by time-of-flight shows combinatorial cytokine expression and virus-specific cell niches within a continuum of CD8+ T cell phenotypes. Immunity 36(1): 142–152
6. Newell EW, Lin W (2014) High-dimensional analysis of human CD8(+) T cell phenotype, function, and antigen specificity. Curr Top Microbiol Immunol 377:61–84
7. Bandura DR et al (2009) Mass cytometry: technique for real time single cell multitarget immunoassay based on inductively coupled plasma time-of-flight mass spectrometry. Anal Chem 18:6813–6822
8. Bendall SC et al (2011) Single-cell mass cytometry of differential immune and drug responses across a human hematopoietic continuum. Science 332(6030):687–696
9. Bodenmiller B et al (2012) Multiplexed mass cytometry profiling of cellular states perturbed by small-molecule regulators. Nat Biotechnol 30(9):858–867
10. Akram A, Inman RD (2012) Immunodominance: a pivotal principle in host response to viral infections. Clin Immunol 143(2):99–115
11. Newell EW et al (2009) Simultaneous detection of many T-cell specificities using combinatorial tetramer staining. Nat Methods 6(7): 497–499
12. Hadrup SR et al (2009) Parallel detection of antigen-specific T-cell responses by multidimensional encoding of MHC multimers. Nat Methods 6(7):520–526
13. Newell EW et al (2013) Combinatorial tetramer staining and mass cytometry analysis facilitate T-cell epitope mapping and characterization. Nat Biotechnol 31(7):623–629
14. Ramachandiran V et al (2007) A robust method for production of MHC tetramers with small molecule fluorophores. J Immunol Methods 319(1-2):13–20
15. Howarth M, Ting AY (2008) Imaging proteins in live mammalian cells with biotin ligase and monovalent streptavidin. Nat Protoc 3(3): 534–545
16. Bakker AH et al (2008) Conditional MHC class I ligands and peptide exchange technology for the human MHC gene products HLA-A1, -A3, -A11, and -B7. Proc Natl Acad Sci U S A 105(10):3825–3830

17. Toebes M et al (2006) Design and use of conditional MHC class I ligands. Nat Med 12(2):246–251
18. Fuss IJ et al (2009) Isolation of whole mononuclear cells from peripheral blood and cord blood. Curr Protoc Immunol Chapter 7:Unit7.1
19. Yokoyama WM, Thompson ML, Ehrhardt RO (2012) Cryopreservation and thawing of cells. Curr Protoc Immunol. Appendix 3:3G
20. Leipold MD, Maecker HT (2012) Mass cytometry: protocol for daily tuning and running cell samples on a CyTOF mass cytometer. J Vis Exp 69, e4398
21. Lissina A et al (2009) Protein kinase inhibitors substantially improve the physical detection of T-cells with peptide-MHC tetramers. J Immunol Methods 340(1):11–24
22. Fienberg HG et al (2012) A platinum-based covalent viability reagent for single-cell mass cytometry. Cytometry A 81(6):467–475

Chapter 10

Imaging and Mapping of Tissue Constituents at the Single-Cell Level Using MALDI MSI and Quantitative Laser Scanning Cytometry

Catherine M. Rawlins, Joseph P. Salisbury, Daniel R. Feldman, Sinan Isim, Nathalie Y.R. Agar, Ed Luther, and Jeffery N. Agar

Abstract

For nearly a century, histopathology involved the laborious morphological analyses of tissues stained with broad-spectrum dyes (i.e., eosin to label proteins). With the advent of antibody-labeling, immunostaining (fluorescein and rhodamine for fluorescent labeling) and immunohistochemistry (DAB and hematoxylin), it became possible to identify specific immunological targets in cells and tissue preparations. Technical advances, including the development of monoclonal antibody technology, led to an ever-increasing palate of dyes, both fluorescent and chromatic. This provides an incredibly rich menu of molecular entities that can be visualized and quantified in cells—giving rise to the new discipline of Molecular Pathology. We describe the evolution of two analytical techniques, cytometry and mass spectrometry, which complement histopathological visual analysis by providing automated, cellular-resolution constituent maps. For the first time, laser scanning cytometry (LSC) and matrix-assisted laser desorption/ionization mass spectrometry imaging (MALDI-MSI) are combined for the analysis of tissue sections. The utility of the marriage of these techniques is demonstrated by analyzing mouse brains with neuron-specific, genetically encoded, fluorescent proteins. We present a workflow that: (1) can be used with or without expensive matrix deposition methods, (2) uses LSC images to reveal the diverse landscape of neural tissue as well as the matrix, and (3) uses a tissue fixation method compatible with a DNA stain. The proposed workflow can be adapted for a variety of sample preparation and matrix deposition methods.

Key words Mass spectrometry, Imaging, Single-cell, Laser scanning cytometry, Fluorescence, In situ protein analysis, Histopathology

1 Introduction

1.1 Flow Cytometry

In recent decades, there has been a paradigm shift in the diagnosis and understanding of disease states. Traditionally, patient samples, either cellular or intact tissues, were stained with dyes, and evaluated by pathologists based on morphological features. The pap-smear screening developed by Papanicolaou had a positive effect in reducing fatalities from cervical cancer, but shortage of technically

Anup K. Singh and Aarthi Chandrasekaran (eds.), *Single Cell Protein Analysis: Methods and Protocols*, Methods in Molecular Biology, vol. 1346, DOI 10.1007/978-1-4939-2987-0_10, © Springer Science+Business Media New York 2015

trained personal necessitated the automation of this procedure [1]. Louis Kamentsky, an image processing scientist at International Business Machines, explored the possibility of using computer-based image processing for screening, but concluded that the computers at the time did not have sufficient processing power for the task [2]. He developed a new model where cells were suspended in a stream of liquid, and flowed through a laser beam illuminated integration zone. As cells flowed through, photomultiplier tubes and photodiode detectors measured the total amount of fluorescence dye and light scattering features on a per cell basis. This was the beginning of flow cytometry, now the gold standard of quantitative cellular analysis [3]. Over the years, key developments included immunostaining, monoclonal antibody technology, green fluorescent proteins (GFP) and related derivatives. Today, research grade cytometers employing multiple lasers and photomultiplier detectors can measure over 30 separate markers per cell. Recently, mass spectrometry (MS)-based CyTOF flow systems have been developed, where antibodies tagged with distinct rare earth elements are used to label cells, and MS is used to quantify their elemental composition [4]. A major limitation of flow cytometry systems is that the cells must be in suspension. This is ideal for blood based cells and bone marrow, but the technology falls short in the analysis of adherent cells and tissue sections.

1.2 Laser Scanning Cytometry

Laser scanning cytometry was developed to address the limitations presented with flow cytometry. The same optical components used in flow cytometers are employed, but with the sample on a fixed substrate. The motion of the interrogating lasers in the *Y* direction, combined with motion of the mechanical stage in the *X*-direction, form two-dimensional arrays of digital photodetector measurements (Q-arrays). These arrays have *X* and *Y* dimensions corresponding to the spatial location within the sample, but unlike camera based images, where there is a one to one correspondence between the pixel size and the sample area. In laser scanning both the diameter of the interrogating laser and the detector field of view are significantly greater than the pixel size, so quantitative data is obtained from the neighborhood of the pixel, in the *X*, *Y*, and *Z* directions. These Q-arrays have the appearance of images, and are segmented to identify events using standard and novel image processing techniques to obtain quantitative data on a per event basis. Q-Arrays obtained from first-generation instruments had poor spatial resolution. If traditional higher resolution images of identified events of interest were desired, the stage was moved to the location of the event, and a camera image was captured. The Deja-Viewer concept allowed for analyzing cells under one set of conditions; for example staining with fluorescent dyes, visualizing the cells, and re-staining them with chromatic dyes. This was the method of choice by cytologists and histopathologists using these systems [5].

As laser scanning technology developed, modifications to the optical system such as reducing the diameter of the interrogating laser beams, and development of Nomarski-like bright field laser imaging, improved the resolution of the Q-arrays to the point where the oculars and cameras of the microscope were no longer needed [6, 7]. A comprehensive software superstructure was developed for both instrument control and data analysis, which extended the capabilities to complex cellular studies [8] as well as analysis of subcellular resolution analysis of tissue microarrays [9].

1.3 Mass Spectrometry Imaging

1.3.1 Evolving MSI Technology

Matrix-assisted laser/desorption ionization (MALDI) is a versatile ionization method that enables mass spectrometry (MS) of a wide variety of analytes, ranging from small molecules to proteins. In MALDI, a small molecule "matrix solution" (such as sinapinic acid described below) is applied to a sample and allowed to co-crystallize. The sample is then ablated by a UV laser, as the matrix heavily absorbs the laser light, leading to ionization of molecules in the sample. The mass-to-charge ratio (m/z) of ions is typically measured with a time of flight (TOF) detector, which permits analysis of large biomolecules such as intact proteins. Application of matrix directly on tissue sections followed by MALDI-TOF-MS repeatedly across a range of grid points enabled mass spectrometry imaging (MSI), whereby the spatial distribution of a given molecule could be mapped based on the intensities of a particular m/z detected across an analyzed region [10].

However, as with cytometry, early single-cell analyses with MALDI mass spectrometry were performed on isolated cells [11]. This was necessary because spatial resolution was limited by the beam diameter of commercially available MALDI lasers and the size of matrix droplets, which were both larger than 100 μm. Amongst the methods used for cell isolation are dissection [12] and immunoaffinity capture [13], although this can lead to loss of spatial information and requires substantially more mechanical disruption of the sample. Alternatively, matrix can be applied only to cells of interest in situ [11, 12], which preserves tissue localization information. However, this can be laborious and results in surrounding cells not being analyzed.

Technological advances in lasers, matrix application methods, and detector technology have enabled cellular-resolution mass spectrometry imaging of protein distributions, in situ. For example, high resolution and high sensitivity MALDI MSI can be achieved with tissue at the subcellular levels down to 5 μm through the use of a Gaussian beam laser [14]. Using a two-dimensional detector to achieve ion microscopy has produced mass spectrometry images of tissue surfaces with a lateral resolution of 4 μm [15]. In addition to advances in instrumentation, advances in sample preparation occurred, including commercially available alternatives for matrix deposition of small spot sizes [16]. It is well established that several aspects of sample preparation, including tissue fixation onto a target, washing, and matrix deposition are keys to good spatial resolution, specificity, and reliability in MSI at the cellular level [16–19].

1.3.2 Combining MALDI with Other Imaging Modalities

MALDI MSI has been combined with light microscopy since its inception, and was first combined with confocal fluorescence microscopy to study the distribution of matrix and analyte throughout a sample after co-crystallization and investigate the impact of this on analyte ionization [20]. Mass spectrometry imaging has been combined with numerous imaging modalities. For example, co-registering MALDI with secondary ion mass spectrometry (SIMS) affords the detection of vitamin E in situ [21]. Three dimensional MSI, achieved by block registration of images obtained from serial sections [22, 23], paired with magnetic resonance imaging (MRI) can enable correlation of the rich and high spatial resolution data provided by MSI with MRI signals obtained in vivo [24]. Raman imaging can also provide information complementary to MSI analysis, helping provide a deeper interpretation of MS data [25]. In general, cross-modality approaches offer many opportunities for deeper interpretation of the large amounts of information obtained in MSI analyses, making this a fertile field of research.

1.3.3 Synergy of Laser Scanning Cytometry (LSC) and MALDI MSI

There are numerous advantages of combining MALDI MSI and LSC within an experimental workflow. In terms of optical configurations, LSC and MALDI are similar, and both technologies provide quantitative data regarding the constituents of cells and tissues. In general, LSC technology detects larger intact structures of the cells and tissues, such as nucleic acids, intact chromatin, cellular proteins, and via immune-staining, specific epitopes and structures on portions of protein molecules. MALDI, on the other hand, detects the (ionized) molecules themselves, often hundreds at a time, including small molecules, lipids, peptides, and proteins. The MALDI ablation process destroys the structural integrity of the tissue, and thus prevents further analysis.

1.4 Experimental Design

Our laboratory is interested in the neurons that are affected by amyotrophic lateral sclerosis (ALS), including layer V cortical motor neurons, and ventral horn spinal cord motor neurons. Transgenic "YFPH" mice with cell-specific overexpression of yellow fluorescent protein (YFP) under control of the Thy1 promoter were used for this study [26]. This protocol describes and compares two tissue fixation methods that preserve spatial localization and fluorescence of YFP during fixation, META-8 (named after the solvents used, *see* below) and ethanol. In addition, META-8 maintains fluorescence throughout the MALDI imaging procedure (Fig. 3). A DNA stain, Hoechst 33342, was incorporated to label cell nuclei, and found to be compatible with both fixation methods (Fig. 5). Two matrix deposition methods were tested: (1)

META-8 which requires no specialized equipment and where matrix solution is deposited across the entire tissue section in one large "pool," and (2) ImagePrep (Bruker Daltonics) that utilizes small droplets generated by acoustic nebulization (Fig. 6). In the course of this study, a new and direct method of using LSC to monitor the efficacy of the manual matrix preparation stages was developed. Using this method, significant differences in homogeneity of the matrix layer and in mass spectral quality were observed, demonstrating the importance of matrix deposition techniques that can evenly coat the tissue with thin matrix layers. In the absence of specialized equipment that help in achieving homogenous matrix deposition, care should be taken in interpreting data and whether changes in molecular distributions observed are actually related to inhomogeneous crystallization yielding a variety of spectra qualities. LSC imaging of the crystallized sample provides one means of assessing this potential.

To demonstrate our current preferred workflows for single-cell protein analysis we analyze specific regions of coronal brain tissue sections in the YFP mice [26] (Fig. 1). To permit detection of all cells within the tissue (i.e., including nonfluorescent cells), we demonstrate a MALDI-MSI compatible method for staining nuclei that does not lead to protein delocalization, an essential aspect to yield high spatial resolution. Localization of fluorescently labeled cells is performed via LSC, followed by MALDI imaging to analyze the defined region. The methods and instruments used promote the detection of proteins, but can be adapted to detect smaller molecules.

2 Materials

2.1 Tissue Slide Preparation

1. Microm HM525 cryostat.
2. Embedded medium for frozen tissue specimens to ensure optimal cutting.
3. HPLC grade water, methanol, ethanol, and acetonitrile.
4. Trifluoroacetic acid.
5. META-8 fixation buffer—prepare solvent containing 8 parts methanol, 8 parts ethanol, 1 part HPLC grade water with TFA, and 1 part acetonitrile (need to switch these so the META acronym makes sense!).
6. Ethanol fixation buffers prepare two solvents, 70 % ethanol for the first step, 90 % ethanol for the second step.
7. WiteOut™ Liquid Paper Pen.
8. Transgenic mice with cell-specific fluorescence overexpressing YFP under control of the Thy1 promoter (Stock No. 003782, The Jackson Laboratories, Bar Harbor, ME, USA) [26]. This

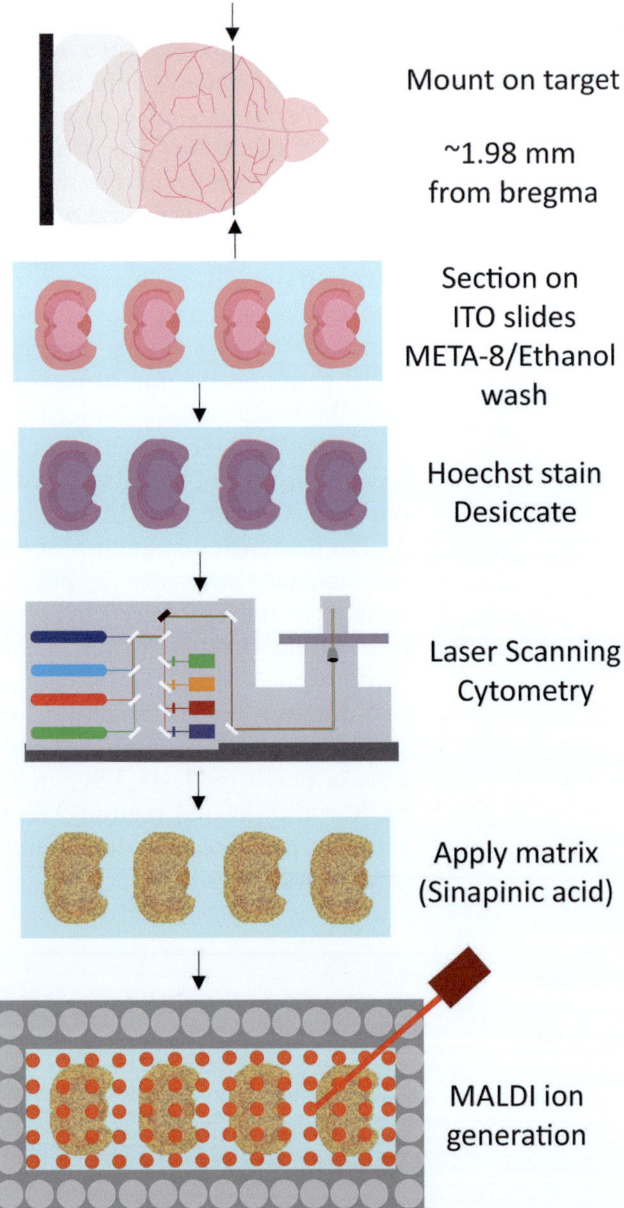

Fig. 1 Outline of the MALDI and LSC imaging of coronal YFP mouse brain tissue. A brain is shown as it should be sectioned and thaw mounted on ITO slides. Prepared tissue slides should be analyzed in a timely fashion, but can be stored desiccated to prevent the accumulation of surface moisture, particularly after removal from the cryostat. The wash step is performed after sectioning. If incorporating the Hoechst stain, it should be combined with the wash/fixative step. Laser scanning cytometry is performed prior to matrix application, although it can be repeated after matrix application if an image of the matrix surface is desired. After matrix application, MALDI-TOF-MS analysis is performed to obtain MS images

method's animal manipulations are approved by the Brandeis University Animal Care and Use Committee and are carried out by the Brandeis University Animal Care Facility in accordance with the federal, local, and institutional guidelines.

9. Indium tin oxide (ITO) coated slides.

2.2 LSC Imaging Sample Preparation

1. Hoechst stain 33342—if used, add to the fixation buffer (META-8 or 70 % ethanol) at a final concentration of 10 μg/mL.
2. iCyte Automated Imaging Cytometer with iGen Software (CompuCyte Corporation, Westwood MA, USA—now marketed by ThorLabs, Inc., Newton, NJ.).

2.3 MALDI Imaging Sample Preparation

1. Just protein calibration standard.
2. Sinapinic acid (SA) ideally recrystallized—Prepare fresh SA at 10 mg/mL in META-8 or 6:4 acetonitrile–water (0.2 % TFA). Sonicate for ~10 min before centrifuging for 10 min at 16,000 ×*g*.
3. ImmEdge™ PEN (Vector Laboratories Inc., Burlingame, CA, USA).
4. ImagePrep (Bruker Daltonics, Billerica, MA, USA).
5. UltrafleXtreme MALDI TOF/TOF with a 1 kHz smart beam laser (Bruker Daltonics, Billerica, MA, USA).
6. flexImaging, flexControl, and flexAnalysis software (Bruker Daltonics, Billerica, MA, USA).
7. SCiLS (SCiLS GmbH, Bremen, Germany).

3 Methods

3.1 Tissue Slide Preparation

1. Embed frozen tissue onto plate using OCT media with the cerebellum facing down. Ideally, stereotaxic techniques are employed. In the tissue used here, the position of bregma and the midline form the basis of an *x*-*y* coordinate system.
2. Slice 14 μm thick coronal sections, ~1.98 mm relative to bregma to better access the motor and sensory neurons [28] (*see* **Note 1**). Thaw mount sections on ITO coated slides by placing your finger on the bottom of the slide under the tissue, as previously described, for <2 s (Fig. 3a, b and e, f) [29]. Use care to have the same orientation and location on multiple slides.
3. ImmEdge™ Pen is used to create a hydrophobic barrier approximately 1 mm away from the edge of the tissue to maintain a constant ratio of matrix solution to tissue volume. Use WiteOut™ liquid paper pen to create a three to four distinct teaching points on the ITO slide as close as possible to high resolution scan region (Fig. 2). It is important to add this

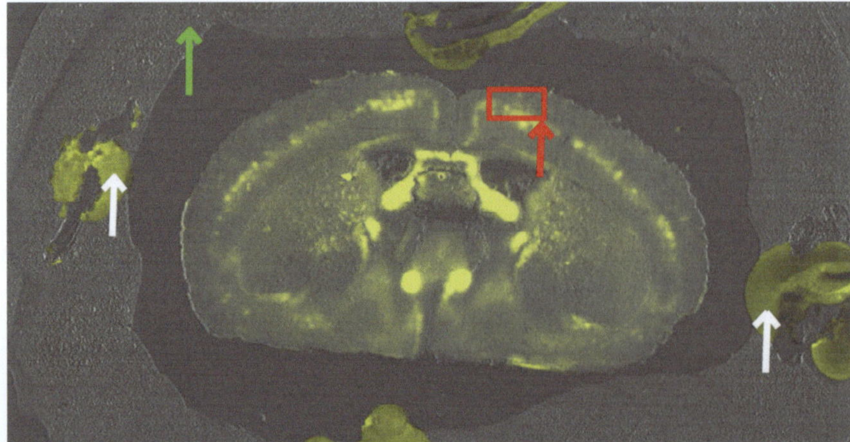

Fig. 2 Laser scan of the slide preparation details necessary for the MALDI MSI. The *green arrow* point to the hydrophobic barrier from the ImmEdge Pen, *white arrows* point to the fiducial markers made with WiteOut™ used for teach points and the *red arrow* points to the area selected for the high resolution scan, which here is also the region we analyze by MSI

first before any LSC scans so that the image obtained from LSC can be used in flexImaging for teaching so that the LSC and MSI images will be co-registered (*see* **Note 2**).

4. Wash and fix tissue using either META-8 (Fig. 3c, d) or 70 % followed by 95 % ethanol wash (Fig. 3g, h) [19]. Since the hydrophobic pen and the teaching points are present pre-tissue washing, the slide should not be submerged in the solvent to wash. Instead, pipette the solvent (~100 μL) directly onto the tissue, and remove using a vacuum and a glass pipette tip until dry. This preserves the tissue and prevents the fiducial markers from washing off (*see* **Note 3**).

5. For the META-8 manual method of matrix application, deposit 100 μL of 10 mg/mL SA, wait for it to dry, and repeat two more times (total of three depositions) [30]. Figure 4 shows LSC scans with varying concentrations of SA, both on and off tissue. This establishes the importance of homogeneity, crystal size, and distribution of matrix.

6. For the ImagePrep matrix deposition method, use the manufacturer's default methods. For this experiment, approximately 80 layers of matrix was added (Fig. 6g, h) (*see* **Note 4**) [16]. It is suggested when using the ImagePrep to coat only half of the slide at a time to achieve more even deposition, covering the half of the slide not receiving matrix at a given time with a coverslip.

3.2 LSC Analysis

1. Add Hoechst stain at a concentration of 10 μg/mL diluted in META-8 or 70–80 % ethanol (*see* **Note 5**). The Hoechst/META-8 fixation not only preserves the tissue on the slide

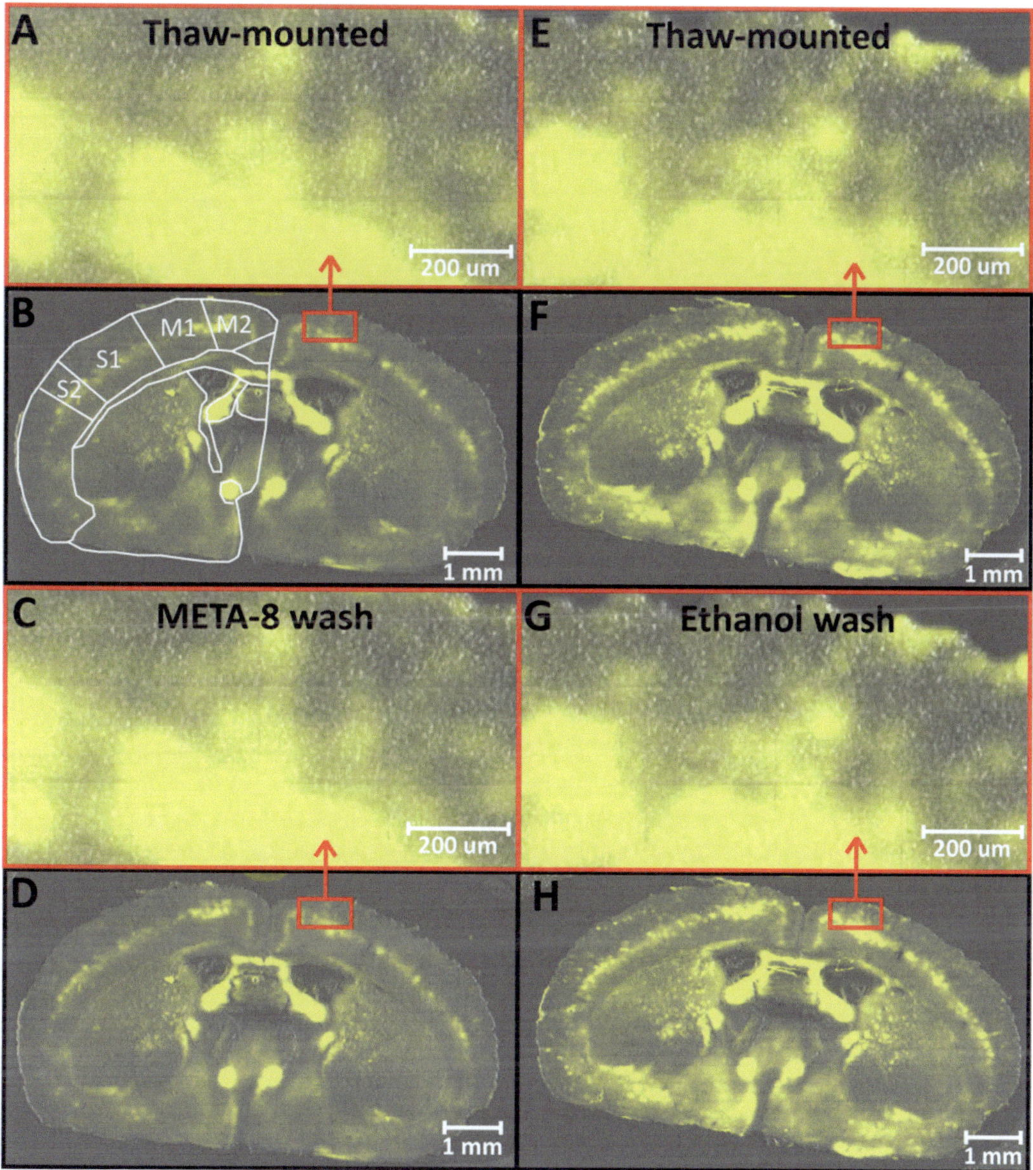

Fig. 3 Laser scanned images of YFP expression on tissue sections before and after fixation. These methods of tissue washing and fixation preserved both fluorescence and protein localization. The regions of the brain, primary motor cortex (M1), secondary motor cortex (M2), primary somatosensory cortex (S1), and secondary somatosensory cortex (S2), are labeled based on a mouse brain atlas with the stereotaxic coordinates of ~0.86 mm relative to bregma. The correlation of the regions of the brain can be done manually but can also be done automatically with specialized software [27]. *Panels* **a**, **b**, **e**, and **f** show the low and high resolution LSC images of the thaw-mounted tissue pre-washing/fixation. The *red boxes* delineate the location of the high resolution scans of the motor cortex. *Panel* **d** shows the low resolution LSC scan of the tissue section (from *panel* **b**) washed with META-8 and the corresponding high resolution scan (**c**). *Panel* **h** shows the low resolution LSC scans of the tissue section (from *panel* **f**) washed with ethanol and the corresponding high resolution scan (**g**)

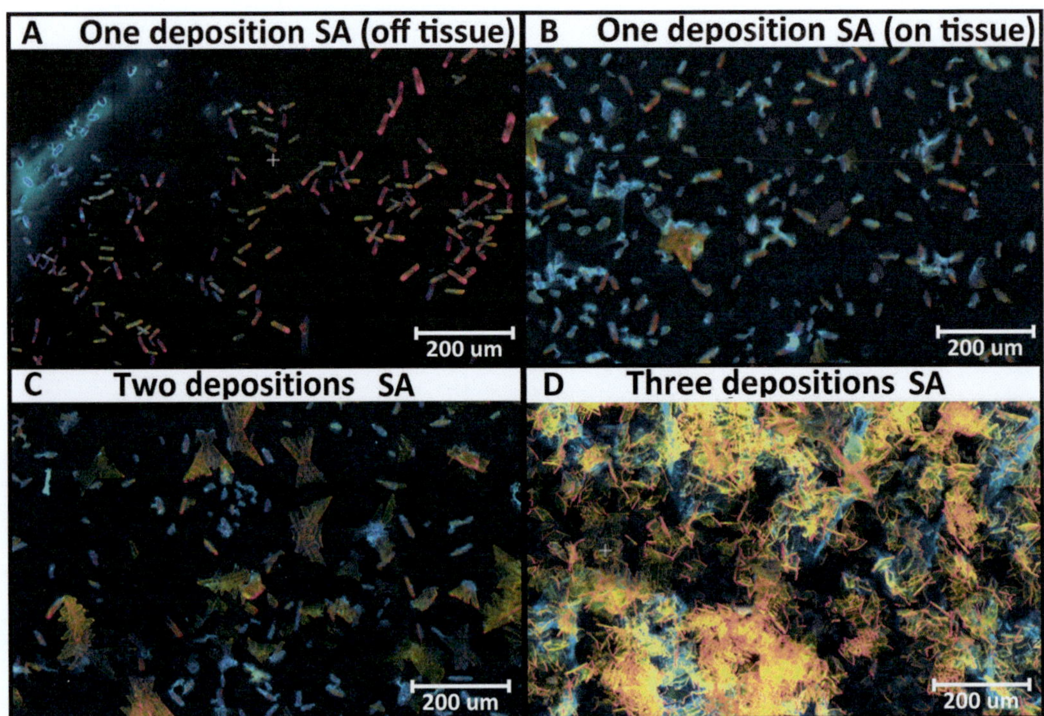

Fig. 4 Monitoring matrix deposition using LSC. High resolution laser scans where blue, yellow, and long red photomultiplier tubes were used with 405 and 488 nm excitation. (**a**) 1 deposition of 10 mg/mL SA off tissue, (**b**) 1 deposition of 10 mg/mL on tissue, (**c**), 2 depositions of 10 mg/mL on tissue (two layers), and (**d**) 3 depositions of 10 mg/mL on tissue (three layers). This shows the homogeneity, distribution and coverage of the matrix at different concentrations. The colors are the result of the iridescence of the crystalline matrix structures

longer but also permits the localization of soluble proteins while staining the nuclei (Fig. 5).

2. Perform laser scanning cytometry in two stages: a low resolution scan for obtaining overviews of the entire tissues, and a high resolution scanning of regions of interest. All laser scanning was performed with a 40× 0.75 N.A. objective lens. Control spatial resolution by varying the spacing of the laser scan lines in the *X* direction. In low resolution scans, the spacing was 10 μm and in the high resolution scans, the spacing was 0.25 μm. By keeping the objective lens constant, it maintains the same analytical sensitivity for both resolutions.

3. For YFP fluorescence evaluation, use a blue (488 nm) laser for excitation. Primary channels were: yellow fluorescence (550/30 nm), and laser light scatter for bright field imaging. For tissues stained with Hoechst 33342, use a violet (405 nm) laser to excite blue fluorescence (440/30 nm), followed by excitation by the blue laser for YFP measurement.

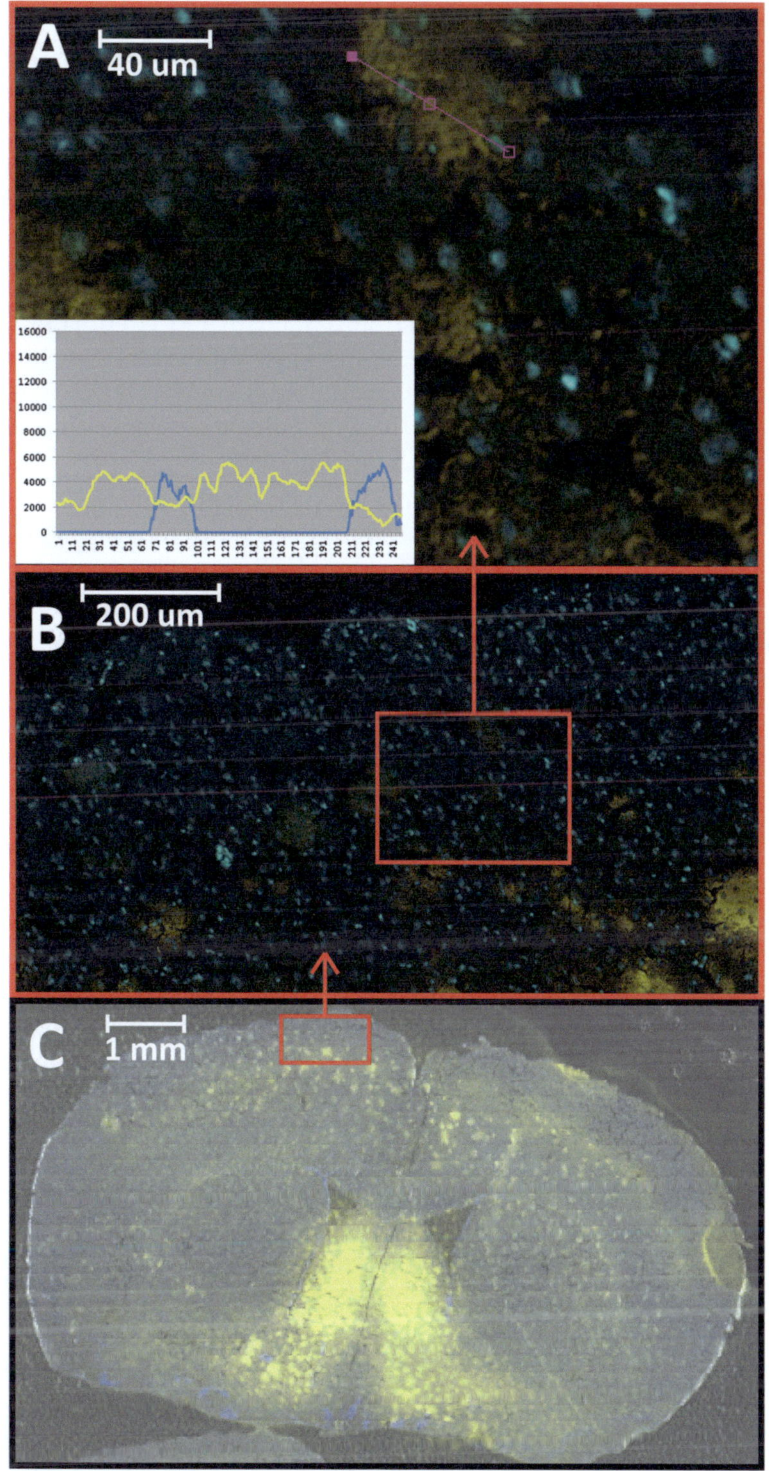

Fig. 5 Low resolution (**c**), high resolution scan region in a tessellated (mosaic) combining all of the scan fields in the region (**b**), and a single scan field view (**a**), LSC images of Hoechst stained, META-8 fixed tissue. Adding Hoechst staining enables the visualization and segmentation of the nuclei of unlabeled and YFP fluorescent cells. *Inset* of **a** shows a plot of the intensity of the nuclei (*blue*) and the YFP (*yellow*) in the cytoplasm of the cells covered by the magenta line

4. For tissues treated with the MALDI matrix, use violet and blue lasers simultaneously, and use photomultiplier tubes for blue, green (510/21 nm), yellow and long red (650 nm LP) to collect fluorescence/iridescence of the crystals (*see* **Note 4**).
5. Use the JPEG images of the tissues with the teaching points clearly visible. Include the rectangles defining the regions of interest for the high resolution scans and export for co-registration with the MALDI instrument.

3.3 MALDI Imaging

1. Perform MALDI imaging with a MALDI-TOF/TOF, such as an UltrafleXtreme, operated with a 1 kHz smart beam laser in linear mode. Define the areas for MALDI analysis on the tissue slide with flexImaging software while using flexControl to coordinate the fiducial markers.
2. As a reference, use the yellow high resolution scan regions from the LSC when choosing the measurement regions for imaging (Subheading 3.2, **step 4**). This makes it easy to correlate the high resolution MALDI with high resolution LSC images to allow the targeted analysis of a single cell or tissue region (Fig. 6).
3. Raster the tissue enclosed in the high resolution areas every 25 μm in the *X* and *Y* directions. If necessary, raster the remaining tissue every 500 μm both with a mass range of 3500–34,000 Da with partial sampling enabled.
4. Typically, tissue sections can no longer generate quality MSI once the laser has ablated the surface. In the studies illustrated here, however, it was possible to obtain two images from the same tissue section with the laser in the "small beam" setting, provided the laser raster positions were shifted ~12 μm from the first sampling. In Fig. 6b and f, the matrix distribution is affected and the raster points are visible on the tissue.
5. High resolution MSI of the entire tissue section is feasible with quality sample preparation, reducing the size of the laser, and having smaller raster points [31].

3.4 Data Analysis

1. In flexImaging, a mass filter list can be made when the masses of the compounds of interest are known, or from the peaks found in individual, or the total average spectrum (*see* **Note 6**).
2. After this analysis, the same regions can be compared between MSI and LSC to determine the extent to which various molecular distributions do or do not spatially correlate with fluorescently labeled cellular populations.
3. Further data preprocessing and statistical analysis can be done in software like SCiLS. For example, classification models can be built based on regions corresponding to fluorescently labeled cells in the LSC images. Unbiased spatial segmentation maps can be evaluated as well to determine cellular subpopulations

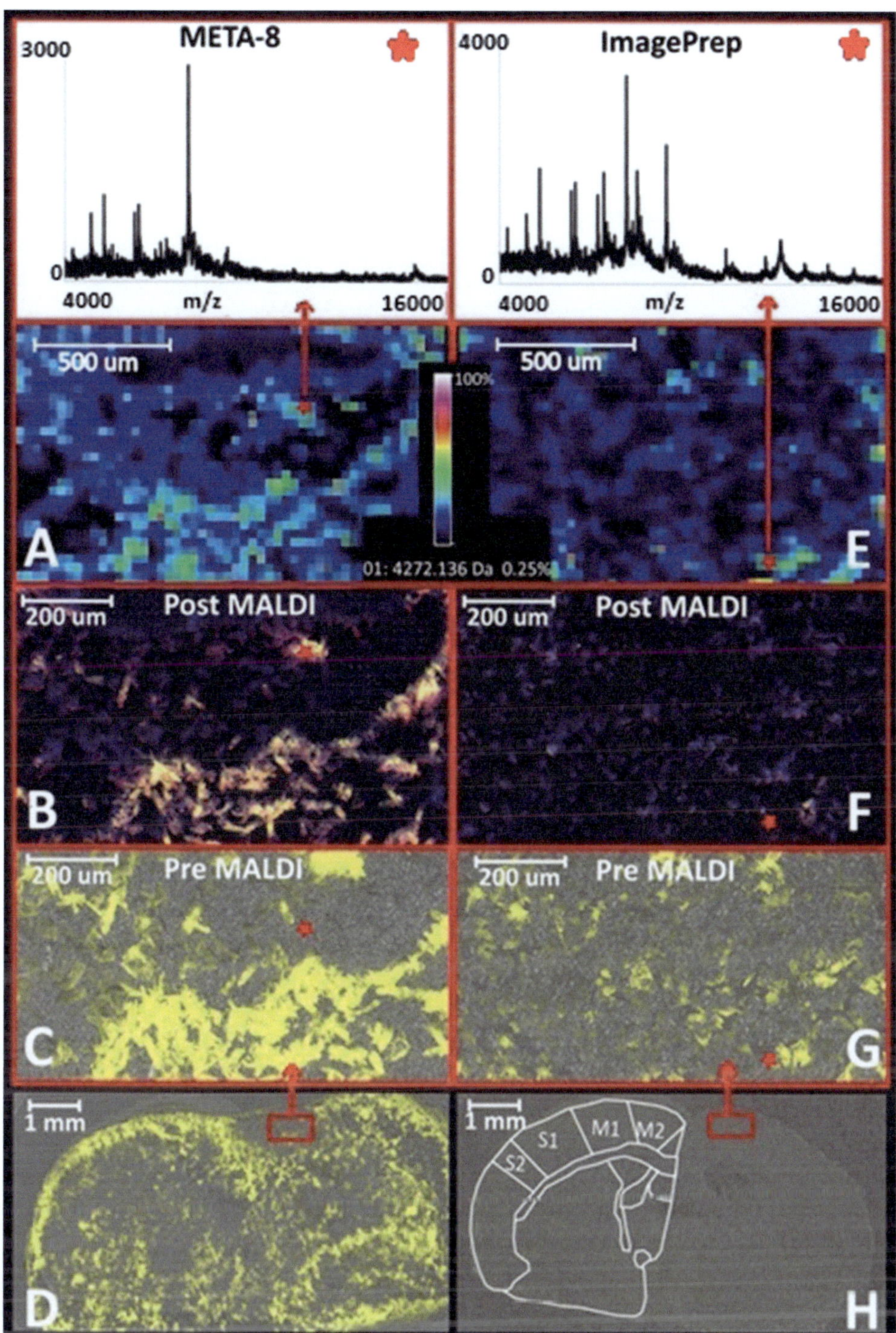

Fig. 6 Co-registered high resolution LSC and MALDI images of the same tissue sections of Fig. 3 following matrix deposition. (**a** and **e**) 25 μm resolution MALDI MS image and a corresponding spectrum following META-8 (**a**) and ImagePrep (**e**) deposition. The "pixel" from which the spectrum was obtained is marked with a *red star*. The *m/z* selected for the MS image correlates theoretically to beta amyloid protein (4512 Da). LSC images illustrate the matrix and tissue landscape before and after MALDI laser ablation. Tissue with 10 mg/mL SA in META-8 applied three times before (**d**, **c**) and after (**b**) MALDI MS. Tissue with 10 mg/mL SA in ACN–H_2O applied by ImagePrep before (**h**, **g**) and after (**f**) MALDI MS. In the post MALDI scans, a violet laser was added to the LSC excitation causing the laser ablated areas to fluoresce blue

within fluorescently labeled regions that could be used to further delineate cell types [32]. A spatial segmentation map divides the MSI based upon similarity of detected ions (i.e., proteins, peptides, lipids, etc.) depending upon the mass range examined. These groups can be viewed in a dendrogram to illustrate the hierarchical relationship of cell types based upon similar distributions of detected analytes. Since the cellular environment around motor neurons contains mainly glia and astrocytes (stained by the Hoechst), these cell types can be distinguished based on the differences in their protein expression shown in the segmentation maps if cellular resolution is achieved. Co-registration of the segmentation map (generated from the MSI data) with the fluorescent LSC image can be used to confirm whether cell types within a particular branch of the dendrogram hierarchy correspond to a fluorescently labeled population. By subdividing the dendrogram branch that corresponds to the fluorescently labeled population, the MSI data could be used to further define cell types by analogous distributions of detected analytes and discover novel cellular subtype molecular markers.

4. Analytes within MSI are represented only by their detected mass-to-charge ratios. In order to validate novel cellular subtype markers, the exact identity of MS-detected analytes that define a particular cellular subpopulation must be determined with a variety of complementary strategies [33]. Databases also have been created to aid in the identification of analytes in MSI that include the identified ions from human and mouse tissue [34]. Once the identity of the putative marker(s) has been determined, the distinct distribution of these analytes within a cell population can be confirmed with immuno-based methods.

4 Notes

1. Typically, thinner tissue sections are better for MALDI imaging (8–20 μm) [29]. 12–14 μm tissue sections were found to be optimal for the robustness of the tissue sections.
2. The teach points communicate to the mass spectrometer the region where data should be collected. WiteOut™ liquid paper is used because it can be easily seen in the MALDI scope while other markers such as black marker are not distinguishable. A high resolution LSC image (JPEG) with marks for co-registration and the yellow regions defined will be compatible with the flexImaging software (Fig. 2). If marks need to be made on the slide, use a diamond pen.
3. The META-8 wash/fixation can show improved image resolution in both MALDI and fluorescence images compared to ethanol and other washes [16]. Proceed with caution ensuring the

application of the META-8 does not cause the tissue to break apart, fold in on itself, or create air bubbles. The ethanol wash does not drastically diminish fluorescence and proves to be just as effective. Carnoy's solution (6:3:1 ethanol–chloroform–acetonitrile) is another effective fixative technique for MALDI-MSI [35], but was shown to decrease fluorescence in LSC images. As seen in Fig. 6, the ImagePrep allows for homogenous and even distribution of matrix across the tissue generating better MSI. Matrix applied via ImagePrep is done by generating small droplets and depositing by acoustic nebulization in an aerosol spray. It detects using optical sensors the amount of light that is scattered to determine the number of layers of matrix to apply. In this workflow, 80 layers was the automated calculation by the instrument. This will vary depending on size, type, and thickness of the sample.

4. Due to the crystalline nature of the matrix, the amount of light scattered diminishes the fluorescence signal of the YFP as seen in Fig. 4. The amount of matrix needed to generate quality MALDI images inhibits the ability to simultaneously resolve the tissue landscape in LSC. There is also the challenge in generating a homogenous distribution of matrix across the sample [36]. Concurrently, small concentration of matrix yields poor MALDI images but unique and descriptive LSC images. In Figs. 4 and 6 the shape of the crystals can be seen and the results of matrix inhomogeneity are visible. We suspect that the loss of fluorescence in the tissue post ImagePrep is due to the water–acetonitrile solvent the SA is diluted in.
5. The Hoechst stain is a fluorescent dye that stains the DNA of cells. Preliminary experiments did not show the Hoechst stain having an effect on the MALDI spectrum [37]. The stain is water soluble and should be added prior to the application of matrix. Keep in mind the stain is light sensitive, keep the vial stored at ~4 °C in a dark, sealed container.
6. A caveat to MALDI imaging is the large amount of data to manage. A mass list is generated for a single raster point and will generate only one aspect of an image as a whole. The selection of peaks is not arbitrary and should be chosen based on established sources. While sample preparation garners a lot of the attention in these kinds of studies, the analysis portion cannot be overlooked. Bias can easily confound or misrepresent the results of a study and analysis should be preceded with proper considerations [32, 38, 39].

Acknowledgements

This work was made possible by grant 1392 from the Amyotrophic Lateral Sclerosis Society of America to J.A. We would like to acknowledge The Barnett Institute for Biological and Chemical

Analysis at Northeastern University as well as the Department of Pharmaceutical Sciences' Core Imaging Facility. We would like to thank the Brandeis University Animal Care Facility for the maintenance of the transgenic mouse colony, Emily Y. Chen and David DeFilippo for their preliminary work in developing the tissue preparation method.

References

1. Traut HF, Papanicolaou GN (1943) Cancer of the uterus: the vaginal smear in its diagnosis. Cal West Med 59:121–122
2. Kamentsky LA, Burger DE, Gershman RJ, Kamentsky LD, Luther E (1997) Slide-based laser scanning cytometry. Acta Cytol 41: 123–143
3. Shapiro HM (2003) Practical flow cytometry, 4th edn. Wiley-Liss, New York
4. Bendall SC, Simonds EF, Qiu P et al (2011) Single-cell mass cytometry of differential immune and drug responses across a human hematopoietic continuum. Science 332: 687–696
5. Kamentsky LA, Thorell B (1970) Cell population identification studies. Acta Cytol 14: 307–312
6. Luther E, Kamentsky L, Henriksen M, Holden E (2004) Next-generation laser scanning cytometry. Methods Cell Biol 75:185–218
7. Pozarowski P, Holden E, Darzynkiewicz Z (2013) Laser scanning cytometry: principles and applications-an update. Methods Mol Biol 931:187–212
8. Darzynkiewicz Z, Smolewski P, Holden E et al (2011) Laser scanning cytometry for automation of the micronucleus assay. Mutagenesis 26:153–161
9. Ananthanarayanan V, Deaton RJ, Amatya A et al (2011) Subcellular localization of p27 and prostate cancer recurrence: automated digital microscopy analysis of tissue microarrays. Hum Pathol 42:873–881
10. Stoeckli M, Farmer TB, Caprioli RM (1999) Automated mass spectrometry imaging with a matrix-assisted laser desorption ionization time-of-flight instrument. J Am Soc Mass Spectrom 10:67–71
11. Li L, Garden RW, Romanova EV, Sweedler JV (1999) In situ sequencing of peptides from biological tissues and single cells using MALDI-PSD/CID analysis. Anal Chem 71:5451–5458
12. Li L, Moroz TP, Garden RW, Floyd PD, Weiss KR, Sweedler JV (1998) Mass spectrometric survey of interganglionically transported peptides in Aplysia. Peptides 19:1425–1433
13. Yew JY, Wang Y, Barteneva N et al (2009) Analysis of neuropeptide expression and localization in adult drosophila melanogaster central nervous system by affinity cell-capture mass spectrometry. J Proteome Res 8:1271–1284
14. Zavalin A, Yang J, Haase A, Holle A, Caprioli R (2014) Implementation of a Gaussian beam laser and aspheric optics for high spatial resolution MALDI imaging MS. J Am Soc Mass Spectrom 25:1079–1082
15. Luxembourg SL, Mize TH, McDonnell LA, Heeren RM (2004) High-spatial resolution mass spectrometric imaging of peptide and protein distributions on a surface. Anal Chem 76:5339–5344
16. Agar NY, Kowalski JM, Kowalski PJ, Wong JH, Agar JN (2010) Tissue preparation for the in situ MALDI MS imaging of proteins, lipids, and small molecules at cellular resolution. Methods Mol Biol 656:415–431
17. Maier SK, Hahne H, Gholami AM et al (2013) Comprehensive identification of proteins from MALDI imaging. Mol Cell Proteomics 12:2901–2910
18. Boggio KJ, Obasuyi E, Sugino K, Nelson SB, Agar NY, Agar JN (2011) Recent advances in single-cell MALDI mass spectrometry imaging and potential clinical impact. Expert Rev Proteomics 8:591–604
19. Seeley EH, Oppenheimer SR, Mi D, Chaurand P, Caprioli RM (2008) Enhancement of protein sensitivity for MALDI imaging mass spectrometry after chemical treatment of tissue sections. J Am Soc Mass Spectrom 19: 1069–1077
20. Dai Y, Whittal RM, Li L (1996) Confocal fluorescence microscopic imaging for investigating the analyte distribution in MALDI matrices. Anal Chem 68:2494–2500
21. Monroe EB, Jurchen JC, Lee J, Rubakhin SS, Sweedler JV (2005) Vitamin E imaging and localization in the neuronal membrane. J Am Chem Soc 127:12152–12153
22. Andersson M, Groseclose MR, Deutch AY, Caprioli RM (2008) Imaging mass spectrometry of proteins and peptides: 3D volume reconstruction. Nat Methods 5:101–108

23. Crecelius AC, Cornett DS, Caprioli RM, Williams B, Dawant BM, Bodenheimer B (2005) Three-dimensional visualization of protein expression in mouse brain structures using imaging mass spectrometry. J Am Soc Mass Spectrom 16:1093–1099
24. Sinha TK, Khatib-Shahidi S, Yankeelov TE et al (2008) Integrating spatially resolved three-dimensional MALDI IMS with in vivo magnetic resonance imaging. Nat Methods 5:57–59
25. Bocklitz TW, Creceius AC, Matthaus C et al (2013) Deeper understanding of biological tissue: quantitative correlation of MALDI-TOF and Raman imaging. Anal Chem 85: 10829–10834
26. Feng G, Mellor RH, Bernstein M et al (2000) Imaging neuronal subsets in transgenic mice expressing multiple spectral variants of GFP. Neuron 28:41–51
27. Abdelmoula WM, Carreira RJ, Shyti R et al (2014) Automatic registration of mass spectrometry imaging data sets to the Allen brain atlas. Anal Chem 86:3947–3954
28. Paxinos G, Franklin KBJ (2004) The mouse brain in stereotaxic coordinates, compact 2nd edition. Amsterdam. Elsevier Academic Press, Boston
29. Goodwin RJ, Lang AM, Allingham H, Boren M, Pitt AR (2010) Stopping the clock on proteomic degradation by heat treatment at the point of tissue excision. Proteomics 10:1751–1761
30. Agar NY, Yang HW, Carroll RS, Black PM, Agar JN (2007) Matrix solution fixation: histology-compatible tissue preparation for MALDI mass spectrometry imaging. Anal Chem 79:7416–7423
31. Chaurand P, Schriver KE, Caprioli RM (2007) Instrument design and characterization for high resolution MALDI-MS imaging of tissue sections. J Mass Spectrom 42:476–489
32. Trede D, Kobarg JH, Oetjen J, Thiele H, Maass P, Alexandrov T (2012) On the importance of mathematical methods for analysis of MALDI-imaging mass spectrometry data. J Integr Bioinform 9:189
33. Mascini NE, Heeren RMA (2012) Protein identification in mass-spectrometry imaging. Trac-Trend Anal Chem 40:28–37
34. McDonnell LA, Walch A, Stoeckli M, Corthals GL (2014) MSiMass list: a public database of identifications for protein MALDI MS imaging. J Proteome Res 13:1138–1142
35. Agar NYR, Yang HW, Carroll RS, Black PM, Agar JN (2007) Matrix solution fixation: histology-compatible tissue preparation for MALDI mass spectrometry imaging. Anal Chem 79:7416–7423
36. Gabriel SJ, Schwarzinger C, Schwarzinger B, Panne U, Weidner SM (2014) Matrix segregation as the major cause for sample inhomogeneity in MALDI dried droplet spots. J Am Soc Mass Spectrom 25(8): 1356–1363
37. Koomen JM, Stoeckli M, Caprioli RM (2000) Mapping of surrogate markers of cellular components and structures using laser desorption/ionization mass spectrometry. J Mass Spectrom 35:258–264
38. Norris JL, Cornett DS, Mobley JA et al (2007) Processing MALDI mass spectra to improve mass spectral direct tissue analysis. Int J Mass Spectrom 260:212–221
39. Alexandrov T (2012) MALDI imaging mass spectrometry: statistical data analysis and current computational challenges. BMC Bioinformatics 13 Suppl 16:S11

Chapter 11

SPLIFF: A Single-Cell Method to Map Protein-Protein Interactions in Time and Space

Alexander Dünkler*, Reinhild Rösler*, Hans A. Kestler, Daniel Moreno-Andrés, and Nils Johnsson

Abstract

Protein interactions occur at certain times and at specific cellular places. The past years have seen a massive accumulation of binary protein-protein interaction data. The rapid increase of this context-free information necessitates robust methods to monitor protein interactions with temporal and spatial resolution in single cells. We have developed a simple split-ubiquitin-based method (SPLIFF) that uses the ratio of two fluorescent reporters as a signal for protein protein interactions. One protein of the pair of interest is attached to the linear fusion of mCherry, the C-terminal half of ubiquitin, and GFP (mCherry-C_{ub}-GFP). The other potential binding partner is expressed as a C-terminal fusion to the N-terminal half of ubiquitin (N_{ub}). Upon co-expression the interaction between the two proteins of interest induces the reassociation of N_{ub} and C_{ub} to the native-like ubiquitin. GFP is subsequently cleaved from the C-terminus of C_{ub} and degraded whereas the red-fluorescent mCherry stays attached to the C_{ub}-fusion protein. We first implemented this method in the model yeast *Saccharomyces cerevisiae*. One fusion protein is expressed in cells of the a-mating type and the complementary fusion protein in cells of the α-mating type. Upon mixing, both cell types fuse and the N_{ub}- and C_{ub}-fusion proteins are free to interact. The red and green fluorescence is monitored by two-channel fluorescence time-lapse microcopy. The moment of cell fusion defines the start of the analysis. The calculated ratio of green to red fluorescence allows mapping the spatiotemporal interaction profiles of the investigated proteins in single cells.

Key words Protein-protein interaction, Split-protein sensor, Split-ubiquitin, Protein complementation assay, Yeast, Single-cell measurement, Fluorescent reporter, Time-lapse analysis

1 Introduction

Split protein sensors comprise a family of analytical tools that based on a common principle allow monitoring protein-protein interactions in vitro and in vivo [1, 2]. The technique relies on the ability of two halves of a fragmented protein to refold into its native-like structure. This reassociation occurs only very ineffectively and

*These authors contributed equally.

Anup K. Singh and Aarthi Chandrasekaran (eds.), *Single Cell Protein Analysis: Methods and Protocols*, Methods in Molecular Biology, vol. 1346, DOI 10.1007/978-1-4939-2987-0_11, © Springer Science+Business Media New York 2015

slowly but can be accelerated upon linking the two fragments to proteins that interact. Upon interaction the local concentration of the two fragments rises dramatically, thus forcing the two polypeptides to reassemble into the original structure of the uncut protein [1, 2]. The various split protein sensors each differs by the activity of the reconstituted protein. The split-ubiquitin (split-Ub) system is based on the reassociation of an artificially fragmented ubiquitin [2, 3]. Once forced into close proximity the reassociation between the N-terminal (N_{ub}) and the C-terminal (C_{ub}) fragment of Ub to its native-like structure will stimulate the release of a reporter protein from the C-terminus of the C_{ub} [2, 3]. The altered activity of the cleaved reporter signals the interaction between the N_{ub}- and C_{ub}-coupled proteins. The split-Ub system, like all other subsequently developed split-protein sensors, is able to detect direct as well as indirect interactions of proteins [1]. Different reporter proteins have been used to transform the interaction between the fusion proteins into a measurable signal. Thus far, the output of the most commonly used reporter proteins for large-scale analyses has been the ability of the cells expressing the pair of interacting proteins to grow on a specifically designed medium [4–7]. This yes/no output ignores the dynamic and spatial aspects of the measured interaction.

We recently introduced a novel fluorescent reporter configuration (SPLIFF) for the split-Ub technique in the yeast *Saccharomyces cerevisiae* (from here on yeast) [8]. Here the C_{ub} is sandwiched between two autofluorescent proteins with different spectral properties (mCherry-C_{ub}-GFP; CCG). After refolding to the native-like Ub the GFP is cleaved from the C_{ub} by the ubiquitin-specific proteases and subsequently degraded. The second fluorescent protein mCherry stays attached to the C_{ub}-fusion protein (Fig. 1). As a consequence, the interaction between N_{ub}-X and Y-CCG is indicated by the decrease in the cellular ratio of green-to-red fluorescence. By expressing the N_{ub}-and the C_{ub}-fusion in haploid yeast cells of different mating types the interaction between the N_{ub}-and C_{ub}-fusion can be initiated by mixing a- and α-cells and observing the fate of the red and green fluorescence by time-lapse microscopy of the obtained diploids (Fig. 1) [8].

Due to the rapid refolding of the fragmented Ub and the instantaneous cleavage at the C-terminus of C_{ub} we never detected a significant delay between the interaction of Y with X and the cleavage of the GFP from Y-CCG [8–10]. However, the system was not yet challenged below the minute scales and it will be important to find out at which time interval the cleavage at the C_{ub} becomes indeed limiting for the temporal resolution of the technique.

The split-Ub technique critically depends on the presence of the Ub-specific proteases [9]. Due to their restricted distribution, SPLIFF can only monitor protein interactions in the cytosol and

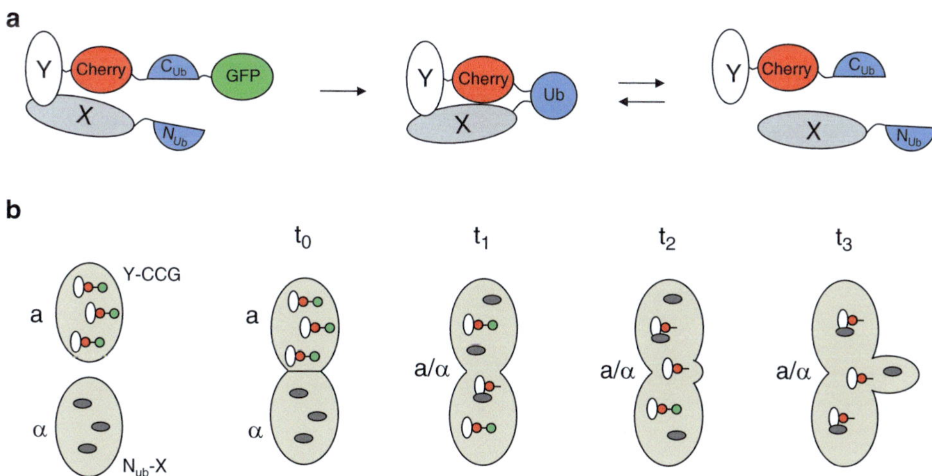

Fig. 1 The SPLIFF method: (**a**) Protein Y is coupled to the mCherry-C_{ub}-GFP module (Y-CCG) and protein X is fused to N_{ub} (N_{ub}-X). Interaction between Y and X stimulates the reassembly of C_{ub} and N_{ub} into a native-like ubiquitin (Ub). The Ub-specific proteases recognize the Ub-fold and cleave off the GFP. The N-terminally exposed arginine leads to the rapid degradation of GFP. (**b**) *MATa* cells of yeast expressing Y-CCG and *MATα* cells expressing N_{ub}-Y fuse to form the zygote (t_0). The cytosols of both cell types mix and Y-CCG is free to interact with N_{ub}-X. The time course of interaction causes the progressive conversion of Y-CCG to Y-CC at t_1, t_2, t_3

the nucleus of the cell but not interactions that occur in the lumen of the different organelles. The interactions between membrane proteins can be recorded provided that the termini that are coupled to N_{ub} or C_{ub} extend into the cytosol of the cell [4, 5] (our unpublished data).

The successful applications of the technique now include the interactions between nucleolar proteins, between proteins of the nuclear pore, between proteins randomly distributed in the cytosol, between proteins localized at the cortex of cells, and between proteins that interact with the growing tip of a microtubule [8].

Two features of the system have to be considered during the interpretation of the measurements:

1. The cleavage of the GFP from Y-CCG and the subsequent shift in the ratio of green-to-red fluorescence indicates time and place of interaction with N_{ub}-X. The sheer accumulation of the red fluorescence records only the localization of the fusion but not its interaction. Consequently, once all Y-CCG fusions are converted to Y-mCherry-C_{ub} (Y-CC) no further measurements are feasible in this cell.
2. The affinity between N_{ub} and C_{ub} disturbs the equilibrium of the interaction between the proteins X and Y in the context of the fusions. The extent depends on the compatibility of the refolded Ub with the structure of the X-Y complex. Quantitative statements about the interaction are thus possible only under certain, restrictive conditions [8]. An irreversible trapping of

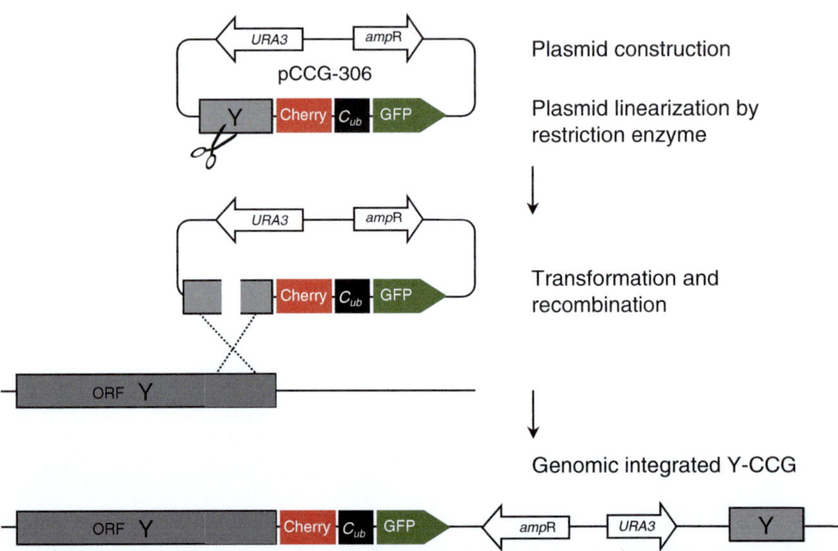

Fig. 2 Manipulation of the yeast genome to integrate Y-CCG fusion genes at their native loci. A PCR fragment of the 3′ end of the ORF of interest (Y) without its stop codon is ligated in front of and in frame to the mCherry-C_{ub}-GFP sequence on a pCCG-306 vector (derivate of pRS306). The fragment harbors a unique restriction site situated a minimal 100 bp from the 5′ and 3′ ends of the amplified sequence. Linearization of the plasmid induces homologous sequences to Y at the 5′ and 3′ ends. Transformation into yeast cells and selection on SD Ura$^-$ medium lead to the C-terminal insertion of the mCherry-C_{ub}-GFP module into the genome by homologous recombination

an interaction upon reassociation of the coupled N_{ub} and C_{ub} was not yet observed (our unpublished data).

The following protocols can be generally applied for performing a SPLIFF experiment in yeast using either heterologous proteins or yeast proteins (Figs. 2, 3, 4, 5). To demonstrate examples of SPLIFF analyses we present three measurements between proteins belonging to the polar cortical domain (PCD) in yeast (Figs. 6, 7, 8, Table 1).

The polar cortical domain is an assembly of more than 100 proteins that forms below the membrane at sites of polar growth [11]. For simplicity, we distinguish between the PCD during the formation of the mating projection and membrane fusion (PCD I); during bud growth at the tip of the bud (PCD II); and at the bud neck during late mitosis and cytokinesis (PCDIII). Many proteins appear at all three manifestations of the PCD whereas others are specific for only one or two. The PCD is highly dynamic and moves to different positions during the cell cycle.

The polarisome is an assembly of several proteins that is part of all three PCDs [12]. We show in a first application the interaction between two members of the polarisome Spa2p and Pea2p (Figs. 6 and 7), and in the second application the interaction between Spa2p and Hof1p (Fig. 7) [8]. Hof1p is involved in cytokinesis and is not permanently associated with the polarisome [13]. As a third application we demonstrate the possibility to monitor with SPLIFF the competition between members of a

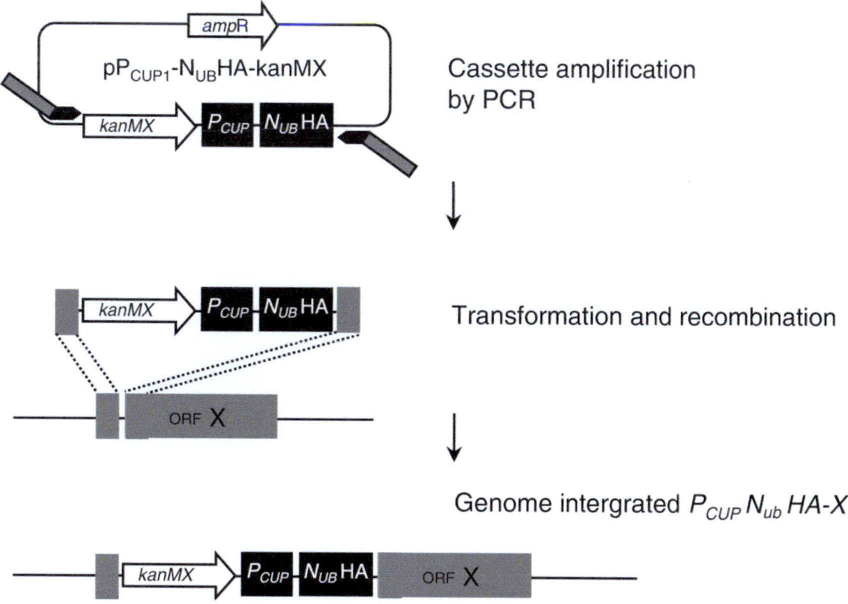

Fig. 3 Genomic integration of P_{CUP1} N_{ub}*HA-X* cassettes. A PCR with two primers annealing at the 5′ and 3′ end of the *kanMX-*P_{CUP1} N_{ub}*HA* cassette yields a fragment harboring beside the N_{ub} module, the P_{CUP1} promoter, and the Geneticin resistance gene a stretch of approximately 45 bp at each of its ends. These stretches are identical to the sequences within the yeast genome, where the integration of the cassette will occur after transformation and selection on Gen^+ medium

protein complex in living cells. Here we quantify the difference in the rate of interaction between Spa2-CCG and N_{ub}-Pea2p in the presence and absence of additional copies of unlabeled Pea2p (Figs. 7 and 8) [8]. Being sensitive enough to measure competition between proteins is important for the analysis of protein interaction networks as the topologies of these networks postulate many hubs of converging interactions that might or might not compete with each other for binding [14, 15].

2 Materials

2.1 *S. cerevisiae* Strains

CCG- and N_{ub}-fusion proteins are generated in cells of different mating types. We routinely transform CCG-fusions in JD47 (*MATa*) and N_{ub}-fusions in JD53 (*MATα*) strains [16]. Any other yeast strain background should be suitable.

1. Yeast JD47: *MATa*, *leu2-3, 112 lys2-801*, *his3-Δ200*, *trp1-Δ63*, *ura3-52*.
2. Yeast JD53: *MATα*, *leu2-3, 112 lys2-801*, *his3-Δ200*, *trp1-Δ63*, *ura3-52*.

2.2 Media

1. Yeast SD minimal medium: 6.7 g/L YNB (yeast nitrogen base) without amino acids (e.g., ForMedia, Hunstanton, UK), 60 mg/L L-isoleucine, 20 mg/L L-arginine, 40 mg/L L-lysine,

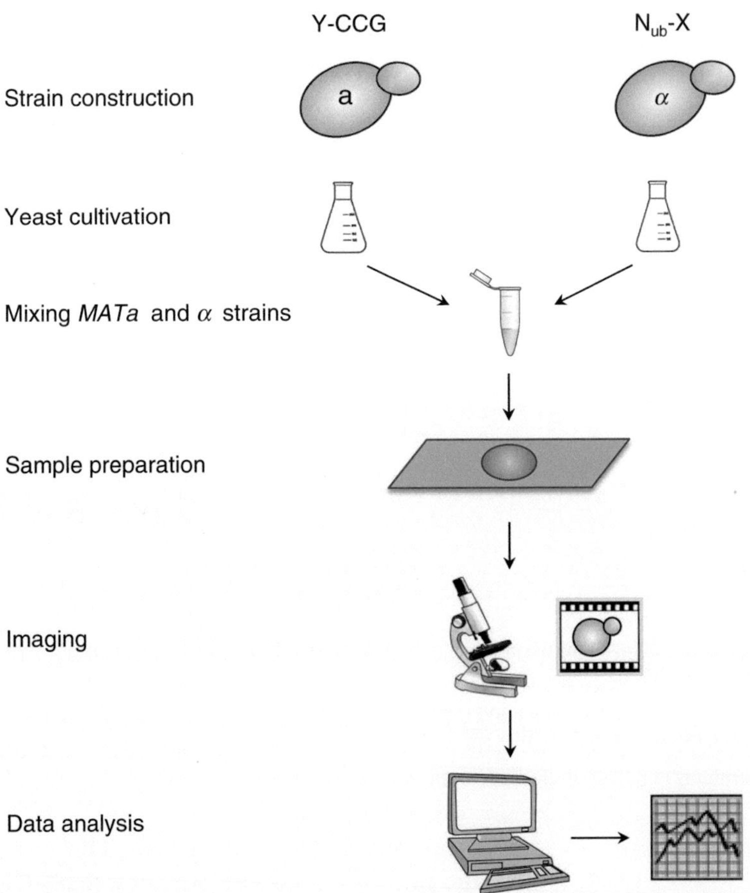

Fig. 4 Schematic flowchart for the application of SPLIFF in yeast cells. See text for details

60 mg/L L-phenylalanine, 10 mg/L L-threonine, 10 mg/L L-methionine, 50 mg/L adenine sulfate, 20 mg/L L-histidine, 60 mg/L L-leucine, 40 mg/L L-tryptophan, 50 mg/L uracil, and 20 g/L glucose. Leave out amino acids or uracil for the selection of the corresponding auxotrophic markers.

2. YPD medium: 10 g/L yeast extract, 20 g/L peptone, 20 g/L glucose.
3. Solid media are obtained from liquid media by addition of 20 g agar per liter medium.
4. The following concentrations of antibiotics were added to YPD and SD plates: *kanMX* marker, 200 μg/ml Geneticin (e.g., Gibco BRL Life Technologies, Grand Island, NY); *natNT2* marker, 100 μg/ml Nourseothricin (ClonNAT, Werner BioAgents, Jena, Germany).
5. For selection of antibiotic markers in SD minimal medium a modified formula is used: 1.9 g/L YNB without ammonium

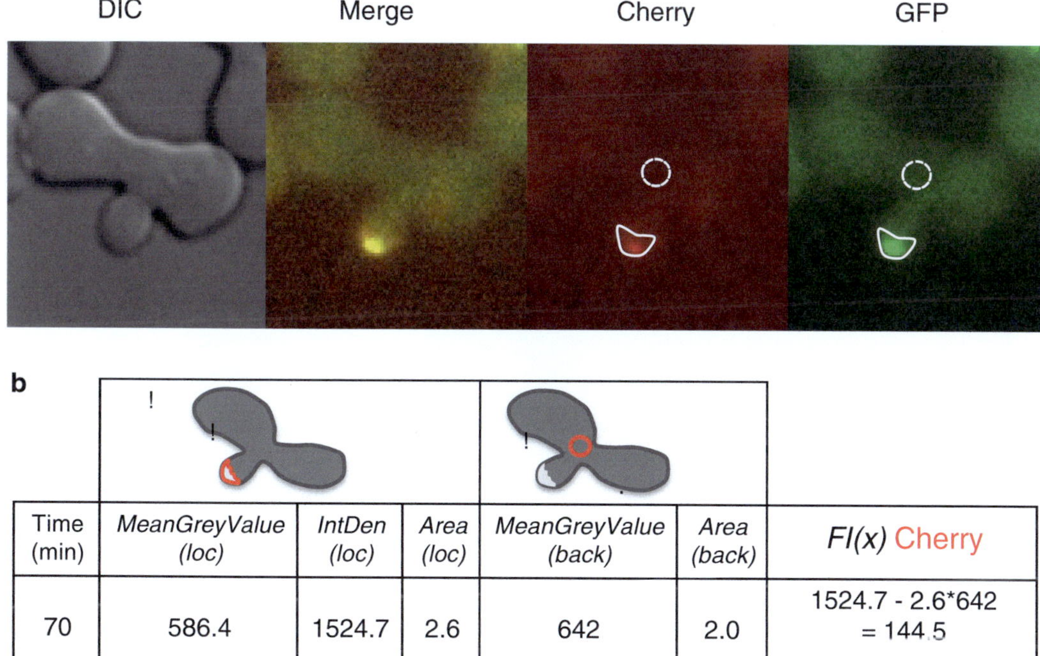

Time (min)	*MeanGreyValue (loc)*	*IntDen (loc)*	*Area (loc)*	*MeanGreyValue (back)*	*Area (back)*	*FI(x)* Cherry
70	586.4	1524.7	2.6	642	2.0	1524.7 - 2.6*642 = 144.5

Fig. 5 Example for the selection of those areas in the cell that are used for the quantification of the fluorescence signals. (**a**) The CCG-fusion protein involved in polar growth localizes during bud growth below the membrane of the growing bud tip. The location of the CCG-fusion (loc) is *circled by a continuous line*, the area used for background subtraction (back) is *circled by a dashed line*. (b) Example for the calculation of fluorescence intensities FI(*x*). Shown are the values of the mCherry channel; the same calculations have to be done for the GFP channel. The Mean Grey Value(back) projected onto Area(loc) has to be subtracted from the *IntDen*(loc) at each time frame to obtain the corrected FI(*x*) of the localized CCG-fusion protein

sulfate and amino acids, 1 g/L asparagine in addition to the amino acids and glucose found in the conventional SD minimal medium.

6. The media for Y-CCG fusions under control of P_{MET17} promoter contain 70 μM or no methionine (highest expression level). The media used for the SPLIFF measurements contain 0 or 100 μM copper sulfate to adjust the expression of the P_{CUP1} promoter-controlled N_{ub}-X fusion.
7. Live cell imaging medium is prepared by mixing in equal ratio a 3.4 % agarose solution and a twofold SD liquid medium containing the required methionine and copper sulfate concentrations.

2.3 Reagents for Yeast Transformation

1. 10× TE buffer (100 mM Tris–HCl pH 8.0, 10 mM EDTA pH 8.0).
2. Lithium acetate solution: 100 mM lithium acetate, 1× TE buffer.

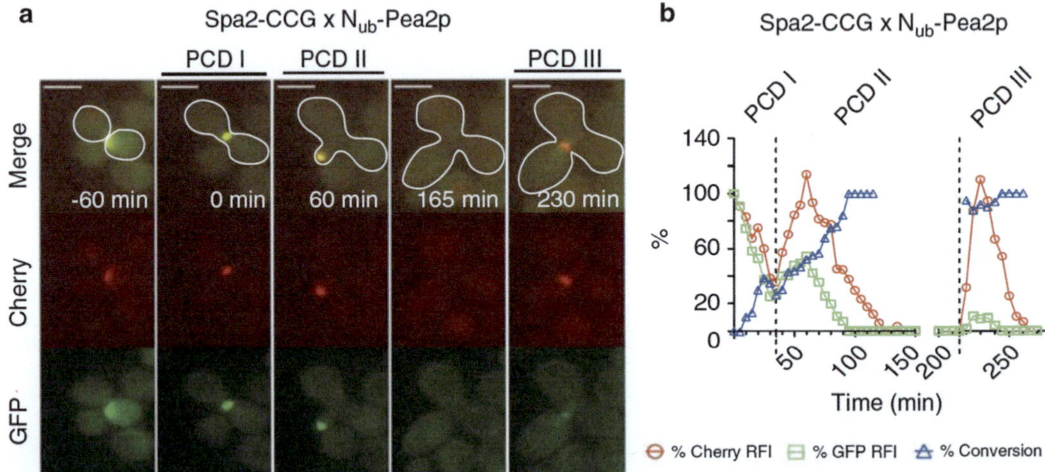

Fig. 6 SPLIFF analysis of the polarisome members Spa2-CCG and N_{ub}-Pea2p [8]. (**a**) Selected frames of the time-lapse analysis of the a-cell-expressing Spa2-CCG before and after mating with the α-cell-expressing N_{ub}-Pea2p (*white frames*). PCDI indicates the region below the membrane of the shmoo tip, PCDII the region below the membrane of the growing bud, and PCDIII the region of cell separation. (**b**) Quantitative analysis of the experiment in (**a**). Time-dependent relative fluorescence intensity (*RFI*) of mCherry is indicated in *red*; *RFI-GFP* is indicated in *green*. The *blue line* indicates the calculated fraction of converted Spa2-CC of the experiment shown in (**a**). *Dashed orthogonal lines* separate the analysis of subsequent PCDs. Time 0 indicates the time point shortly before fusion of the haploid cells. Note that the interaction between Spa2p and Pea2p occurs during all three phases of the PCD. Scale bar: 5 μm

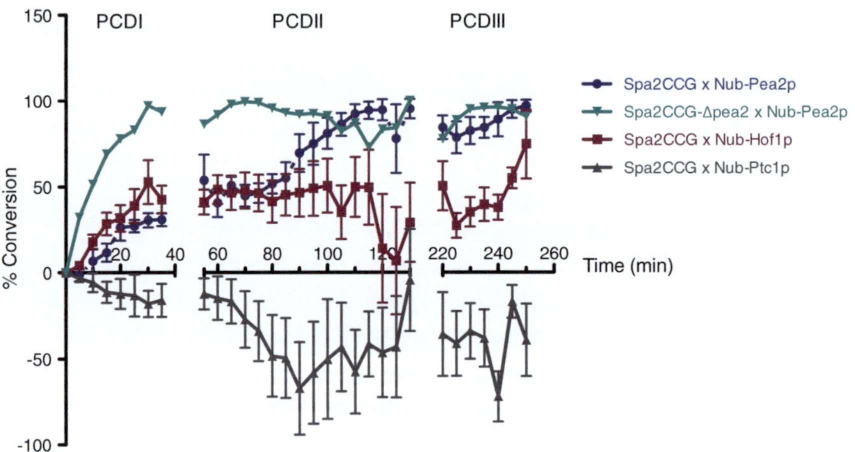

Fig. 7 Time-dependent Spa2-CCG conversion induced by the N_{ub}-Ptc1p (*grey*), N_{ub}-Hof1p (*purple*) and N_{ub}-Pea2p in cells expressing (*blue*) and lacking (*green*) the native copy of Pea2p [8]. The identities of the three PCDs are indicated above the corresponding time axis intercept; values of the transition between the PCDs are not shown. The cytosolic protein N_{ub}-Ptc1p serves as a negative control for the experiment. The averages of $n = 6$ independent matings (error bars, s.e.) are shown. Note that Hof1p interacts with Spa2p during PCDI and PCDIII but not during PCDII

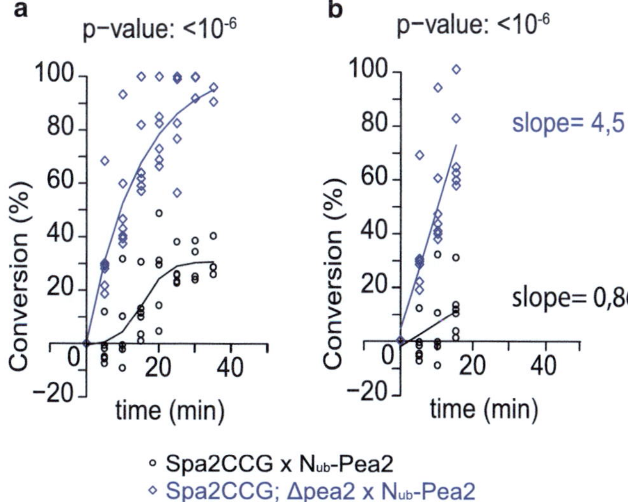

Fig. 8 Pairwise comparisons of the time-dependent conversion of Spa2-CCG by N_{ub}-Pea2p in the presence and absence of additional copies of unlabeled Pea2p [8]. Spa2-CCG was expressed in diploid cells derived from either wild-type a-cells (*black symbol and line*) or a-cells lacking *PEA2* (*blue symbol and line*). (**a**) Lines show best fits for each interaction profile selected via the Akaike Information Criterion after fitting multiple models. *P* values are calculated from the *F*-test [8, 19]. (**b**) Linear regression lines and slopes for the first 15 min of Spa2-CCG conversion induced by N_{ub}-Pea2p in the presence and absence of native *PEA2*. Note that the presence of Pea2p reduces the rate of interaction between Spa2-CCG and N_{ub}-Pea2p. R script is available: http://sysbio.uni-ulm.de/soft/CuCompare

3. PEG solution: 40 % PEG3000, 100 mM lithium acetate, 1× TE buffer.
4. Single-stranded carrier DNA 2 mg/mL: Dissolve 200 mg of salmon sperm DNA (sodium salt) in 100 ml sterile 1× TE buffer and heat to approximately 50 °C while vigorously stirring. Aliquots are boiled for 10 min at 95 °C, cooled in an ice-water bath, and stored at −20 °C. Boiling and cooling may be repeated before transformation to increase transformation efficiency.

2.4 CCG-Gene Fusion Expression Plasmids

Yeast shuttle plasmids were adapted to the SPLIFF system. The CCG (mCherry-C_{ub}-GFP) gene fusions (baits) were constructed either directly in yeast by homologous recombination into the genome or by inserting the full-length open-reading frame (ORF) or a fragment of the ORF between the P_{MET17} promoter and the CCG reporter on a yeast shuttle plasmid, followed by transformation into the *MATα* strain JD47.

1. pCCG-306, an integrative plasmid for construction of full-length CCG-fusions into the yeast genome: The 3′ ends of yeast genes are amplified by PCR and inserted in frame 5′ to the CCG cassette. After cloning, a single-restriction site in the

Table 1
Spa2-CCG is converted to Spa2-CC through interaction with N_{ub}-Pea2p. Shown is an example for the calculation of the conversion rate *FD(x)* of *x*. Characteristic time points during the cell cycle are specifically annotated in the *left column*. *FI(0)*, the fluorescence intensity just before cell fusion, is taken as the reference value for the calculation of *RFI(x)* of the following time frames (*RFI(start)* = 100 %). Time frames without a localized signal in the reference channels are excluded from the *FD(x)* calculation

	Red		Green		$FD(x) =$
Time (*x*)	FI(*x*)	RFI(*x*)	FI(*x*)	RFI(*x*)	$\frac{RFI_{RED} - RFI_{GREEN}}{RFI_{RED}} \times 100$
0 *FI*(start)	129.0	100	230	100	0
10	86.4	67	110.4	48.3	35.8
20	54.2	42	55.2	24	42.8
30	40.0	31	34.7	15.1	51.2
40 – *no localized signal*	20.6	16	27.6	12	N.A.
50 (bud emergence)	65.8	51	53.1	23.1	54.7
60	83.9	65	65.8	28.6	56
70	144.5	112	108.8	47.3	57.7
80	104.5	81	35.4	15.4	80.1
90	60.6	47	3.22	1.4	97.0
100	39.9	31	0	0	100
120	28.4	22	0	0	100
140	15.5	12	0	0	100
160 – *no localized signal*	4.1	3.2	0	0	N.A.
180 – *no localized signal*	0	0	0	0	N.A.
200 – *no localized signal*	1.4	1.1	0	0	N.A.
220	46.4	36	1.15	0.5	98.6
240	138.0	107	5.29	2.3	97.9
260	29.7	23	0	0	100

amplified gene fragment is used for plasmid linearization. After transformation, the homologous recombination between the chromosomal gene copy and the introduced CCG gene fusion fragment leads to the formation of the full-length CCG gene fusion at the chromosomal locus (Fig. 2).

2. pP_{MET17}CCG-316, a centromere-based vector which drives the ORF-CCG fusion of interest by the regulatable P_{MET17} promoter: This vector is suitable for heterologous gene expres-

sion. Stable, low-copy-number maintenance of the plasmid is provided by a *CEN6* and *ARSH4* element. Introduction of the ORF is achieved by a multiple cloning site between the sequence of the P_{MET17} promoter and the CCG cassette.

3. Both plasmids contain a *URA3* marker gene to allow selection in yeast and an antibiotic resistance gene for the selection against ampicillin in *E. coli*.
4. pYM-N35 is used for genomic integration of the P_{MET17} promoter in front of CCG-fusion genes by PCR-based gene targeting (*natNT2* and *amp* marker) [17].

2.5 N_{ub} Fusion Expression Plasmids

The N_{ub}-fusion genes are expressed under the control of the copper-inducible P_{CUP1} promoter. Fusions were obtained by homologous recombination of PCR-amplified P_{CUP1}-*Nub* cassettes or by cloning of the full-length ORFs behind the P_{CUP1} promoter [16, 18]. N_{ub}-fusion-expressing strains are generated in the *MATα* strain JD53 to allow mating with the CCG fusion-expressing strains.

1. p$P_{CUP1}N_{ub}$HA-kanMX is used as a template for the PCR amplification and site-specific integration of $P_{CUP1}N_{ub}HA$ sequences where the sequence *HA* codes for the HA epitope. Using primers composed of sequences up- and downstream of the start codon (usually 45 bp) of the respective yeast gene and sequences annealing with the *kanMX*-$P_{CUP1}N_{ub}HA$ cassette (N_{ub}-FW, N_{ub}-RV; *see* primers in Subheading 2.6) a PCR product is obtained that will integrate in front or within the gene of interest via homologous recombination (Fig. 3). Successful recombination is verified in Geneticin-resistant strains by a diagnostic PCR using primers annealing with the N_{ub} sequence and sequences 300–600 bp downstream of the integration site within the respective gene. The PCR product is sequenced to reveal any errors that might have occurred during the procedure to generate the fusion.
2. p$P_{CUP1}N_{ub}$HA-kanMX CEN is a centromere-based vector (*CEN6* and *ARSH4* element) with the option of heterologous expression. Common restriction sites are used to insert the ORF of interest behind the $P_{CUP1}N_{ub}HA$ cassette. Expression of the N_{ub}-fusion in yeast is obtained after transformation of the plasmid and selecting for its presence by Geneticin [16, 18].

2.6 Primers

1. N_{ub}-SEQ: CATTGGAAGTTGAATCTTCCG.
2. N_{ub}-FW: 45 bp upstream of the start codon of gene X + GCATAGGCCACTAGTGGATC.
3. N_{ub}-RV: 45 bp downstream of the start codon of gene X + GGTCGACCCCGCATAGTCAGG.

2.7 Technical Equipment

1. For time-lapse imaging, a wide-field fluorescence microscope equipped with bright-field, GFP (470/525 nm) and mCherry (572/632 nm) channel is required. In all experiments a 100× NA 1.4 UPlanSApo oil immersion objective (Olympus) was used. To reduce bleaching and phototoxicity, imaging with an electron-multiplying charge-coupled device (EMCCD) camera is beneficial and capture time has to be adapted to the intensity of mCherry and GFP signal for every construct.
2. To record yeast cell growth for several hours the cavities of the microscope slides were fully filled with the required solid media and incubated at 30 °C with a steady-state heating chamber under the microscope.

3 Methods

3.1 Single-Cell Protein-Protein Interaction Analysis by SPLIFF

We recommend the following general protocol for performing a successful SPLIFF analysis (Fig. 4).

1. Construct the CCG- and N_{ub}-fusion proteins in the required strain backgrounds.
2. Cultivate Y-CCG and N_{ub}-X yeast strains in appropriate liquid media.
3. Mix logarithmically grown cells of both strains in liquid media.
4. Prepare the microscope slide with the required solid media and trap the cells.
5. Image cell growth of fusing cells using a microscope and record the fluorescence intensities of mCherry and GFP.
6. Measure the fluorescence intensities and plot the data to visualize interaction kinetics.

3.2 Introducing CCG and P_{CUP1}-N_{ub}-HA Cassettes into Yeast

1. Inoculate a fresh culture from a plate with wild-type yeast (JD47 or JD53) in YPD medium.
2. To obtain a mid-log-phase culture on the day of transformation, inoculate 50 mL fresh YPD medium with the overnight culture to an OD_{600} of 0.3 and incubate at 30 °C with shaking to a final OD_{600} of 0.6–0.8 (approximately 4 h).
3. Spin down culture for 5 min at 1500 ×*g* and wash the cell pellet once with 50 mL water.
4. Centrifuge suspension and resuspend the pellet in 1 mL water.
5. Pellet cells again, remove the supernatant, and resuspend yeast in 1.5 mL lithium acetate solution.
6. Mix 100 μL of the prepared cell suspension with 10 μL single-stranded carrier DNA and 10 μL linearized pY-CCG-306 plasmid or $P_{CUP1}N_{ub}$HA-kanMX cassette (each with a total DNA amount of 0.5–2 μg).

7. Add to the complete transformation mix 600 μL PEG solution, vortex for 30 s, and incubate for 30 min at 30 °C with gentle agitation.
8. Heat shock samples for 15 min in 42 °C water bath and afterwards let them adjust to room temperature.
9. Remove supernatant, resuspend cells in 300 μL water, and plate on selective media immediately when using auxotrophic markers for selection (e.g., SD Ura- for pY-CCG306-transformed cells). In case of $P_{CUP1}N_{ub}$*HA-kanMX* transformation resuspend in 300 μL YPD, incubate for 1 h at 30 °C to allow expression of the antibiotic resistance marker *kanMX*, and plate afterwards on selective media (YPD Gen$^+$).
10. Incubate plates for 2–3 days at 30 °C and check for colony growth.
11. Verify positive clones by diagnostic PCR.

3.3 Sample Preparation

A variety of techniques exist for live cell imaging. There is an option between commercially available cell chambers and the more traditional way of preparing microscope slides for imaging. A glass slide with a cavity fully filled with media is sufficient for SPLIFF analysis and is described below.

1. Incubate the CCG and N_{ub} fusion-expressing a- and α-cells separately in 3 mL selection medium (SD Ura$^-$ and YPD Gen$^+$) at 30 °C overnight.
2. Dilute the cells into 3 mL fresh medium without copper sulfate and let them grow to an OD_{600} of 0.6–1. Low N_{ub}-X expression levels can be compensated by including 100 μM copper sulfate.
3. Pour the hot live cell imaging media mixture into the cavity of the microscope slide and pull a cover slip over it to obtain a bubble-free, flat surface of the agarose pad. Let it cool down for 10 min.
4. Mix equal amounts of the a- and α-cells (between 0.5 and 0.75 ml each), spin the culture immediately down, and resuspend in 50 μL of SD medium.
5. Remove the cover slip and immobilize 3.5 μL of the prepared cell suspension on the agarose pad by pushing back and forth a new cover slip, afterwards fix the cover glass with parafilm strips.
6. Incubate the slide at 30 °C under the microscope. After 45–75 min of incubation usually the first zygote formations become visible. To obtain enough single-cell measurements for the analysis select 8–16 positions of cells on the slide that will mate but have not started to fuse and collect them in a point list of the microscope software.

3.4 Cell Imaging by Time-Lapse Experiment

1. Pictures have to be taken in three channels: bright field, GFP, and mCherry. Wavelengths of emission and excitation have to be suitable to obtain a good signal of the fluorophores.
2. Autofocus should be enabled to gain pictures in focus during 3–6 h of image acquisition despite shifting due to cell growth.
3. To receive a good and stable fluorescence signal during all phases of the cell cycle (PCDI-III), a four-section z stack with intervals of 0.5 μm at each position is recommended. Pictures should be taken at each z-level for all three channels before moving to the next z-level to obtain the most accurate overlay of mCherry and GFP.
4. Camera capture time needs to be adapted to the intensity of GFP and mCherry signal in every construct to reduce bleaching and phototoxicity.
5. Two- to five-minute intervals are recommended for time-lapse measurements. Of course, intervals have to be adjusted to the kinetics of the measured interaction and the experimental question.

3.5 Quantification

Fluorescence intensity measurements of the obtained images were processed and analyzed with ImageJ 1.47 software (US National Institutes of Health; http://rsbweb.nih.gov/ij/).

1. Split each raw file generated by the microscope software from hyperstacks to stacks (4 slices/3 channels/*x* time frames) and create cutouts of the files showing one mating event. Run a sum projection of the z-series to obtain two-dimensional images, and then split the channels (Fig. 5a).
2. Accurately select regions of interest (ROI) with the Polygon Tool in the mCherry channel for each time frame starting before the mating begins and save them in the ROI-Manager. Double-check, and if necessary correct the ROIs in the GFP channel.
3. Measure the intensity of the selected ROIs in both channels (measuring parameters: Area, *IntDen*, Stack position) and save the results in an Excel worksheet. (loc) represents the position of the selected ROIs and integrated density (*IntDen*(loc)) specifies the fluorescence signal intensity in the corresponding areas (Fig. 5b):

$$IntDen(\text{loc}) = \text{Area}(\text{loc}) \times \text{Mean Grey Value}(\text{loc})$$

4. Select ROIs within the cell in each time frame that could serve as background for the quantification. Measure the intensity of the selected background ROIs in both channels and save the values in Excel. (back) specifies the signal of the background areas.

5. To obtain the fluorescence intensities (FI) of each channel subtract the background values, adjusted to the Area(loc) from the integrated density values at each time frame. In time-dependent interaction experiments (x) represents time (Fig. 5b):

$$\mathrm{FI}(x) = \mathit{IntDen}(\mathrm{loc}) - (\mathrm{Area}(\mathrm{loc})) \times \mathrm{Mean\ Grey\ Value}(\mathrm{back})$$

6. FI(start) represents the fluorescence intensities at the time point shortly before the fusion of the cells. Calculate the relative fluorescence intensities (RFI(x)) by

$$\mathrm{RFI}(x) = \mathrm{FI}(x) \times 100 / \mathrm{FI}(\mathrm{start})$$

Mark the time points of FI(start), as well as characteristic cell cycle events such as bud emergence and cytokinesis. Indicate time points where the localization of the CCG-labeled proteins could not be recorded (Table 1, Fig. 6).

7. Calculate the conversion (FD) of the CCG fusion to the CC fusion by:

$$\mathrm{FD}(x) = (\mathrm{RFI}_{\mathrm{RED}} - \mathrm{RFI}_{\mathrm{GREEN}}) / \mathrm{RFI}_{\mathrm{RED}} \times 100$$

If the reference signal is too weak at some time points ($\mathrm{RFI}_{\mathrm{RED}} < 10$), do not use them for the following calculations.

8. Do **steps 1–8** for at least six to eight matings. As not all cells grow equally well, create a table and summarize the FD(x) values of all matings. The values of FD(start) should be listed in one row. If the monitored CCG-fusion proteins alter their positions during the cell cycle, organize the values of their first appearance in one row (here: bud emergence (PCDII), cytokinesis (PCDIII)).
9. Calculate the mean and standard error (s.e.) for each row. Create a graph with x = time and cell cycle phase, y = FD(x) and error bars (±s.e.) (Fig. 7).
10. Perform statistical evaluations and curve fittings of the experimental data when comparing two or more data sets with each other (Fig. 7). Software for statistical evaluation and curve fitting is available from http://sysbio.uni-ulm.de/soft/CuCompare. The webpage contains source code (R language) and guides through an example workflow. Statistical evaluation is performed by minimizing the least squares deviation of a set of regression functions to the data. The best fitting function is determined by the Akaike information criterion. For statistical comparison of two groups, represented as curves, the sum of the residual sum of squares of the individual regression processes is then compared to that of the regression for the combined data of both groups using the F-test [19, 20] (Fig. 8).

4 Notes

1. Attaching N_{ub} or CCG to a yeast protein involved in mating can inhibit cell fusion and consequently SPLIFF analysis. Expressing the fusion protein from a centromeric plasmid and leaving the genomic copy untouched should solve this problem (*see* Subheadings 2.4 and 2.5).
2. A fast interaction might completely convert Y-CCG into Y-CC already during mating and therefore prevent the SPLIFF analysis during the later stages of the cell cycle. We propose two approaches: (1) expressing the Y-CCG fusion under control of the P_{MET17}-promoter and thereby increasing the expression of the Y-CCG fusion and (2) downregulating the N_{ub} fusion or expressing it during later phases of the cell cycle through the proper choice of a promoter that is preferentially active only during these times (*see* Subheading 2.5).
3. Always run a control experiment under the same conditions as the Y-CCG/N_{ub}-X combination of interest. If available, a test with a N_{ub} fusion of a known binding partner of Y will place the results of the investigated interaction in the context of what is already known about Y and X. A measurement between Y-CCG and a N_{ub} fusion known not to interact with Y allows to better judge the significance of the measured interaction (*see* Subheading 3.1).
4. A low intensity of the mCherry signal sometimes limits the application of SPLIFF. Increasing the level of the Y-CCG fusion by expressing it under the control of the P_{MET17}-promoter often alleviates this problem. The methionine concentration in the media for an optimal Y-CCG expression, laser power, emission time, and camera gain for the time-lapse experiments have to be carefully adjusted to achieve the optimal yield of fluorescence signals with low background fluorescence and minimal phototoxic effects (*see* Subheading 3.4).
5. Phototoxicity is an issue when measuring interactions throughout one or more complete cell cycles. As a general rule we leave at least 2 min between two measurements. Try to minimize the stress on the cells by reducing laser power/emission time (*see* Subheading 3.4).
6. The maturation time of the two fluorophores in the Y-CCG fusion differs (mCherry matures slower, GFP faster). When the rate of Y-CCG conversion is much slower than the synthesis of new Y-CCG fusion during the cell cycle a negative FD(x) might be encountered (*see* Subheading 3.5).
7. In experiments where the Y-CCG fusion is confined to a distinct region of the cell, the background fluorescence should be measured within the Y-CCG-expressing cell before mating,

and later in the diploid cell. Make sure not to choose an area with an aberrant fluorescence in either the mCherry or the GFP channel and avoid the vacuole as region of reference. In experiments where the CCG fusion is randomly distributed throughout the cell, a section from a haploid N_{ub} fusion-expressing cell occupying the same optical field should be used as reference (*see* Subheading 3.5).

Acknowledgement

This work was supported by the BMBF Initiative SysTec (0315690B) and by the BMBF Initiative GerontoSys2 (SyStaR, 03158 94A).

References

1. Stynen B, Tournu H, Tavernier J, Van Dijck P (2012) Diversity in genetic in vivo methods for protein-protein interaction studies: from the yeast two-hybrid system to the mammalian split-luciferase system. Microbiol Mol Biol Rev 76:331–382
2. Müller J, Johnsson N (2008) Split-ubiquitin and the split-protein sensors: chessman for the endgame. Chembiochem 9:2029–2038
3. Johnsson N, Varshavsky A (1994) Split ubiquitin as a sensor of protein interactions in vivo. Proc Natl Acad Sci U S A 91:10340–10344
4. Stagljar I, Korostensky C, Johnsson N, te Heesen S (1998) A genetic system based on split-ubiquitin for the analysis of interactions between membrane proteins in vivo. Proc Natl Acad Sci U S A 95:5187–5192
5. Wittke S, Lewke N, Müller S, Johnsson N (1999) Probing the molecular environment of membrane proteins in vivo. Mol Biol Cell 10:2519–2530
6. Tafelmeyer P, Johnsson N, Johnsson K (2004) Transforming a (beta/alpha)8—barrel enzyme into a split-protein sensor through directed evolution. Chem Biol 11:681–689
7. Ear PH, Michnick SW (2009) A general life-death selection strategy for dissecting protein functions. Nat Methods 6:813–816
8. Moreno D, Neller J, Kestler HA, Kraus J, Dünkler A, Johnsson N (2013) A fluorescent reporter for mapping cellular protein protein interactions in time and space. Mol Syst Biol 9:647
9. Johnsson N, Varshavsky A (1994) Ubiquitin-assisted dissection of protein transport across membranes. EMBO J 13:2686–2698
10. Wittke S, Dünnwald M, Johnsson N (2000) Sec62p, a component of the endoplasmic reticulum protein translocation machinery, contains multiple binding sites for the Sec-complex. Mol Biol Cell 11:3859–3871
11. Gao JT, Guimera R, Li H, Pinto IM, Sales-Pardo M, Wai SC, Rubinstein B, Li R (2011) Modular coherence of protein dynamics in yeast cell polarity system. Proc Natl Acad Sci U S A 108:7647–7652
12. Sheu YJ, Santos B, Fortin N, Costigan C, Snyder M (1998) Spa2p interacts with cell polarity proteins and signaling components involved in yeast cell morphogenesis. Mol Cell Biol 18:4053–4069
13. Lippincott J, Li R (1998) Dual function of Cyk2, a cdc15/PSTPIP family protein, in regulating actomyosin ring dynamics and septin distribution. J Cell Biol 143:1947–1960
14. Kiel C, Serrano L (2012) Challenges ahead in signal transduction: MAPK as an example. Curr Opin Biotechnol 23:305–314
15. Vidal M, Cusick ME, Barabasi AL (2011) Interactome networks and human disease. Cell 144:986–998
16. Dünkler A, Müller J, Johnsson N (2012) Detecting protein-protein interactions with the split-ubiquitin sensor. Methods Mol Biol 786:115–130
17. Janke C, Magiera MM, Rathfelder N, Taxis C, Reber S, Maekawa H, Moreno-Borchart A, Doenges G, Schwob E, Schiebel E, Knop M (2004) A versatile toolbox for PCR-based tagging of yeast genes: new fluorescent proteins, more markers and promoter substitution cassettes. Yeast 21:947–962

18. Hruby A, Zapatka M, Heucke S, Rieger L, Wu Y, Nussbaumer U, Timmermann S, Dunkler A, Johnsson N (2011) A constraint network of interactions: protein-protein interaction analysis of the yeast type II phosphatase Ptc1p and its adaptor protein Nbp2p. J Cell Sci 124:35–46

19. Motulsky HJ, Ransnas LA (1987) Fitting curves to data using nonlinear regression: a practical and nonmathematical review. FASEB J 1:365–374

20. Lomax RG, Hahs-Vaughn DL (2012) Statistical concepts: a second course. Routledge Chapman & Hall, New York

Chapter 12

Microfluidic Proximity Ligation Assay for Profiling Signaling Networks with Single-Cell Resolution

Matthias Blazek, Günter Roth, Roland Zengerle, and Matthias Meier

Abstract

The proximity ligation assay (PLA) is a technique that can be used to characterize proteins, protein–protein interactions, and protein modifications at the single-cell level. Image-based in situ detection of proteins using PLA is a quantitative method with a high degree of sensitivity and specificity. The miniaturization and parallelization of the PLA onto a microfluidic chip and concurrent use of an automated cell-culture system increase the throughput of this technology. Here, we describe the performance of PLA on a microfluidic chip. We provide protocols for on-chip cell culture, time shifted cell stimulation and fixation, PLA implementation, and computational image analysis in order to achieve single-cell resolution. As a proof of concept, we studied the phosphorylation of Akt in response to stimulation with platelet-derived growth factor.

Key words Proximity ligation assay, Microfluidics, Polydimethylsiloxane, Single-cell segmentation, Cell culture, Cell signaling

1 Introduction

Single-cell analytics are important for identifying rare events and biological variance and for characterizing low-abundance cell types in valuable or small-volume patient samples [1]. Single-cell resolution is equally important for cell-signaling studies [2, 3]. Specifically, differences in cell size, protein quantities, culture density, cell cycle, and microenvironment can cause variations in cell signaling, signal transduction, and termination [4]. Technical advances in biological engineering have yielded analytical methods for obtaining single-cell genomic information. However, few assays are available for proteomic analysis at the single-cell level [5].

The proximity ligation assay (PLA) is an in situ technique for quantifying proteins and identifying protein–protein interactions and posttranslational modifications in cells [6]. The PLA reaction utilizes two nucleotide-labeled antibodies to detect the protein target. When the bound antibodies are in close proximity, the labeled

Anup K. Singh and Aarthi Chandrasekaran (eds.), *Single Cell Protein Analysis: Methods and Protocols*, Methods in Molecular Biology, vol. 1346, DOI 10.1007/978-1-4939-2987-0_12,

oligonucleotides are hybridized and ligated. Subsequently, the ligation product serves as a template for isothermal rolling-circle amplification (RCA). The amplified DNA is stained with probe molecules and can be detected as polymer dots using a standard fluorescence microscope. Proximity events from a single cell can be obtained by image analysis, including cell segmentation, fluorescence dot counting, and subcellular localization.

The use of PLA for quantitative cell-signaling studies is limited by the complexity of the biochemical steps and the mode of endpoint detection. Previously, we demonstrated that these drawbacks could be overcome by miniaturization and parallelization of the PLA and concurrent use of an automated cell-culture system on a microfluidic chip [7, 12]. The microfluidic device and a representative image obtained after PLA processing are shown in Fig. 1. The PLA chip consists of two polydimethylsiloxane (PDMS) layers, the control and flow layers, and a glass substrate [8, 9]. Both PDMS layers are manufactured by standard rapid prototyping [10]. The flow layer contains 128 cell-culture chambers arranged in a 16 × 8 matrix. Each cell chamber is 400 × 400 μm^2 in size with a volume of 8 nL and has a capacity of 100–200 cells (Figs. 1 and 2). A thin PDMS membrane separates the control layer from the flow layer. The control layer has microchannels with a fluid inlet but no outlet. Pneumatic actuation of these dead-end channels leads to deflection of the thin membrane between the flow layer and control layer. At cross sections between the channels in the control layer and flow layer, the membrane acts as a pneumatic valve. These valves guide and control the flow of all reagents required for cell culture and integration of the PLA.

Each column or row of cell-culture chambers can be addressed separately. Thus, the cell-culture array enables the investigation of temporal variations in cell signaling. Upon fixation of the cell cultures, the molecular state of the cell-signaling process is maintained. A fully integrated chip-based PLA is then performed in order to visualize temporal changes in proteins among thousands of cells. Furthermore, the PLA can be multiplexed for more than one target by using cell-culture replicates arranged and addressable in column or row direction. Thus, the chip-based PLA allows for the measurement of two independent parameters (e.g., stimulation time and PLA targets).

Here, we provide a protocol for using the microfluidic PLA chip to measure changes in protein phosphorylation after cell stimulation. First, we delineate methods for culture and stimulation of adherent cells on the chip. Then, we provide detailed methods for performing on-chip PLA. Finally, we describe the retrieval of single-cell data from microscopic PLA images using computational image segmentation. As a proof of concept, we stimulated NIH-3 T3 mouse fibroblasts with platelet-derived growth factor (PDGF) and measured subsequent phosphorylation of Akt at serine 473.

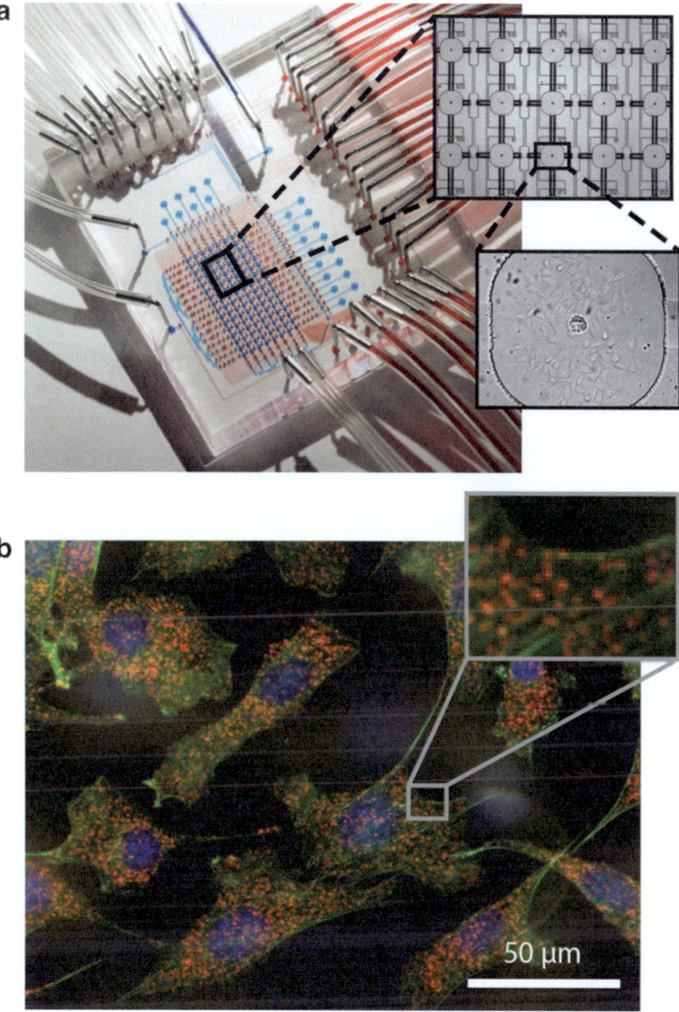

Fig. 1 (**a**) The microfluidic PLA chip. For illustration, microchannels within the PDMS flow layer and control layer are filled with *blue*- and *red-colored* fluids, respectively. The external pressure supply and the chip's fluid channels are connected with plastic tubing and metal pins. *Upper inset*: Enlarged view of the matrix of cell-culture chambers on the chip. *Lower inset*: Enlarged view of a single cell-culture chamber filled with cells. (**b**) Representative image of NIH-3 T3 mouse fibroblasts cultured and automatically processed using the microfluidic PLA chip. *Red staining* represents the DNA polymer obtained after detection of the phosphorylated Akt (p-Ser-473). The nuclei and cytoskeleton (phalloidin) are counterstained with DAPI (*blue*) and Atto 488 (*green*), respectively

2 Materials

2.1 Technical Components

1. CAD file of the microfluidic PLA chip (*see* **Note 1**).
2. Master molds and/or PDMS replica of the microfluidic PLA chip (Stanford, Foundry Service).
3. Microfluidic pressure-control system: Detailed instructions for building a microfluidic pressure-control system are described

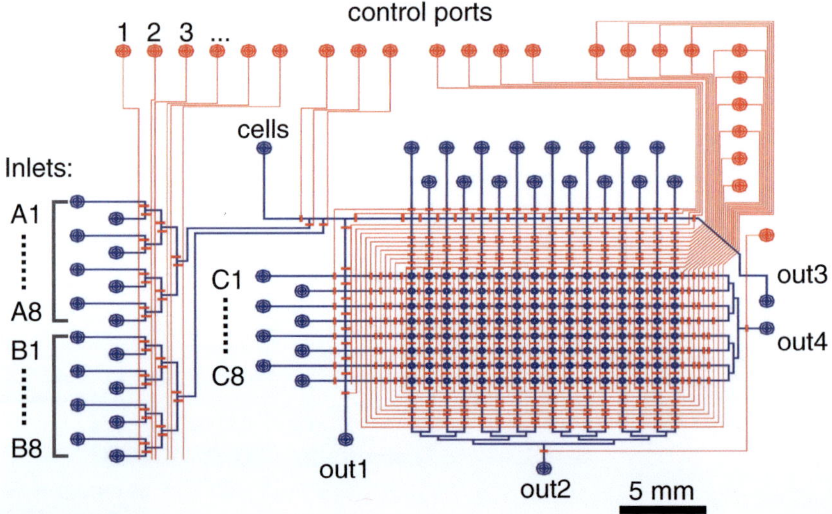

Fig. 2 Layout of the microfluidic PLA chip. The microchannel network of the PDMS control and the flow layer are shown in *red* and *blue*, respectively. The inlet ports for the microchannels in both layers are indicated with a *circle* (control layer, *red*; flow layer, *blue*). Flow-inlet ports labeled A1–A8 are used for cell culture, stimulation, and preparation of reagents; those labeled B1–B8 are used for PLA reagents; those labeled C1–C8 are used for primary antibodies, and that labeled D is used for cell suspensions. The PLA chip contains four flow-outlet ports. The unlabeled inlet ports are not used in this protocol. The control inlet ports 1–24 are connected to the pressure-control unit and filled with water

at http://www.stanford.edu/group/foundry/ or https://sites.google.com/site/rafaelsmicrofluidicspage/home.

4. Control software (graphical user interface, GUI) and assay-automation files (*see* **Note 1**).
5. Microsoft Excel files pla01.xlsx—pla07.xlsx (*see* **Note 1**).
6. Tygon tubing (inner diameter, 0.02 in.; outer diameter, 0.06 in.; Saint Gobain, Paris, France): Each assay requires approximately 8 m in length.
7. Metal pin connectors (outer diameter, 0.025 in.; inner diameter, 0.013 in.; New England Small Tube, Litchfield, NH).
8. 1.5-mL glass bottles sealed with septum (e.g., Mueller, Fridolfing, Germany).
9. Metal forceps.

2.2 Laboratory Equipment

1. Inverted fluorescence microscope equipped with a 5× and 20× objectives for visualization of the chip during the assay and image acquisition, respectively.
2. Fluorescence filters for DAPI (excitation, 365 nm; beamsplitter, 395 nm; emission, 445 nm), green fluorescence protein (GFP) (excitation; 470 nm; beamsplitter, 495; emission,

525 nm), and rhodamine (excitation, 550 nm; beamsplitter, 570 nm; emission, 605 nm) channels.

3. Cell incubator for an inverted microscope for temperature and CO_2 control.
4. Desktop PC (minimum RAM, 4 GB).
5. Matlab version 2008 or later (Mathworks, Natick, MA, USA).

2.3 Chemicals

1. 0.05 % Fibronectin from human plasma in phosphate-buffered saline (PBS).
2. Fixation reagent: 4 % paraformaldehyde in PBS.
3. Growth hormone: 100 μg/mL Platelet-derived-growth-factor-BB (PDGF-BB) from mouse recombinant produced in *E. coli* (ProSpec) in sterile filtered 100 mM acetic acid and 0.1 % BSA.
4. Permeabilization reagent: 0.05 % Tween-20 in water.
5. Blocking reagent: 1 % bovine serum albumin (BSA) in PBS.
6. Primary antibodies (*see* **Note 2**): In this work, we used anti-pAkt-(Ser473) procured from Cell Signaling (Product Number: 4060).
7. PLA detection starter kit (Olink, Uppsala, Sweden), including host-specific PLUS- and MINUS-PLA probes (secondary antibodies), wash buffer A, and wash buffer B, Duolink Ligation stock (5×), Duolink Ligase, Duolink Amplification stock, and Duolink polymerase. The ligation stock contains all oligonucleotides to form a circular DNA template for the RCA reaction in cases that the two PLA probes are in close proximity. Note that the amplification stock contains the fluorescently labeled detection and thus amplification of the RCA template and staining are performed in one step. For detailed description see the Olink product information.
8. Cell-staining solution: 0.5 vol.% Phalloidin-Atto 488 and 0.1 vol.% 4′,6-diamidino-2-phenylindole (DAPI) in PBS.
9. TrypLE Express (Gibco, Life Technologies, Carlsbad, CA).

2.4 Cell Culture

1. Cells: NIH-3 T3 (ATCC® CRL-1658™).
2. Cell-culture medium: DMEM containing 4.5 g/L glucose, 0.58 g/L L-glutamine, 0.11 g/L pyruvate supplemented with 10 % fetal bovine serum (FBS), and 1 % penicillin–streptomycin.
3. Cell starvation medium: DMEM containing 4.5 g/L glucose, 0.58 g/L L-glutamine, 0.11 g/L pyruvate supplemented with 0.05 % FBS, and 1 % penicillin–streptomycin.
4. Cell stimulation medium: 100 ng/mL PDGF-BB (from the stock solution **item 3** of Subheading 2.3) in cell starvation medium.

2.5 Image Segmentation

1. Image segmentation software: Open-source software CellProfiler [11] (*see* **Note 3**).

3 Methods

3.1 Microfluidic Framework

1. Place an order for the molds of the microfluidic PLA chip from any open clean room facility. The CAD file for the PLA chip can be downloaded (*see* **Note 1**).
2. Build a USB pressure-control system for 48 solenoid valves by following the instructions provided by either the Stanford Microfluidic Foundry or the indicated Internet resource. Both control systems have equal efficacy.
3. Connect the microfluidic pressure-control system to a desktop PC via USB.
4. Install Matlab on the desktop PC and start the GUI for the PLA chip (*see* Fig. 3). The control panel is divided into two sections. The first area allows for manual control (buttons A–E) of the PLA chip. The second area allows for automated

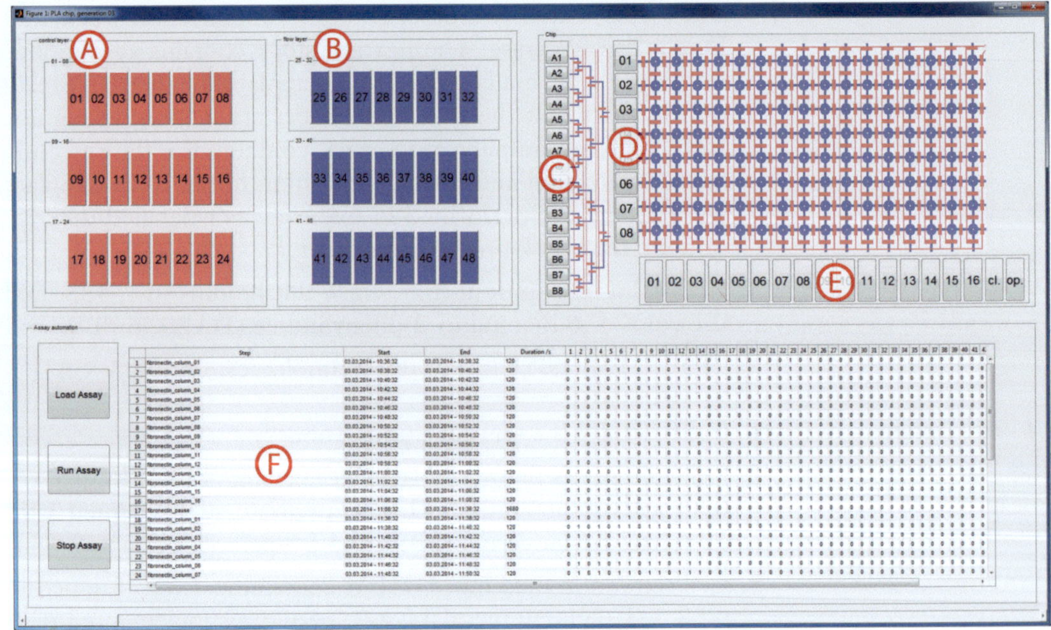

Fig. 3 Matlab-based GUI for controlling reagent flow on the PLA chip. *Red letters* (*A*) through (*E*) indicate the key components for manual control of the chip; (*F*) indicates the area for automated control of the assay. The *red* and *blue* buttons control the control-layer and flow-layer features on the PLA chip, respectively. *See* **step 4** of Subheading 3.1 for a detailed explanation

control (button F) of the PLA chip. Buttons within the manual section are divided into flow controllers (blue), which control the reagents, and static controllers (red), which open and close the PDMS membrane valves. The allocation of the reactions to the inlet port of the chip (blue circles) and static (red circles) controllers is shown in Fig. 2. The combined use of the buttons in areas A and B allows for the manual control of all possible fluid routes. However, shortcuts for the most often used flow routes have been programmed and can be accessed using the buttons in control areas C, D, and E. Use of the buttons in area C opens all control valves to allow selective flow through inlet ports A1–8 and B1–8; use of the buttons in areas D and E routes the fluids through the cell-culture chambers within a specified column or row of the matrix. The lower section of the GUI contains file-handling operations for automating the flow processes on the chip. A program that defines the switch state of all valves on the chip can be programmed in the Microsoft Excel files. The corresponding load, run, and stop functions for the files are labeled in the GUI window.

5. Connect the PLA chip to pressure lines 1–24 of the microfluidic control system by connecting the metal pins and water-filled tygon tubing (Fig. 1a). The metal pins can be bent to 90° with tweezers. For port allocation, follow the instructions in Fig. 2.
6. Set the control-valve pressure on the PLA chip to 100–200 kPa and ensure that all valves are closed.
7. Mount the connected chip to the microscope inside the on-stage cell incubator.
8. Confirm that all PDMS membrane valves on the chip can open and close by manually clicking the corresponding buttons in area A of the GUI.

3.2 PLA Overview and On-Chip Implementation

In this section, we outline the operations for seeding, culturing, stimulating, fixing, and permeabilizing cells on the PLA chip. This sample-preparation process creates an array of cell cultures that encode the cell-signaling transduction process with temporal resolution. Here, we demonstrate how to perform the on chip PLA on the cell-culture array. Table 1 shows the main assay steps and the associated automation protocols, processing times, reagent-inlet-port assignments, and environmental conditions. In general, all solutions with volumes less than 50 μL are stored in tygon tubing, while those with volumes greater than 50 μL are stored in 1.5-mL glass vials. The flow pressure applied to the vials or tubing in all steps is 25 kPa, which corresponds to a volume flow rate of 0.65 μL/min.

Table 1
Integrated processing steps on the PLA chip

Assay step	Excel file	Duration	Flow inlet allocation	Figure	Conditions
Fibronectin coating	pla01.xlsx	3 h	A1: fibronectin A2: full medium	4a	37 °C 5 % CO_2
Cell seeding	Manual	~1 h	Cell inlet: cells	4b	37 °C 5 % CO_2
Cleaning	Manual	10 min	Cell inlet: trypsin-EDTA	4c	37 °C 5 % CO_2
Cell culture Starvation	pla02.xlsx	Up to 96 h	A2: full/starvation medium	4d	37 °C 5 % CO_2
Stimulation Fixation Permeabilization Blocking	pla03.xlsx	3 h	A3: stimulation medium A4: fixation reagent A5: perm. reagent A6: blocking reagent	4d	37 °C 5 % CO_2
Primary antibodies	pla04.xlsx	8 h	C1 to C8: antibody dilutions	4e	20 °C
Secondary antibodies	pla05.xlsx	4 h	B1: wash buffer A B2: secondary antibodies	4f	20 °C
Ligation	pla06.xlsx	1.5 h	B3: ligation dilution	4f	32 °C
Amplification Staining	pla07.xlsx	4 h	B4: amplification dilution B5: wash buffer B B6: staining dilution	4f	32 °C

3.3 Fibronectin Surface Treatment of the PLA Chip

1. Start the cell incubator to establish cell-culture conditions of 37 °C and 5 % CO_2. Due to the gas permeability of PDMS, the chip equilibrates with the external gas environment within minutes.
2. Connect the fibronectin solution and cell culture medium to inlet ports A1 and A2, respectively.
3. To coat the glass surface of the cell-culture chambers with fibronectin, load and start protocol “pla01.xlsx” in the GUI. The fluid route on the chip for this step is shown in Fig. 4a. At the end of the 2-h protocol, the fibronectin solution is replaced by the cell-culture medium.

3.4 Cell Loading

1. Harvest the cells from cell-culture flasks using a trypsin solution. Centrifuge the cell suspension for 5 min at 1000 × g. Resuspend the cells with preheated cell-culture medium to a concentration of 10^6 cells/mL.
2. Load the cell suspension into tygon tubing, and connect the tubing to the cell-inlet port of the chip (Port: Cells in Fig. 2).

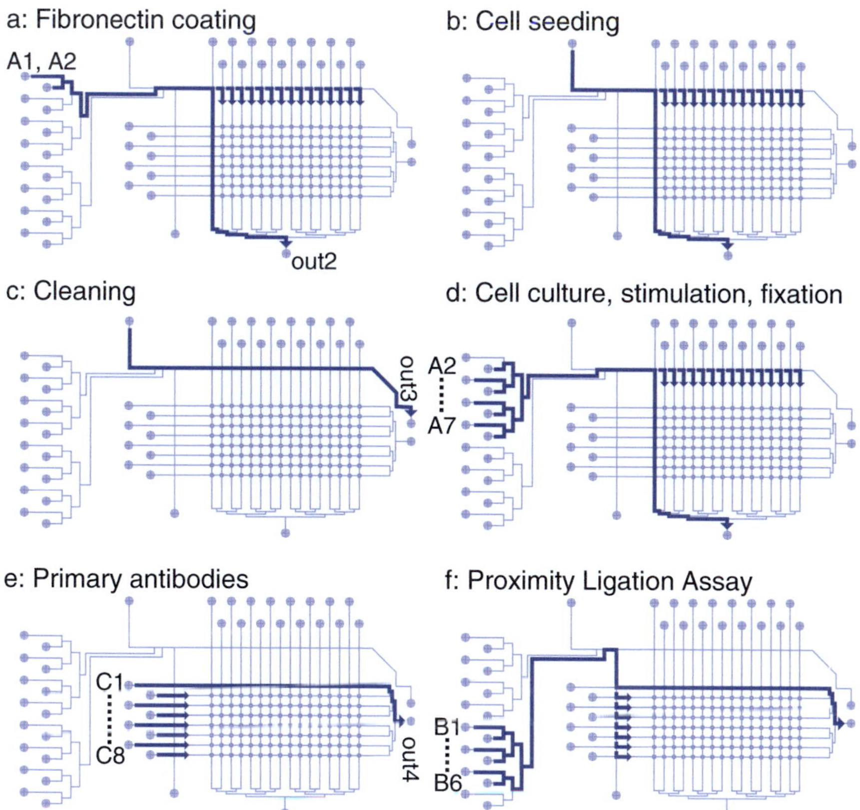

Fig. 4 Routing of the reagents through the microfluidic chip during the main assay. The *highlighted* channels and *arrowheads* indicate the flow direction on the chip. During steps (**a**) through (**d**) the corresponding reagents are flushed through columns of cell-culture chambers; during steps (**e**) and (**f**), the reagents are flushed through rows

The cells are flushed onto the chip manually. To perfuse cell-culture chambers in column elements of the matrix, use the buttons in area E of the GUI, and monitor the cell density in real time with the microscope. The desired confluence within the cell-culture chamber is 80 %. Wait for 10 min to allow the cells adhere to the fibronectin coated glass surface of the PLA chip. If the cell density is insufficient, repeat the cell-seeding procedure.

3. Disconnect the tubing containing the cell suspension, and connect the TrypLE express solution to the cell-inlet port. Manually flush the TrypLE express solution through the inlet line for 10 min in order to remove cells and cell debris (Fig. 4c). Alternately stopping and starting the flow helps to remove cell debris from the channels.

3.5 Cell Culture

1. Load and start the automated cell-culturing protocol “pla02. xlsx.” Each column of cell-culture chambers is flushed with cell culture medium for 2 min at an interval of 90 min and a volume flow rate of 0.65 μL/min. Reduction or prolongation

of the feeding interval below 60 and 120 min, respectively, will reduce the cell viability. Cell cultures can be maintained for at least 96 h (*see* **Note 4**).

2. (Optional) Cell starvation: For assays requiring a starvation step, replace the cell-culture medium in inlet A2 with cell starvation medium. Run protocol "pla02.xlsx" for the desired starvation period.

3.6 Cell Stimulation, Fixation, and Permeabilization

1. Connect the cell stimulation medium to flow inlet A4, the fixation reagent to flow inlet A5, the permeabilization reagent to flow inlet A6, and the blocking reagent to flow inlet A7.
2. Load and run protocol pla03.xlsx from the GUI. This protocol implements a cell stimulation time series for 16 time points. For this, cell cultures in the column elements 1–15 of the PLA chip are stimulated for 120, 90, 75, 60, 50, 40, 30, 20, 15, 10, 8, 6, 4, 2, and 1 min, respectively. Column 16 is used as a no-stimulation control (*see* **Note 5**). The volume of the cell-culture chambers is replaced multiple times, resulting in a homogeneous concentration of stimulation reagent across all cell-culture chambers.
3. To fix the cell cultures after a given stimulation interval, the fixation reagent is flushed into the columns in reverse order, starting at column 16 and ending at column 1. The fixation pulse is included in the flush protocol "pla03.xlsx."
4. The permeabilization and blocking steps are the final steps in the flush protocol "pla03.xls." The flow path of the flush steps is indicated in Fig. 4d.

3.7 PLA Assay

1. The mLSI chip allows to multiplex up to eight primary antibody pairs or single antibodies. Each antibody solution has to be prepared fresh at appropriate dilution.
2. Aspirate approximately 25 μL of each antibody solution into a tygon tube and connect the tubes to the flow-inlet ports C1–C8 (*see* Fig. 4e). Manually apply the flow pressure to the antibody channels (GUI buttons 33–40). Dead-end filling of the inlet channels prevents cross contamination of the antibody solutions. Therefore, on-chip valves of the antibody input channels remain closed during the fill process, whereas the air within the inlet channels is released through the PDMS. Ensure that the antibody inlet channels are free of air before proceeding to the next step.
3. Load and start protocol "pla04.xlsx" within the GUI. Antibody solutions are flushed in by row through the cell-culture chambers. Each flush/incubation cycle lasts for 2 h and is repeated three times. The flush protocol can be modified if longer incubation times are required.

4. Use a 1.5-mL glass container to connect wash buffer A (from the PLA kit) to inlet B1, as indicated in Fig. 4f.
5. Freshly prepare the secondary antibody at the appropriate dilution, and connect it to flow-inlet port B2. For a total volume of 30 μL, dilute 5 μL of PLUS probe and 5 μL of MINUS probe into 20 μL of the antibody diluent from the Olink PLA kit. Select secondary antibodies with host specificity corresponding to that of the primary antibodies.
6. Load and run the flush protocol "pla05.xlsx." This protocol starts by clearing the primary antibody solutions with wash buffer A. Then, the secondary antibody solution is perfused over the cell cultures followed by another washing step with buffer A.
7. Set the cell incubator to 32 °C, and turn off the CO_2 regulation.
8. Prepare the ligation solution, and connect it to flow-inlet port B3 (*see* Fig. 4f). For a total reaction volume of 20 μL, dilute 4 μL of Duolink Ligation stock (5×) in 16 μL high-purity H_2O. Then, add 0.5 μL Duolink ligase.
9. Load and run protocol "pla06.xlsx" to flush, incubate, and clear the ligation solution.
10. Prepare the amplification solution, and connect it to flow-inlet port B4. For a total reaction volume of 20 μL, dilute 4 μL of Duolink Amplification stock (5×) in 16 μL high-purity H_2O. Then, add 0.75 μL of Duolink polymerase. The amplification stock contains the fluorescently labeled probe oligonucleotide for the RCA product and thus no further flush step for staining is required.
11. Connect buffer B to flow-inlet port B5.
12. Connect the cell staining solution for counterstaining the F-actin and the cell nucleus to flow-inlet port B6.
13. Load and run protocol "pla07.xlsx" to flush, incubate, and clear the amplification mix, staining solution, and wash buffer B. Protect the chip from light during the entire procedure.

3.8 Image Acquisition

1. To obtain optimal fluorescence PLA signals, the chip should be imaged after completion of the assay procedure; however, the chip can be stored at 4 °C and protected from light for up to 2 weeks.
2. Acquire three images per position corresponding to the fluorescence channels for DAPI (nuclei), GFP (cytoskeleton), and rhodamine (PLA signal). We recommend obtaining a bright-field image at each position for orientation on the chip. The bright-field image is not used for image analysis. Use identical camera integration times for the entire imaging procedure.

3. Save all images as separate files in a non-compressed file format (e.g., .bmp or .tif). Use appropriate nomenclature for subsequent image analysis (e.g., "c1," "c2," "c3,") in order to identify the three fluorescence channels.

3.9 Image Analysis with Single-Cell Resolution

Depending on the cell line, cell density on the chip, and magnification used for image acquisition, a single image contains 10–200 cells. For cell segmentation and isolation of single-cell PLA dot information, use the free image-analysis software, CellProfiler [11]. The following steps summarize the procedure for obtaining single-cell PLA dot-count information from NIH 3 T3 cells using CellProfiler.

1. Collect all images in a single folder. Each channel must be separated and encoded under its own filename (e.g., "c1" for DAPI, "c2" for GFP, and "c3" for rhodamine).
2. Install and start the CellProfiler software.
3. Set the default input folder to the location where the files are stored.
4. Initialize a pipeline (i.e., a workflow) for image analysis. Pipelines can be saved and reused for similar experiments.
5. Add the module "LoadImages" to the pipeline. Add the three channels with the appropriate filename identifiers (e.g., "c1," "c2," and "c3") corresponding to the fluorescence channel for nuclei, cytoskeleton, and PLA. Name each channel accordingly (e.g., "dapi," "gfp," and "rhod").
6. Segmentation of PLA dots: Add the module "Identify PrimaryObjects" to the pipeline. Select the identifier "rhod" as the input image. Name the primary objects to be identified (e.g., "pladots"). Enter the typical diameter in pixels of the PLA dots in the images (*see* **Note 6**). Set the thresholding method to "MoG Adaptive." Set the "Approximate fraction of image covered by objects" field to 0.01. Deselect the option "Speed up by using lower-resolution image to find local maxima." Keep the default settings for all other parameters.
7. Segmentation of cell nuclei: Add the module "Identify PrimaryObjects" to the pipeline. Select the identifier "dapi" as the input image. Name the primary objects to be identified (e.g., "nuclei"). Enter the typical diameter in pixels of the PLA dots in the images (*see* **Note 7**). Set the thresholding method to "Otsu Global." Keep the default settings for all other parameters. Figure 5a illustrates nuclei segmentation.
8. Segmentation of cells: Add the module "IdentifySecondary Objects" to the pipeline. This module uses primary objects as seeds for identifying secondary objects. In this case, each nucleus is considered a seed for one cell. Select the identifier

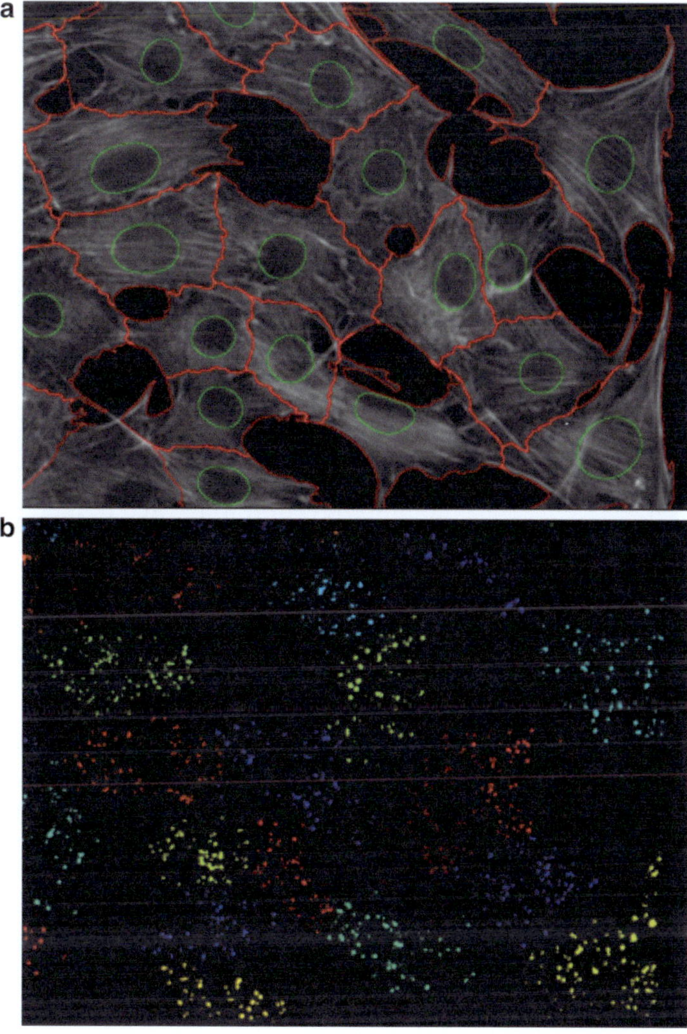

Fig. 5 Cell-segmentation results using CellProfiler to analyze a PLA image of NIH-3 T3 mouse fibroblasts. (**a**) *Green* and *red outlines* correspond to nuclei (primary objects) and cell boundaries (secondary objects), respectively. (**b**) *PLA dots* are identified and assigned to single cells

"gfp" as the input image. Select "nuclei" as the input objects. Name the objects to be identified (e.g., "cells"). Use the method "Propagation" and the thresholding method "Otsu global." Keep the default settings for all other parameters. Figure 5a illustrates cell segmentation from nuclei seeds.

9. Add the module "RelateObjects" to the pipeline in order to relate each identified PLA dot to a segmented cell. Define "pladots" as child objects and "cells" as parent objects. Figure 5b shows an example of a PLA-dot assignment.

10. Add the module "ExportToSpreadsheet" to the pipeline. Select "pladots" in the field "Data to export." Each identified PLA dot will result in a row in the spreadsheet, including its position and the cell's identifier. This spreadsheet can be used for further analysis, such as quantifying PLA dot counts in a single cell.
11. Define the default output folder.
12. Run the pipeline with a test dataset; check the segmentation result of each module, and optimize the parameters.
13. Run the pipeline for the entire dataset.

3.10 Example: Single-Cell Microfluidic PLA Dataset

We ran the device using the above protocol to generate a representative single-cell microfluidic PLA dataset. First, we stimulated NIH-3 T3 cells with 100 ng/mL PDGF. Cell-culture chambers in column direction were used to increment the cell stimulation time from 1 to 15 min. The stimulation time course was used as a standard in the file "pla03b.xlsx." The eight remaining cell cultures in column 16 were used for control experiments and were not stimulated with PDGF. After cell fixation, the PLA protocol was performed with a single primary antibody against Akt phosphorylated at Ser-473. In general to increase the specificity of phospho-antibodies the PLA is build up with a second primary antibody against an epitope on the target protein different than its phosphorylation site. For the here used Akt-Ser-473 antibody the specificity was previously demonstrated by knockdown of the corresponding kinase, i.e., mTOR/rictor complex [5], and thus we used only the phosphor-antibody.

The cell cultures in the other eight rows of the matrix were used as replicates. Figure 6 shows the PLA dot count per cell and mean Akt phosphorylation signal in response to PDGF across the entire time course. The cell-to-cell variability in Akt phosphorylation over the stimulation time course was obtained by plotting the PLA dot count per cell against cell number. The resulting histograms are shown in the lower panel of Fig. 6.

4 Notes

1. Supplementary information can be downloaded from "http://www.imtek.de/professuren/anwendungsentwicklung/forschung/microfluidic-lsi/microfluidic-and-biological-engineering."
2. The PLA chip can be used to determine the functionality and appropriate dilution factors of primary antibody by varying both parameters within separate PLA chip runs. High antibody concentrations within a PLA experiment cause an optical overlap between adjacent PLA events in the cell and the resultant loss of quantitative information. In contrast, dilute antibody

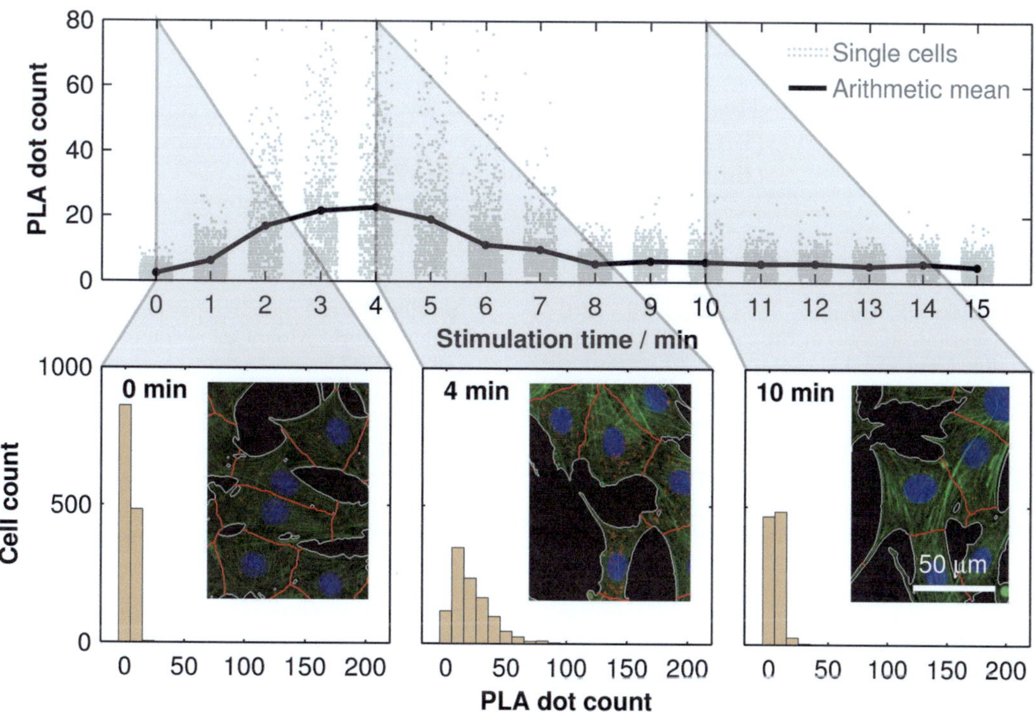

Fig. 6 PLA dot quantification of Akt phosphorylation at residue Ser-473 upon PDGF stimulation. *Gray dots* indicate PLA dot count per cell. The *black line* indicates the arithmetic means of the PLA dot counts at a given stimulation time. During the first 8 min, there was a strong phosphorylation response of p-Akt (Ser-473). The single-cell PLA data were evaluated and presented as a histogram for three stimulation conditions (no stimulation, 4 min, and 10 min). The *insets* show representative PLA images of the corresponding stimulation conditions. *Red lines* indicate single-cell segmentation obtained using CellProfiler with the pipeline described in Subheading 3.9. The cells were starved for 12 h with 0.1 % FBS before stimulation

solutions result in low PLA dot counts per cell, which may be insufficient for statistical analysis.

3. The software can be downloaded at www.cellprofiler.org. The software is based on Matlab but is distributed as a stand-alone version.
4. For a cell culture period of 24 h, approximately 1.5 mL of cell medium is required at the recommended flow pressure of 25 kPa.
5. Stimulation times can be adjusted by changing the corresponding pause times in the Microsoft Excel file "pla03.xlsx."
6. These parameters depend on the magnification used during image acquisition. For a 20× objective, min = 1 and max = 10 are appropriate starting parameters.
7. These parameters depend on the magnification during image acquisition and the cell type. For a 20× objective and typical fibroblasts, min = 40 and max = 100 are appropriate starting parameters.

Acknowledgements

This study was supported by the German Excellence Initiative (BIOSS-Project C7) and by the German Research Foundation (Emmy-Noether Grant ME3823/1-1).

References

1. Kalisky T, Blainey P, Quake SR (2011) Genomic analysis at the single-cell level. Annu Rev Genet 45:431–445
2. Sachs K (2005) Causal protein-signaling networks derived from multiparameter single-cell data. Science 308:523–529
3. Shi Q, Qin L, Wei W et al (2012) Single-cell proteomic chip for profiling intracellular signaling pathways in single tumor cells. PNAS 109:419–424
4. Snijder B, Pelkmans L (2011) Origins of regulated cell-to-cell variability. Nat Rev Mol Cell Biol 12(2):119–125
5. Bendall SC, Simonds EF, Qiu P et al (2011) Single-cell mass cytometry of differential immune and drug responses across a human hematopoietic continuum. Science 332(6030):687–696
6. Fredriksson S, Gullberg M, Jarvius J et al (2002) Protein detection using proximity-dependent DNA ligation assay. Nat Biotechnol 20:473–477
7. Blazek M, Betz C, Hall MN et al (2013) Proximity ligation assay for high-content profiling of cell signaling pathways on a microfluidic chip. Mol Cell Proteomics 12(12): 3898–3907
8. Unger MA, Chou HP, Thorsen T et al (2000) Monolithic microfabricated valves and pumps by multilayer soft lithography. Science 288 (5463):113–116
9. Gomez-Sjöberg R, Leyrat AA, Pirone DM et al (2007) Versatile, fully automated, microfluidic cell culture system. Anal Chem 79(22): 8557–8563
10. Duffy DC, McDonald JC, Schueller OJA, Whitesides GM (1998) Rapid prototyping of microfluidic systems in poly (dimethylsiloxane). Anal Chem 70(23):4974–4984
11. Carpenter AE, Jones TR, Lamprecht MR et al (2006) Cellprofiler: image analysis software for identifying and quantifying cell phenotypes. Genome Biol 7(10):R100
12. Blazek M, Silva Santisteban T, Zengerle R, Meier M (2015) Analysis of fast protein phosphorylation kinetics in single cells on a microfluidic chip. LabChip 15:726–734

Chapter 13

Dynamics and Interactions of Individual Proteins in the Membrane of Single, Living Cells

Stephen Anthony, Amanda Carroll-Portillo, and Jerilyn Timlin

Abstract

Total internal reflection fluorescence (TIRF) microscopy is a powerful technique for interrogating protein dynamics in the membranes of living single cells. Receptor–ligand interactions are of particular interest for improving our understanding of cell signaling networks in a variety of applications. Here, we describe methods for fluorescently labeling individual receptors and their ligands, conducting single-molecule TIRF microscopy of receptors and ligands in single, living cells, and importantly, performing image analysis on the resulting time sequence of images to extract quantitative dynamics. While we use Toll-like receptor 4 and its ligand lipopolysaccharide as a specific example, the methods are general and readily extendable to other receptor–ligand systems of importance in cellular biology.

Key words Single-molecule methods, Single-particle tracking, Fluorescence microscopy, Superresolution

1 Introduction

Understanding the proteome, protein networks, and protein modifications at the single-cell level is critical to advancing current understanding in widespread applications in human health and disease, environmental sciences, and renewable energy [1]. While static information about protein abundance and function is useful and cannot be replaced, it has been shown that the location in time and space of many proteins is highly regulated. Cellular response to external stimuli or internal cues is governed by complicated sets of changes in the location and abundance of proteins. Thus, the addition of spatially resolved protein dynamics and the understanding of heterogeneity in time and space of the dynamic response across multiple scales are necessary for building a more complete understanding of protein networks. At a coarse level (several minutes to hours), spatial-temporal response of proteins can be deduced from time-lapse fluorescence [2] and electron microscopy [3]. Faster

Anup K. Singh and Aarthi Chandrasekaran (eds.), *Single Cell Protein Analysis: Methods and Protocols*, Methods in Molecular Biology, vol. 1346, DOI 10.1007/978-1-4939-2987-0_13, © Springer Science+Business Media New York 2015

spatial-temporal dynamics (milliseconds to minutes), can be probed with live cell, single-molecule fluorescence microscopy [4].

In this chapter, we detail the method of total internal reflection fluorescence (TIRF) microscopy for the detection and tracking of individual proteins in the membranes of living cells. TIRF microscopy offers the advantage of restricting the laser excitation volume to very near the membrane (<100 nm), thus reducing the effect of interfering autofluorescence from the rest of the cellular milieu, and permitting sensitive detection of single molecule dynamics in the plasma membrane [5]. TIRF microscopy has been employed extensively to determine protein dynamics in living cells in a variety of applications [6–12], including Toll-like receptor 4 (TLR4) and its ligand lipopolysaccharide (LPS), which we use here as our example system [13, 14].

TLR4 is a member of the Toll-like receptor family, a family of pathogen recognition receptors which respond to external stimuli to initiate a phosphorylation cascade that elicits cellular response [15]. Specifically, TLR4 recognizes LPS, a component of gram-negative bacterial membranes, and responds through dimerization and subsequent phosphorylation to generate a signaling response. TLR4 response can lead to innate immune responses as well as other cellular modifications depending on the specific signaling pathway [16].

Often protein labeling, fluorescence microscopy, and quantitative image analysis are considered independent tasks, with little thought given to the analysis of the results until after the experiment is complete. Unfortunately this approach can result in substandard results, because successful single-molecule analysis requires specific signal-to-noise and density criteria be met in the images. Iterative optimization of labeling techniques and imaging parameters with analysis will lead to greater success in the extraction of reliable, quantitative results from the experiment. Therefore, we have chosen to detail the entire method: labeling the proteins, performing the microscopy, and basic analysis of the resulting data to obtain single-molecule locations and track their positions with respect to time. The methods detailed herein are applicable and easily extended to a variety of proteins in the cellular membrane.

2 Materials

Prepare all solutions using ultrapure water (prepared by purifying deionized water to attain a sensitivity of 18 MΩ cm at 25 °C) and analytical grade reagents unless otherwise noted. All solutions should be prepared at room temperature and stored at 4 °C unless otherwise noted. Follow all appropriate regulations when disposing of waste materials.

2.1 Reagents for Labeling LPS

1. 0.5 % triethylamine (TEA) in diH_2O.
2. Neuraminidase.
3. Galactose oxidase.

4. Sodium borohydride.
5. Dye-conjugated hydrazide.
6. Purified lipopolysaccharide (LPS).
7. Phosphate-buffered saline (PBS) (137 mM NaCl, 1.5 mM KH_2PO_4, 2.7 mM KCl, 8 mM Na_2HPO_4, pH 7.4).
8. G25 Sephadex spin columns.

2.2 Reagents for Labeling Antibody

1. Anti-TLR4 monoclonal antibody.
2. Dye-conjugated succinimidyl ester (Alexa 488 or similar).
3. Sodium bicarbonate.
4. PBS.
5. G25 Sephadex spin columns.
6. 10 kDa MWCO centrifugal concentrators, 500 μL maximum volume.

2.3 Reagents for Cleaving Antibody

1. Fluorescently labeled antibody.
2. 2-Mercaptoethylamine•HCl (2-MEA).
3. Ethylenediaminetetraacetic acid (EDTA).
4. Iodoacetamide.
5. 10 kDa MWCO centrifugal concentrators.

2.4 Reagents for Live Cell Imaging (See Note 1)

1. Cells of interest grown or seeded on cover slips.
2. High Tolerance #1.5 cover slips (Harvard Apparatus, Holliston, MA, USA).
3. Live cell imaging chamber, Warner QE-1 Quick Exchange Platform (Harvard Apparatus, Holliston, MA, USA) or similar.

2.5 Microscope Imaging Setup (See Note 2)

1. Laser excitation TIRF microscope (objective-based TIRF with excitation preferred, *see* **Note 3**).
2. High readout rate detector with single-molecule sensitivity such as an EMCCD or scientific CMOS detector (*see* **Note 4**).
3. Laser capable of providing at least 10 mW of the desired excitation wavelength (*see* **Note 5**).
4. Dichroic filter to project the laser to the sample and the fluorescence emission to the detector (*see* **Note 6**).
5. Notch filter to prevent the laser wavelength(s) from reaching the detector or the eyepiece (*see* **Note 7**).

2.6 Analysis Software

1. Matlab Software with the Image Processing Toolbox (Mathworks, Natick, MA, USA) (*see* **Note 8**).

3 Methods

Protein labeling: In order to visualize proteins in the membrane of living cells, a fluorescent label must be conjugated to the protein of interest. In the special case of receptor and ligand interactions the receptor (through use of a conjugated antibody or genetic expression), the ligand, or both can be labeled, depending on the application. Here we present strategies for labeling an antibody to the TLR4 receptor and its ligand (LPS) and for demonstration purposes we show data of TLR4 receptor trafficking on the plasma membrane in live mouse macrophage cells. A variety of organic dyes with reactive moieties are available from different companies for the purpose of labeling proteins or ligands. These include dyes that react with primary amines (succinimidyl esters), thiols (maleimides), and modified carbohydrate residues (hydrazides). Quantum dots are also quite popular because of their superior photophysical properties, but do have a disadvantage of relatively large size [17]. Care should be taken to verify that protein function is maintained post-labeling, regardless of label (*see* **Note 9**). All procedures should be conducted at room temperature unless otherwise noted.

3.1 Protein Labeling: Ligand Labeling via Carbohydrate Residues

This section is based on a protocol by Triantoflou et al. [18]. This protocol modifies carbohydrate residues within the ligand (LPS) to aldehyde groups allowing for reactivity toward dye-conjugated hydrazide molecules. It can be extended to other ligands with carbohydrate residues.

1. Resuspend lyophilized, purified LPS in 0.5 % TEA at a 2 mg/mL concentration and sonicate for 15 min on ice to generate a monomeric, homogenous solution of LPS.
2. Remove 100 μL of this solution for immediate use and store the remainder in aliquots at −20 °C or −80 °C depending on the length of storage (*see* **Note 10**).
3. To the 100 μL, add 0.2 U/mL neuraminidase and 20 U/mL galactose oxidase and bring the final volume to 1 mL with PBS, pH 7.4 (*see* **Note 11**).
4. Incubate with gentle agitation (on a rocking mixer such as nutator or orbital shaker at low speed) for 5 min at 37 °C.
5. Add 100 μL of 1 mg/mL fluorescently conjugated hydrazide. Resuspension of the dye can be done into PBS or DMSO (*see* **Note 12**).
6. Incubate at 37 °C in the dark for 1 h and 25 min.
7. Stop reaction with 100 μL of 2 mM sodium borohydride and mix with gentle inversion.
8. Separate the unlabeled dye from the labeled LPS by loading the sample onto a centrifugal G-25 column following the manufacturer's protocol. One column should allow for purification

of ~100 μL of sample and retains the ability to assume protein concentration based on dilution of the original sample (*see* **Note 13**).

9. Quantitation of label: The concentration of dye in the final preparation is determined spectrophotometrically using the dye extinction coefficient (*see* **Note 14**).

Concentration of Dye (M) = Absorbance of sample at emission maximum of the dye/extinction coefficient of dye × path length of sample (this value is usually 1).

3.2 Protein Labeling: Fluorescent Dye Conjugation to Whole Antibody

This procedure will label the whole antibody with a fluorescent dye through a reaction of the primary amines located on the lysine residues within the structure of the antibody with fluorescent succinimidyl esters. While for some applications receptors still may function normally for dynamic studies, there is a strong potential that the bivalency of antibodies could result in artificial crosslinking of proteins (*see* **Note 9**). Therefore we recommend following this procedure with that described in Subheading 3.3 to generate monovalent versions of the antibody for live cell studies. The fluorescent dye-conjugated whole antibody is sufficient for labeling receptors on fixed cells; however this is not covered in this protocol.

1. Dilute antibody to a concentration of 1 mg/mL in PBS, pH 7.4 (*see* **Note 15**).
2. Add sodium bicarbonate to the antibody solution such that the final concentration is 100 mM. This raises the pH to between 7.5 and 8.5 where succinimidyl esters are most reactive, allowing for optimal conjugation.
3. Add the reactive dye to the antibody such that the degree of labeling (DOL) is around 6 through use of a 10× excess of dye (*see* **Note 16**).
4. Incubate the reaction in the dark at room temperature for 20 min, with gentle agitation.
5. Upon completion of the reaction, load sample into a 10 kDa MWCO centrifugal concentrator and follow the manufacturer's protocol to remove the unbound dye and concentrate the sample.
6. Calculate the protein concentration and DOL, by measuring the absorbance at 280 nm (A280) and the emission maximum of the dye (Adye) of the sample and then using these values in the following equations:

$$\text{Molar concentration of protein} = \left[\text{A280} - (\text{Adye} - \text{correction factor for the dye})\right] / \text{extinction coefficient of IgG}\left(210{,}000\ \text{M}^{-1}\text{cm}^{-1}\right)$$

$$\text{DOL} = \text{Adye} / \text{extinction coefficient of the dye} \times \text{molar concentration of the protein} \left(\textit{see}\ \textbf{Note 14}\right)$$

3.3 Protein Labeling: Cleaving Dye-Conjugated Antibody for Live Cell Labeling

As described above and in **Note 9** the bivalency of antibodies could result in artificial cross-linking of proteins. This protocol cleaves the dye-conjugated antibody from Subheading 3.2 to generate monovalent versions. Cleavage of the antibody also allows for the option of conjugation to sulfhydryls with fluorescent maleimides.

1. Labeled antibody from Subheading 3.2 is incubated with 2-MEA (50 mM final concentration per 10 mg of antibody) with 1-10 mM EDTA for 90 min at 37 °C in the dark with gentle agitation to cleave dithiol bonds (preferentially in the neck region of the antibody where the heavy chains join [19]).
2. At the end of incubation, 20 mM iodoacetamide is added to the reaction to block the free sulfhydryl groups, thus preventing reformation of dithiol bonds. Incubate this reaction in the dark for 1 h at 4 °C with gentle agitation.
3. Purify the cleaved antibody by running the sample through a 10 kDa MWCO concentrator following the manufacturer's protocol. This step removes the 2-MEA and the iodoacetamide as well as concentrates the sample.
4. Add PBS to the concentrate to bring the sample volume up to the starting volume. Concentration of protein should now be double that of the whole antibody used in the cleavage reaction (*see* **Note 17**).

3.4 Microscopy: Labeling Live Cells

1. Assemble cover slip with cells into live cell imaging chamber as per the manufacturer's instructions (*see* **Notes 18** and **19**).
2. The amount of antibody for labeling cells at a desirable density needs to be empirically determined and is variable with antibody concentration and DOL. Testing several dilutions ranging from 1:50 to 1:500 is usually sufficient to indicate at what concentration the antibody should be used (*see* **Note 20** and **21**).
3. For studies with LPS stimulation: Add 100 ng/mL of labeled or unlabeled LPS to the chamber and place on microscope.

3.5 Microscopy: Image Acquisition

Image acquisition parameters will be dependent on the specific microscope manufacturer. Here we provide general concepts for collecting single-molecule data and the reader will need to relate these to their specific software parameters. Care must also be taken to (1) ensure sufficient signal-to-noise for single-molecule detection while keeping the intensities in the linear range of the detector for the most reliable performance (*see* **Note 22**), (2) avoid laser power levels that result in photobleaching of more than 25 % (*see* **Note 23**), and (3) control the variation of acquisition parameters between images of an experiment (*see* **Note 24**). **Steps 1–9** below describe a process to empirically determine values for laser power, detector gain, and acquisition time (along with the concentration of antibody (Subheading 3.4, **step 2**)) with a test sample perhaps a few days prior to actual experiment. TIRF microscope alignment and TIRF illumination optimization should be performed as per the manufacturer's or imaging facility's standard protocol.

1. Set acquisition time. This value determines the temporal resolution of the data and thus the dynamics you want to resolve should be considered (*see* **Note 25**).
2. Set detector gain to a reasonable starting value for single-molecule imaging based on the manufacturer's recommendation. For the EMCCD on our microscope this is a gain setting of 100 at a temperature of −100 °C 100. This is extremely manufacturer and technology dependent.
3. With no sample on the microscope and no excitation light, acquire a dark image and estimate the background signal simply by looking at the intensities. This is your approximate background level (*see* **Note 26**).
4. Place an aliquot of labeled antibody or labeled LPS onto a cover slip inside an imaging chamber.
5. Wait 3–5 min for settling to occur and add sufficient PBS to fill the chamber.
6. Place chamber on the microscope and focus onto the cover slip surface while collecting data in software's "live image" or "focus mode." Bright single-molecule spots should be visible. Aim for approximately 10–40 spots per field of view. If spots are too sparse or dense, repeat with a fresh cover slip and different dilution until appropriate density is achieved.
7. Adjust laser power to achieve brightness greater than the estimated background and lower than the maximum linear range of the detector (*see* **Note 27**).
8. Take a sequence of 500–1000 images and plot the mean intensity of the thresholded spots (*see* Subheadings 3.6 and 3.7 for methods to calculate this) over time.
9. If the initial intensities are too dim, too bright, or significant (>25 %) photobleaching occurs in the image sequence, adjust laser power, then adjust gain and/or acquisition time, and repeat.
10. Place the chamber containing the cell sample on the microscope stage.
11. Starting with parameters determined in **steps 1–9**, acquire a few single images, and verify that parameters will provide sufficient signal-to-noise for single-molecule detection. Adjust if necessary following recommendations in **steps 1–8** and corresponding notes.
12. Acquire sequences of images.

3.6 Quantitative Single-Molecule Analysis (See Note 28): Preprocessing

1. Assemble all files in the movie into a single directory, as the only files in that directory, and input that directory name into Matlab's command prompt (*see* **Note 29**).

```
>>directory = 'C:\ExampleDirectory\'
```

2. Verify that Matlab will read the files in the appropriate order for the movie (*see* **Note 30**).

```
>>files = dir(directory);
>>files = files(3:end);
>>files.name
```

3.7 Quantitative Single-Molecule Analysis: Identifying Single Molecules (See Note 31)

1. Select a representative frame of the movie to determine the appropriate threshold (*see* Fig. 1a). In this example the 17th file is used (*see* **Note 32**).

```
>>image = double(imread(fullfile(directory,files(17).name)));
>>imagesc(image); set(gca,'DataAspectRatio',[1 1 1])
```

2. Determine which pixels are greater than or equal to their surrounding pixels in the image (*see* **Note 33**).

```
>>kernel = ones(5);
>>[~,maxima] = findExtrema(image,kernel,false);
```

3. Apply a simple thresholding algorithm to keep only those maxima whose intensity is greater than the threshold (*see* Fig. 1b). Repeat, varying the threshold value, to find an optimum threshold value which should be recorded to use later (*see* **Note 20**).

```
>>threshold = 350
>>[row,col] = find(maxima & (image>threshold));
>> imagesc(image); set(gca,'DataAspectRatio',[1 1 1])
```

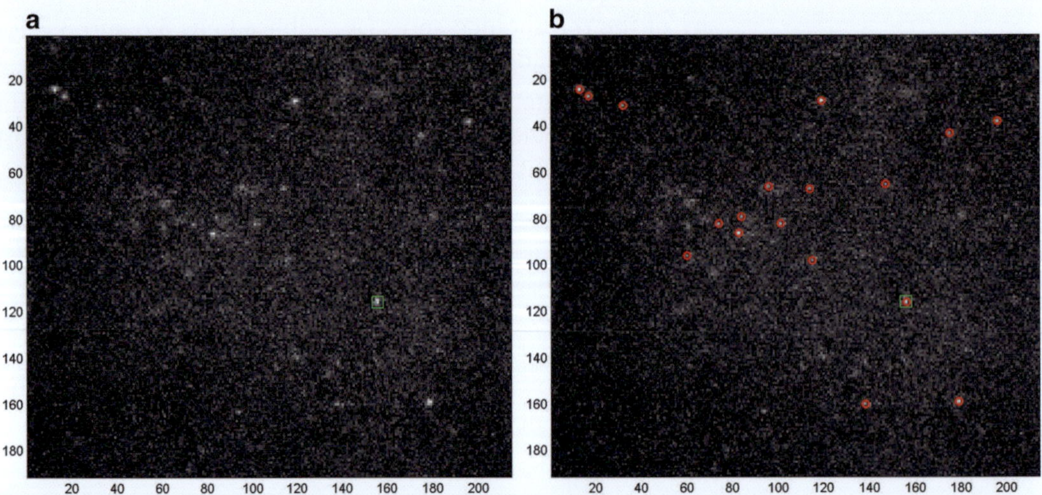

Fig. 1 Illustration of spot detection and fitting algorithms. (**a**) Raw image frame from an image sequence of TLR4 receptor dynamics on RAW 264.7 mouse macrophage cells. (**b**) Results of spot detection. *Red circles* indicate spots (single TLR4 receptors) detected above the threshold

```
>>colormap('gray')
>>hold on; plot(col,row,'ro','LineWidth',2);hold off
```

3.8 Quantitative Single-Molecule Analysis: Calculating Position at or Below Diffraction Limit

Once the threshold has been run and the single molecules detected, each detected position needs to be localized to subpixel resolution. The algorithm employed requires that the background offset be subtracted and an estimate made of the noise level of the background.

1. Estimating the background using a robust estimate of the standard deviation of the image (*see* **Note 34**).

```
>>image = image-median(image(:));
>>sigmaEst = robustStandardDeviation(image)
```

2. Fluorophores which are too close to the edge of the image cannot be accurately located, so those detections should be discarded. In this example detections are required to be at least 3 pixels in from the edge.

```
>>nearE= row<3|row>size(image,1)-2|col<3|col>size(image,2)-2;
>>col = col(~nearE); row = row(~nearE);
```

3. Each detected fluorophore must now be localized by fitting the region around each detection to a 2D-Gaussian (*see* **Note 35**). The following code performs a 2D-Gaussian fitting to the 15th detected fluorophore using a region ±2 pixels from the pixel detected earlier (*see* Fig. 2). The next step automates this process.

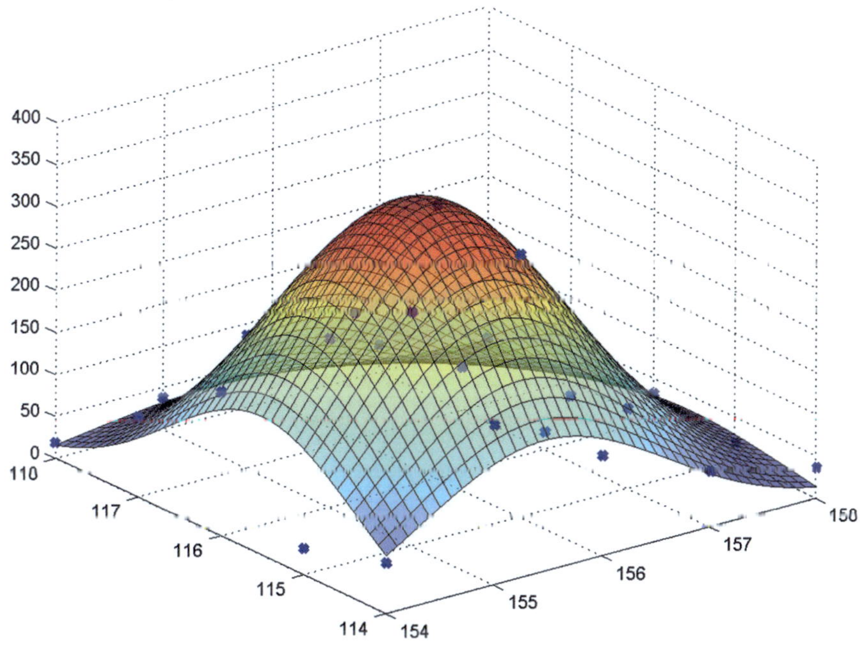

Fig. 2 Illustration of subpixel localization. Image pixel intensities are shown as *blue x*'s, while the fitted Gaussian is the mesh overlay. Pixel shown is indicated by the *green box* in Fig. 1b

```
>> [xOffsets,yOffsets] = meshgrid(-2:2,-2:2);
>>x=xOffsets(:)+col(15);y=yOffsets(:)+row(15);
>> lindex = sub2ind(size(image),y,x);z= image(lindex);
>>[xc,yc]=gauss2dcirc(z,x,y, sigmaEst);
```

4. Automating single-molecule detection and localization. In practice, once an appropriate threshold value is determined, the process must be automated as there may be thousands of frames and more than 100,000 detected fluorophores. The following code performs all the steps in Subheading 3.8, **steps 1–3** in an automated fashion using the previously determined threshold value (*see* **Note 36**).

```
>>positions = singleMoleculeAnalysis(directory,threshold,true);
```

3.9 Quantitative Single-Molecule Analysis: Connecting Tracks (for Dynamics)

1. The desire is to connect observations in one frame to observations in the next frame. To begin, we need to extract the positions in two adjacent frames, in this case the 5th and 6th frames (*see* **Note 37**).

```
>>frame = 5;
>>sources = positions(:,3)==frame;
>>sinks = positions(:,3)==frame+1;
>>sourcePos = positions(sources,1:2);
>>sinkPos = positions(sinks,1:2);
```

2. Next, we need to determine the displacement between each observation in the selected frame and each observation in the following frame. At the same time, it can be helpful to verify that the average displacements are not so large that tracking may be unreliable (*see* **Note 21**).

```
>>displacements = calcDisplacement(sourcePos,sinkPos);
```

3. The maximum allowed displacement (in pixels) must be specified (*see* **Note 38**).

```
>>maxDisp = 2;
```

4. Any displacements which are greater than the maximum allowed displacement must be set to be excluded.

```
>> displacements(displacements> maxDisp) = NaN;
```

5. Among the remaining possible connections, the best matching needs to be found, where here the best matching is defined as the one which minimizes the total displacement over all particles (*see* **Note 39**).

```
>>pairs = callLapjv (displacements);
>>lindex = sub2ind(size(displacements),pairs(:,1),pairs(:,2));
>>keep = ~isnan(displacements(lindex));
>>pairs = pairs(keep,:);
```

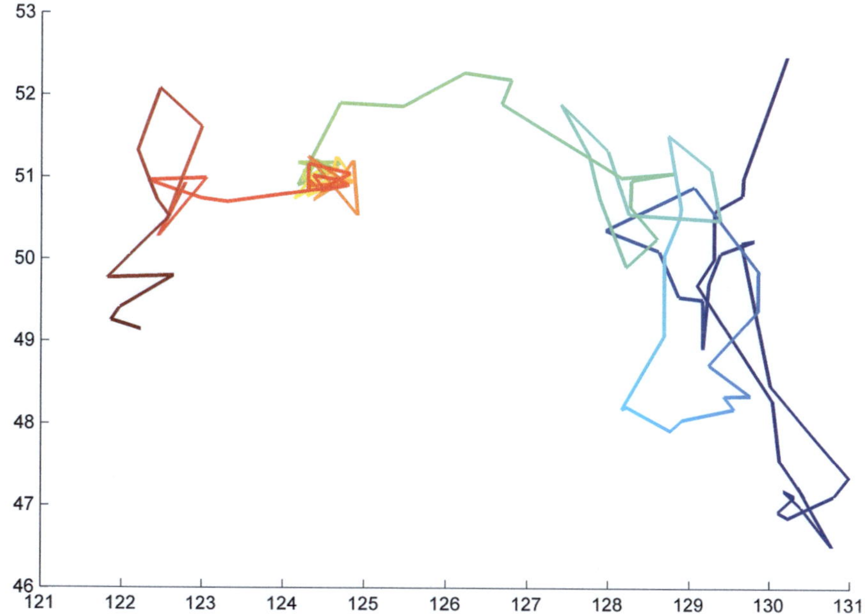

Fig. 3 Track of the single TLR4 receptor highlighted by *green box* in Fig. 1b. The color transitions from *blue* to *red* corresponding to the beginning and end of the track. Distinct diffusion regimes are visible, ranging from random motion, directed motion, and corralled behavior. The trajectory consists of 121 sequential frames spanning just under 12 s

6. Visualizing the tracking can be extremely helpful (*see* **Note 40**).

   >> *visualizeF2Ftrack*

7. **Steps 1–6** illustrate tracking from one frame to the next frame. Obtaining trajectories requires performing these steps over all pairs of frames and stringing together the connections from the different pairs of frames. While conceptually simple, a fair amount of code would be required. This step automates the process to operate on all frames of the movie (*see* Fig. 3). The code provided requires only the positions calculated earlier and the maximum displacement. If the directory containing the images of the frames is also provided and the final input set to true, then the code will pause after each pair of frames and display the tracking results for those frames (*see* **Note 41**).

   >> *[trajectories,rho] = f2fTracking(positions,maxDisp,directory, true)*

4 Notes

1. For the experiments detailed RAW 264.7 mouse macrophage cells were grown on piranha-cleaned High Tolerance Round #1.5 cover slips following standard cell culture methods. Live

cell imaging chambers vary greatly in functionality, price, and ease of assembly. For the inexperienced user, sufficient practice assembling the chamber should be conducted prior to use. Though high-tolerance cover slips are not required, we have found that the use of a high-tolerance cover slip (such as 0.17 ± 0.01 mm vs. the ±0.02 mm variation that is standard) makes switching between samples on a TIRF microscope easier/faster. The smaller variation in cover slip thickness requires less optimization of the TIRF excitation and this is advantageous when probing fast dynamics at early time points.

2. For this work we used a custom objective-based TIRF excitation microscope built on an Olympus IX71 platform with a 60×, 1.45 NA PlanApo TIRFM objective (Olympus America, Center Valley, PA, USA). This system is capable of simultaneous, dual-color detection of single molecules using an image splitter (OptoSplit II, Cairn Research, Kent, UK) which splits the image into two color channels prior to projecting onto the high-sensitivity EMCCD (iXon 887, Andor Technologies, Belfast, Northern Ireland). While for the studies presented in this work only single color imaging was used, for many experiments and systems the capability to image multiple colors simultaneously is advantageous. Systems with similar characteristics are commercially available.

3. We have found that TIRF microscopes with lamp-based excitation generally do not have sufficient sensitivities for single-molecule tracking.

4. For single-molecule sensitivity, low readout noise is essential in order to achieve sufficient signal-to-noise. While using different mechanisms, both EMCCDs and scientific CMOS cameras effectively achieve low readout noise.

5. Generally, excitation powers of 5–20 mW are suitable for objective-based TIRF. For this work, approximately 7 mW of 488 nm light (Cyan, Spectra-Physics, Santa Clara, CA, USA) was used to excite Alexa 488-labeled anti-TLR4. When purchasing a laser, the capability to provide 20 mW rather than the bare minimum of 5 mW is highly preferred as this allows adjustment of the laser intensity based upon experimental needs.

6. The dichroic filter should reflect greater than 95 % of the laser light and transmit greater than 90 % throughout the region of peak fluorescence emission of the selected fluorophore, where higher is better. For microscopes capable of using multiple laser wavelengths for excitation, it can be helpful to use multi-edge dichroic filters. For setups using an image splitter to allow simultaneous multicolor imaging, a second dichroic will be needed to split the emission into two spectral bands. For the

work described here, a quad-edge dichroic was installed in the microscope filter cube as our custom setup is equipped with four different excitation lasers and a single-edge dichroic was used in the image splitter (Di01-R405/488/532/635 and DI02-R635, Semrock, Rochester, NY, USA).

7. Unless a notch filter is employed, the signal from single fluorophores is likely to be swamped by reflections from the laser(s), which may also damage the detector. More importantly, an appropriately positioned notch filter also protects the operator's eyes from damage by preventing the dangerous laser power from reaching the microscope eyepiece. The notch filter should have an optical density (OD) of at least four for all laser wavelengths. In our case, a quad-notch filter is employed to attenuate the four laser wavelengths with which our setup is equipped (NF01-405/488/532/635, Semrock, Rochester, NY, USA).
8. While the code provided here will likely run in most versions of Matlab, version R2013b was the only version specifically tested.
9. Conjugation of a fluorophore to a ligand or antibody has the potential to disrupt functionality; therefore, it is best to run assays to determine whether the addition of the fluorophore has been detrimental to function. For the conjugation strategies pertinent to this chapter, these assays would include an LPS binding assay (for ligand) to determine if signaling through TLR4 is maintained, and labeling of cells with TLR4 antibodies to assess that antibody binding is as expected.
10. Store LPS aliquots at −20 °C for up to 6 months in aliquots to avoid freeze-thaw cycles. Store at −80 °C for long-term storage.
11. Some enzyme preparations are defined as number of units (U) or number of active units per a given weight. One should use active units to determine how much should be added to a reaction rather than a general weight in order to achieve the best results.
12. Use of DMSO allows for freezing of small aliquots of dye for longer periods than PBS as it maintains the reactive group in an organic solvent.
13. Purify volumes of labeled LPS from stock as needed for future experiments. As the concentration of the ligand is known at the beginning of the reaction (2 mg/mL), after final purification through the spin column, the concentration of the labeled LPS can be assumed to be 0.167 mg/mL based on the known dilutions. There are alternative methods for calculating LPS concentration more accurately including use of a KDO assay [20] or mass spectrometry which are outside of the scope of this chapter.
14. Information for many common fluorescent dyes can be found at www.metabion.com/downloads/fluorescentdyes.pdf. The

correction factor for a particular dye is typically provided by the company.

15. The presence of sodium azide, a component in many commercial antibody buffers, is inhibitory to amine conjugation chemistry. A buffer exchange using spin columns (such as the G25 columns already mentioned in this protocol) should be carried out if azide is present.
16. We find that a final DOL at the end of the conjugation between 3 and 6 is optimal for fluorescence microscopy. At DOLs above six labeled antibodies commonly aggregate leading to inconsistent labeling and nonspecific binding. The protocol can be varied to change the DOL by: increase/decrease of dye added, incubation for shorter/longer time periods, and incubation at 4 °C (to slow the reaction).
17. This assumes that no protein was lost during the purification step and that cleavage only occurred within the neck region. DOL calculations would require assumption that the extinction coefficient of the cleaved antibody is exactly half of the whole antibody. A native gel can be run on samples to determine what the population of cleaved antibody consists of (cleaved at neck vs. cleaved at where the heavy and light chains join) if desired.
18. For the Warner Quick exchange sample platform we apply a thin ring of vacuum grease (Dow-Corning) with a small tip paintbrush to the top surface of the bottom chamber and the bottom surface of the top chamber. We then sandwich the cover slip in between and rotate the top chamber about a ¼ of a turn to seal. This trick prevents leakage in most situations. The bottom of the cover slip (that which is closest to the objective) is cleaned gently with ethanol to remove precipitated protein from cell media.
19. TIRF microscopy illuminates the sample only in the region very near (100–300 nm) the cover slip. For strongly adherent cells such as macrophages and dendritic cells, the accessibility of the fluorescent label to the membrane nearest this surface needs to be verified. To determine if a label has access to the membrane at the attached surface, a solution of fluorescent dye consistent with the microscope filter sets can be used. The dye should be non-conjugated and non-reactive on its own to avoid potential interactions that would skew the results. The dye can be added at a dilute concentration to cell samples on cover slips and then imaged in TIRF. The dye should diffuse under the cells leading to uniform or nearly uniform concentrations around and beneath the cells. If the adherent properties of the cells inhibit access of the dye to the membrane surface at the cover slip interface, the images will show voids (areas of no signal) under the cells (supplementary Fig. S5 of Ref. 21). If the dye cannot access the surface readily, the label will need to be added in solution

and the cells deposited on the cover slip after. This precludes the possibility of imaging early dynamics post treatment, but in many cases may be sufficient. Another alternative approach is the use of supported lipid bilayers which allows for cellular binding to antigens in the bilayer and imaging at the bilayer-cellular interface without the concern for exclusion of dye seen with cells directly on a cover slip surface [11].

20. The provided code will display a grayscale image of the frame being considered with superimposed red circles indicating where maxima (*see* Subheadings 3.7). By varying the threshold value, the minimum threshold value (*minThreshold*) high enough that background pixels are not mistakenly detected as signals (false positives) should be determined. Similarly, the maximum threshold value (*maxThreshold*) low enough that no fluorophores are visible but not detected (false negatives) should also be determined. Ideally, the former (*minThreshold*) should be lower than the latter (*maxThreshold*). In this case, the optimum threshold would be a value intermediate between the two, where the greater the separation from both bounds, the more likely the threshold will perform well for all similar images. In the event that there is no threshold value above *minThreshold* and below *maxThreshold*, a decision on the desired threshold must be made taking into consideration whether false positives or false negatives are more problematic. If there is no acceptable compromise, either a more sophisticated detection scheme which produces usable results must be found or better sample images must be acquired (Subheading 3.5, **steps 1–9**).

21. Two important criteria for gauging the feasibility of tracking are the average nearest-neighbor distance (*avgNearNeighbor*) and the average displacement of particles from frame to frame (*avgDisplacementF2F*). A useful guideline is that for typical particle tracking algorithms to perform well, the ratio *ρ=avgDisplacementF2F/avgNearNeighbor* (*see* **Note 42**) should be well below 0.5 [32]. Both values can be readily estimated. To estimate the average displacement, first determine the minimum displacement for each observation from one frame to the next. A useful estimate (estA*vgDisplacementF2F*) is obtained by taking the mean of these values (*see* **Note 43**).

```
>>nearestNext = min(displacements);
>> estAvgDisplacementF2F = mean(nearestNext);
```

To estimate the average nearest neighbor distance, it is first necessary to calculate the displacement between all pairs of particles within a single frame. To avoid mistakenly calculating the distance to the nearest neighbor as zero due to the zero distance between each particle and itself, it is necessary to increase the value of all the terms on the diagonal. Then, the distance to the nearest neighbor can easily be calculated for each particle, at which point an estimate (*estAvgNearNeighbor*) of the average nearest neighbor distance can be obtained by taking the mean.

```
>>distSingleFrame = calcDisplacement(sourcePos,sourcePos);
>> distSingleFrame = distSingleFrame + eye(size(distSingleFrame))*1e10;
>>nearestDist = min(distSingleFrame);
>> estAvgNearNeighbor = mean(nearestDist);
```

Using these two values, it is then possible to estimate ρ, where the larger the value of ρ, the more concern there needs to be whether the tracking results are reliable.

```
>> rhoEstimate = estAvgDisplacementF2F/ estAvgNearNeighbor
```

For cases where ρ is too high, the experiment may need to be rerun, and either *avgNearNeighbor* increased or *avgDisplacementF2F* decreased. The former is typically accomplished by reducing the number of labels or the labeling density. Increasing *avgDisplacementF2F* typically requires acquiring the images at a faster frame rate. Unfortunately, this frequently also results in reduced signal-to-noise ratios (SNR) in each frame, and is not always feasible, since at low SNR, particle detection and localization become difficult or impossible [33].

22. It is very important to fully characterize the detector prior to any imaging experiments. Detector linearity is the range over which the detector responds linearly when additional photons have been detected. This will vary from detector to detector and is not simply equivalent to the pixels' saturation levels. An assessment of the linear range can be made by taking an out-of-focus image of an object. Plotting the variance of each pixel vs. its mean intensity will give a graph from which the linear range can be deduced [22]. The reader is strongly encouraged to work with the microscope vendor or the coordinator at the imaging facility to determine this range appropriately for the system being used.
23. Every fluorescent label is susceptible to photobleaching. If photobleaching is severe, it will reduce the signal-to-noise below the detection limit and thus shorten the duration over which dynamics can be determined. Though the microenvironment of the dye can alter the photobleaching rates (on cells, rather than buffer alone), we have found the process described in Subheading 3.5, **steps 1–9** is a fairly reliable estimate of photobleaching. Photobleaching of 5–15 % from first to last frame is generally acceptable. Most vendors of fluorescent dyes provide photobleaching information; however due to the differences in microscopes it is recommended that procedures in Subheading 3.5, **steps 1–9** are conducted to

determine the settings for a particular instrument/dye/sample combination.

24. If quantitative intensity information is desired, laser power, acquisition time, and detector gain should be kept constant for every image in an experiment. Some image acquisition software will have an "autogain" or similar option that automatically adjusts the detector gain, laser power, or acquisition time for each image to strive for better image quality. In general, this is not recommended because it hinders robust quantitative analysis post acquisition.
25. If the acquisition time is too slow, fast-moving dynamics will not be resolved. However, for a given laser power and detector gain, acquisition time is inversely proportional to signal-to-noise; therefore, faster acquisition times will require higher gain or laser power to achieve detectable spots. We typically use 30–50 frames per second with the EMCCD on our microscope.
26. Take care to reduce extraneous light to the detector. The background levels will vary with detector acquisition time and gain, so they should be estimated again when these change.
27. For example, if the linear range of the detector is 0–8000 counts with dark images showing background of 600 counts, aim for spot intensity of 1500–4000 counts above background.
28. Sample code with extensive comments is available for download. Once posted, a link will be made available on the corresponding author's website (http://www.sandia.gov/bioenergy-biodefense/timlin.html), and the code can also be obtained by emailing the corresponding author. Example commands to be entered into Matlab are also provided on separate lines following the Matlab command prompt (>>). Note that once the desired *threshold* value has been determined, Subheading 3.8, **step 4** automates and replaces Subheading 3.8, **steps 1–3**. Similarly, once the desired *maxDisp* value has been determined, Subheading 3.9, **step 7** automates and replaces Subheading 3.9, **steps 1–6**.
29. The file format of the data will depend upon the specific camera and software employed for image acquisition. Some software will save the entire movie as a single file, in which case it is necessary to determine how to load the individual frames of the movie. The example provided here assumes that each frame of the movie is saved as a separate, sequentially numbered file, in an image format which can be read by Matlab's imread function (such as the .tiff file format), with a file extension matching the file type. While different file formats are acceptable, only uncompressed file formats or ones using lossless compression should be employed, and the file format should support and save data with the full bit depth of the camera.

30. The provided command will display the files to be analyzed in the order they will be analyzed. A common naming scheme is to use a file naming scheme where each file is named *"movieName_XXX.tiff"* where for a movie with 100 frames, for the first frame *XXX* would be *001* (e.g., *"movieName_001.tiff"*), for the second frame *XXX* would be *002*, and so on. By writing the first frame as *001* instead of *1*, the computer will sort all the frames in the appropriate order, whereas otherwise frame *10* would be read before frame *2*.
31. The method employed here to detect single molecules (simple thresholding of local maxima) was chosen for its conceptual and computational simplicity. Advanced algorithms, which are more robust and capable of operating on noisy images or images with nonuniform backgrounds, are available [23, 24].
32. This method assumes that the images from all frames of the movie appear comparable to each other such that a single threshold can be used for all movies. It is frequently a good idea to examine a sampling of images throughout the movie to ensure that all images appear comparable. In cases where the observed intensity of the fluorophores attached to the proteins varies with time (such as changes in intensity due to photobleaching, *see* **Note 23**), it may be necessary to vary the threshold with different frames. Further, this method assumes that the background of the image is relatively flat and uniform (as is often the case for TIRF). In instances where the background is more complicated, spatial filtering techniques may need to be employed for background subtraction [25].
33. The code can be run with any positive odd number for the size of the kernel. For instance, if

    >>kernel=ones(3)

    were used, the code would select the local maxima, only those points which are equal to or greater than all eight of their neighboring pixels. Due to the stochastic nature (inherent noisiness) of single-molecule imaging, it is not uncommon for the image of a fluorophore to correspond to two peaks separated by an intervening pixel. In order to avoid detecting one fluorophore multiple times, a slightly larger search neighborhood is recommended, detecting only those pixels which are at least as bright as all other pixels which are no more than two pixels away. Better detection can also be obtained by first convolving the image with a Gaussian kernel of dimensions approximating the point spread function of the fluorophores [26].
34. The method employed [27] makes use of the median and should provide an accurate estimate so long as the signal density is sufficiently sparse. The quality of the estimates of the

offset and the standard deviation can be evaluated by displaying a histogram of the pixel intensities and ensuring that after the correction, the peak giving the distribution of the noise is approximately centered at 0 and has a standard deviation matching *sigmaEst*.

35. Subpixel localization requires that the point spread function be wide enough that the fluorescent intensity is distributed over multiple pixels. The size of the region selected (±2 pixels) was chosen based upon the width of the point spread function such that the selected region contained the vast majority of the detected photons. The gauss2dcirc Gaussian fitting routine employed [28] was selected because it is freely available, fast, and particularly easy to use. At extremely low signal-to-noise ratios, however, gauss2dcirc has limited precision and an alternate algorithm may be preferable [29]. Using the alternate algorithm is more challenging, though, as it involves GPU computing and requires additional inputs. Additionally, the use of 2D-Gaussian fitting is itself an approximation to the true point spread function, albeit a common, computationally helpful one. Alternate methods using more accurate point spread functions also exist [30]. A particularly valuable resource providing more background and comparing a range of single-molecule localization software is available [31]. Similarly, if the fluorophores are too close to each other, such that their point spread functions overlap or nearly overlap, an alternate algorithm may need to be employed.

36. When run with the third input *true*, the code will pause after each frame, displaying a grayscale image of the frame with superimposed red circles indicating where fluorophores were detected and localized. Switching the third output to *false* will eliminate this behavior. The code is also extensively commented.

37. The scheme employed here represents a frame-by-frame tracking method. Frame-by-frame tracking is conceptually and computationally simple, but has significant limitations. In particular, a fluorophore (such as a quantum dot) blinking off for a frame (or otherwise not being detected for a frame) will result in trajectory fragmentation. There is no capability to connect observations which are separated by more than one frame when there are not intervening observations. An excellent overview of tracking algorithms including a listing of many alternate algorithms and additional tips and tricks is available [32].

38. The *maxDisp* does not need to be an integer. A reasonable number to use for the maximum displacement is three times the estimated average displacement. This value is large enough to be greater than the vast majority of real displacements (assuming 2-dimensional Brownian motion, this would con-

tain ~99.9 % of all displacements) while restrictive enough to exclude most false assignments.

```
>>nearestNext = min(displacements);
>> estAvgDisplacementF2F = mean(nearestNext);
>> maxDisp = estAvgDisplacementF2F*3;
```

Recall (*see* **Note 43**) that *estAvgDisplacementF2F* depends upon the assumption that most observations in one frame can be matched with a corresponding observation in the next frame. When imaging 3-dimensional motion with TIRF, fluorophores will frequently enter and exit the field of view, both limiting the length of trajectories and resulting in an overestimation of est*AvgDisplacementF2F*. Alternatively, fluorophores sometimes cannot be detected in adjacent frames even when physically present, particularly when either the fluorophores are prone to blinking such as many quantum dots (blinking suppressed quantum dots are beginning to be produced) [34] or the signal-to-noise ratio is low enough that detection is unreliable.

39. The problem being solved is known both as a linear sum assignment problem and as finding the optimal matching for a weighted bipartite graph. At first glance, due to the combinatorial number of possible solutions, finding the optimum solution would appear computationally prohibitive. Fortunately, these problems are well known in computer science, and efficient algorithms are available [35–37]. While minimizing the total displacement is intuitive, alternative weightings (or cost matrices) such as the square of the distance [38] or probability-based costs may provide improved results.
40. The provided code will display a grayscale image of the second (sink) frame of the pair, with the observations from the first (source) frame superimposed as red x's and the observations from the second frame superimposed as blue circles. Green connecting lines indicate observations between the frames which were found to be paired.
41. The output variable, *trajectories*, is a matrix where each row corresponds to one observation in a particular frame. The first two columns correspond to the *xy*-coordinates of the observation in units of pixels, while the third column corresponds to the number of the frame where it was observed. The fourth column provides the ID number of the trajectory, unless it is 0. Rows whose fourth column values are 0 correspond to observations which do not belong to any trajectory, but are instead isolated points. For convenience, the code also computes *rhoEstimate*, an estimate of ρ (*see* **Note 21**), using information from the entire movie. The larger the value of *rhoEstimate*, particularly as it approaches 0.5, the more concern there needs

to be whether the combination of particle density and displacement is such that the tracking results may be unreliable

42. The use of the average displacement and average nearest neighbor distance assumes that the images are relatively homogeneous, which is not always the case for the imaging of cells. Frequently, fluorophores will be locally concentrated in certain regions of the cell and less abundant in other regions. In such cases, when tracking feasibility is evaluated using ρ, on average the label density may be sufficiently sparse, but the regions of the cell with greater label density may preclude tracking.

43. Bear in mind that est*AvgDisplacementF2F* should not be expected to be an unbiased estimate. Observations in one frame which do not correspond to an observation in the next frame would tend to bias est*AvgDisplacementF2F* to overestimation of *avgDisplacementF2F*. This would tend to result in an overestimation of ρ. Therefore, this method requires assuming that that most observations in one frame can be linked to a corresponding observation in the next frame. In this case, the larger the value of ρ, the more est*AvgDisplacementF2F* would be expected to underestimate *avgDisplacementF2F*. The reason for this underestimation is that the shorter the nearest neighbor distance, the more likely that a smaller displacement can be obtained by incorrectly linking the observed particle in one frame to one of its nearby neighbors in the next frame. The same rationale governs why higher values of ρ tend to produce incorrect tracking results. An alternate method of examining the frame-to-frame displacement distribution is to examine the histogram of displacements.

```
>> hist(nearestNext,20)
```

Examination of the histogram as well as visual examination of the tracks between frames are useful in revealing whether the above-mentioned assumption that most observations in one frame correspond to observations in the next.

Acknowledgments

This study was supported in part by the National Institutes of Health Director's New Innovator Award Program, 1-DP2-OD006673-01 (probe development and experiments), as well as the Department of Energy's Laboratory Directed Research and Development (LDRD) program (compilation and writing). Sandia National Laboratories is a multi-program laboratory managed and operated by Sandia Corporation, a wholly owned subsidiary of Lockheed Martin Corporation, for the US Department of Energy's National Nuclear Security Administration under contract DE-AC04-94AL85000.

References

1. Wu M, Singh AK (2012) Single-cell protein analysis. Curr Opin Biotechnol 23(1):83–88
2. Giepmans BNG, Adams SR, Ellisman MH et al (2006) The fluorescent toolbox for assessing protein location and function. Science 312(5771):217–224
3. Osafune T, Schwartzbach S (2007) Intracellular protein localization by immunoelectron microscopy. In: van der Giezen M (ed) Protein targeting protocols, vol 390, Methods in Molecular Biology™. Humana, New York, NY, pp 407–416
4. Cognet L, Leduc C, Lounis B (2014) Advances in live-cell single-particle tracking and dynamic super-resolution imaging. Curr Opin Chem Biol 20:78–85
5. Axelrod D (2001) Total internal reflection fluorescence microscopy in cell biology. Traffic 2(11):764–774
6. Anantharam A, Onoa B, Edwards RH et al (2010) Localized topological changes of the plasma membrane upon exocytosis visualized by polarized TIRFM. J Cell Biol 188(3): 415–428
7. Martinière A, Lavagi I, Nageswaran G et al (2012) Cell wall constrains lateral diffusion of plant plasma-membrane proteins. Proc Natl Acad Sci 109(31):12805–12810
8. Mashanov, GI, Molloy, JE (2007) Automatic detection of single fluorophores in live cells. Biophys J 92: 2199–2211
9. Weigel AV, Tamkun MM, Krapf D (2013) Quantifying the dynamic interactions between a clathrin-coated pit and cargo molecules. Proc Natl Acad Sci 110(48):E4591–E4600
10. Spendier K, Carroll-Portillo A, Lidke K et al (2010) Distribution and dynamics of RBL IgE receptors (Fc∑RI) observed on planar ligand-presenting surfaces. Biophys J 99:388–397
11. Carroll-Portillo A, Spendier K, Pfeiffer J et al (2010) Formation of a mast cell synapse: FcεRI membrane dynamics upon binding mobile or immobilized ligands on surfaces. J Immunol 184(3):1328–1338
12. Aaron JS, Greene A, Kotula PG et al (2011) Advanced optical imaging reveals dependence of particle geometry on interactions between CdSe quantum dots and immune cells. Small 7(3):334–341
13. Aaron JS, Carson BD, Timlin JA (2012) Characterization of differential Toll-like receptor responses below the optical diffraction limit. Small 8(19):3041–3049
14. Shawkat S, Karima R, Tojo T et al (2008) Visualization of the molecular dynamics of lipopolysaccharide on the plasma membrane of murine macrophages by total internal reflection fluorescence microscopy. J Biol Chem 283(34):22962–22971
15. Takeda K, Kaisho T, Akira S (2003) Toll-like receptors. Annu Rev Immunol 21(1):335–376
16. Pålsson-McDermott EM, O'Neill LAJ (2004) Signal transduction by the lipopolysaccharide receptor, Toll-like receptor-4. Immunology 113(2):153–162
17. Jaiswal JK, Goldman ER, Mattoussi H et al (2004) Use of quantum dots for live cell imaging. Nat Methods 1(1):73–78
18. Triantafilou K, Triantafilou M, Fernandez N (2000) Lipopolysaccharide (LPS) labeled with Alexa 488 hydrazide as a novel probe for LPS binding studies. Cytometry 41(4):316–320
19. Sun MMC, Beam KS, Cerveny CG et al (2005) Reduction–alkylation strategies for the modification of specific monoclonal antibody disulfides. Bioconjug Chem 16(5):1282–1290
20. Lee C-H, Tsai C-M (1999) Quantification of bacterial lipopolysaccharides by the purpald assay: measuring formaldehyde generated from 2-keto-3-deoxyoctonate and heptose at the inner core by periodate oxidation. Anal Biochem 267(1):161–168
21. Andrews NL, Lidke KA, Pfeiffer JR et al (2008) Actin restricts Fc[epsiv]RI diffusion and facilitates antigen-induced receptor immobilization. Nat Cell Biol 10(8):955–963
22. Lidke KA, Reiger B, Lidke DS et al (2005) The role of photon statistics in fluorescence anisotropy imaging. IEEE Trans Image Process 14(9):1237–1245
23. Olivo-Marin JC (2002) Extraction of spots in biological images using multiscale products. Pattern Recogn 35(9):1989–1996
24. Izeddin I, Boulanger J, Racine V et al (2012) Wavelet analysis for single molecule localization microscopy. Opt Express 20(3):2081–2095
25. Weeks AR (1996) Fundamentals of electronic image processing, SPIE/IEEE series on imaging science & engineering. SPIE Optical Engineering Press; IEEE Press, Bellingham, WA; New York, NY
26. Crocker JC, Grier DG (1996) Methods of digital video microscopy for colloidal studies. J Colloid Interface Sci 179(1):298–310
27. Sadler BM, Swami A (1999) Analysis of multiscale products for step detection and estimation. IEEE Trans Inform Theor 45(3):1043–1051
28. Anthony SM, Granick S (2009) Image analysis with rapid and accurate two-dimensional gaussian fitting. Langmuir 25(14):8152–8160

29. Smith CS, Joseph N, Rieger B et al (2010) Fast, single-molecule localization that achieves theoretically minimum uncertainty. Nat Methods 7(5):373–U352
30. Ober RJ, Ram S, Ward ES (2004) Localization accuracy in single-molecule microscopy. Biophys J 86(2):1185–1200
31. Single molecule localization microscopy. http://bigwww.epfl.ch/smlm/. Accessed August 26 2014
32. Meijering E, Dzyubachyk O, Smal I (2012) Methods for cell and particle tracking. Methods Enzymol 504:183–200
33. Smal I, Loog M, Niessen W et al (2010) Quantitative comparison of spot detection methods in fluorescence microscopy. IEEE Trans Med Imag 29(2):282–301
34. Dennis AM, Mangum BD, Piryatinski A et al (2012) Suppressed blinking and auger recombination in near-infrared type-II InP/CdS nanocrystal quantum dots. Nano Lett 12(11):5545–5551
35. Jonker R, Volgenant A (1987) A shortest augmenting path algorithm for dense and sparse linear assignment problems. Computing 38(4):325–340
36. Jonker R, Volgenant A (1999) Linear assignment procedures. Eur J Oper Res 116(1):233–234
37. Cao Y. LAPJV – Jonker-Volgenant algorithm for linear assignment problem V3.0. http://www.mathworks.com/matlabcentral/fileexchange/26836-lapjv-jonker-volgenant-algorithm-for-linear-assignment-problem-v3-0. Accessed August 26 2014
38. Jaqaman K, Loerke D, Mettlen M et al (2008) Robust single-particle tracking in live-cell time-lapse sequences. Nat Methods 5(8):695–702

Chapter 14

Microfluidics-Enabled Enzyme Activity Measurement in Single Cells

Cinzia Tesauro, Rikke Frøhlich, Magnus Stougaard, Yi-Ping Ho, and Birgitta R. Knudsen

Abstract

Cellular heterogeneity has presented a significant challenge in the studies of biology. While most of our understanding is based on the analysis of ensemble average, individual cells may process information and respond to perturbations very differently. Presented here is a highly sensitive platform capable of measuring enzymatic activity at the single-cell level. The strategy innovatively combines a rolling circle-enhanced enzyme activity detection (REEAD) assay with droplet microfluidics. The single-molecule sensitivity of REEAD allows highly sensitive detection of enzymatic activities, i.e. at the single catalytic event level, whereas the microfluidics enables isolation of single cells. Further, confined reactions in picoliter-sized droplets significantly improve enzyme extraction from human cells or microorganisms and result in faster reaction kinetics. Taken together, the described protocol is expected to open up new possibilities in the single-cell research, particularly for the elucidation of heterogeneity in a population of cells.

Key words Human topoisomerase I, Microfluidics, Single-molecule detection, Single-cell analysis, Rolling circle amplification, Enzyme activity

1 Introduction

Single-cell analysis is becoming increasingly significant in molecular biology offering new possibilities in comprison to the traditional ensemble average measurements where the individual cellular behavior is masked. This is particularly true in notoriously heterogeneous populations of cancer cells. Drastic heterogeneity has been demonstrated not only in morphology, but also in the functionality of cancer cells, including the enzymatic activities.

The DNA-modifying enzyme topoisomerase I (hTopI) relaxes supercoiled DNA, and is essential in nearly all processes of DNA metabolism where tension in DNA is accumulated, such as replication, transcription, recombination, and chromosomal segregation [1]. Clinically, hTopI has also attracted great interest being the sole

Anup K. Singh and Aarthi Chandrasekaran (eds.), *Single Cell Protein Analysis: Methods and Protocols*, Methods in Molecular Biology, vol. 1346, DOI 10.1007/978-1-4939-2987-0_14,

target of the chemotherapeutic drugs of the camptothecin (CPT) family [2]. Several studies have demonstrated a correlation between hTopI expression levels, hTopI activity, and tumor progression [3, 4] and stressed the need for highly sensitive methods to detect its activity in crude extracts from a few cells or even from a single cell.

We have previously developed a powerful tool to detect the activity of hTopI, based on a technique termed, rolling circle-enhanced enzyme activity detection (REEAD) [5] (Fig. 1). The REEAD assay relies on the conversion of a linear dumbbell-shaped DNA molecule S(hTopI) into a circular product mediated by the cleavage-ligation reaction of hTopI. The circular product is subsequently subjected to isothermal rolling circle amplification (RCA) resulting in ralling circle products (RCP), which are visualized at the single-molecule level by hybridization of fluorescent-labeled DNA probes. This results in the appearance of a fluorescent spot as detected in a fluorescence microscope, which represents a single-DNA molecule that has been circularized by a single catalytic reaction of hTopI and, hence, the method is directly quantitative in nature. Combined with microfluidics, we have shown detection of enzymatic activities from one or a few cells using this technique [6].

Presented in this protocol is detection of hTopI activity at the single-cell level; however, this approach can be easily adapted for assaying other enzymatic activities. In fact, in addition to the assaying of human enzyme activities, the microfluidics-based enzyme extraction combined with REEAD detection has recently proven its value for highly sensitive identification of hard-to-break microorganisms, such as the malaria-causing parasite, *Plasmodium falciparum* [7, 8], for disease-relevant diagnostics.

2 Materials

2.1 REEAD

1. 5′ amino primer: 5′-amine-CCAACCAACCAACCAAATAAG CGATCTTCACAGT-3′.
2. S(hTopI): 5′-AGAAAAATTTTTAAAAAAACTGTGAAGATC GCTTATTTTT TTAAAAATTT TTCTAAGTCT TTTA GATCCC TCAATGCTGC TGCTGTACTA CGATCTAAAA GACTTAGA-amine-3′ (*see* **Note 1**).
3. Fluorescent probe: 5′ fluorophore- CCTCAATGCT GCTGCT GTACTAC-3′.

Fig. 1 (continued) to a glass surface to enable amplification of the circularized products directly on the surface. (**c**) The hTopI-circularized substrates are hybridized to the 5′-amino primer attached to the glass slide. Upon amplification with the Phi29 polymerase a tandem repeat of up to 10^3 complementary copies of the s(hTopI) is generated. Then fluorescent DNA probes, complementary to the amplified products, are added and the single RCPs visualized using a fluorescence microscope

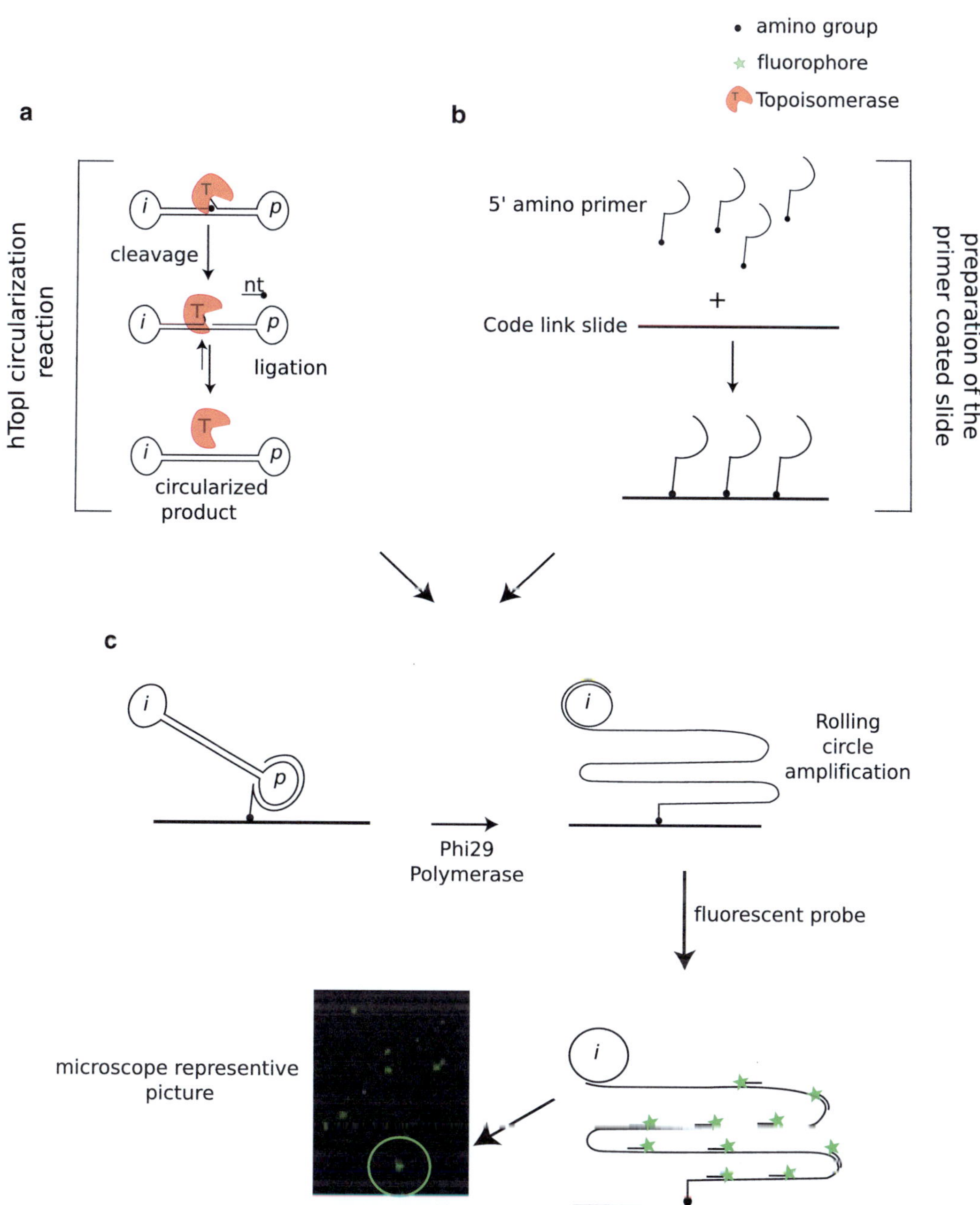

Fig. 1 Representation of the REEAD assay. (**a**) Steps of the circularization reaction. The hTopI substrate, s(hTopI), self-assembles into a dumbbell shape with two loops with a double-stranded sequence in between. One loop contains a sequence for the hybridization to a complementary primer (loop *p*), the second loop contains an sequence for the hybridization with fluorescent oligonucleotides (loop *i*), and the double-stranded stem region contains the preferred hTopI cleavage sequence. hTopI binds to the substrate and cleaves three bases upstream to the 3′ end releasing a short amino-oligo (nt). Subsequently the enzyme religates the 5′ OH end of the substrate, and the open dumbbell substrate is converted into a closed circular molecule that can be used as template in the RCA mediated by the Phi29 polymerase. (**b**) Preparation of the primer-coated slides. The 5′-amino-coupled primer (complementary to the sequence in the loop p of the s(hTopI)) is covalently attached

4. Cell resuspension buffer: Phosphate-buffered saline (1× PBS), 1 % Pluronic F-68, 0.1 % BSA.
5. Lysis buffer: 20 mM Tris–HCL pH 7.5, 0.5 mM EDTA, 1 mM DTT, 1 mM PMSF, 0.2 % Tween-20.
6. Nuclear extraction buffer: 0.5 M NaCl, 20 mM HEPES, pH 7.9, 20 % glycerol.

2.2 HTopI Circularization Reactions

1. HTopI reaction buffer: 10 mM Tris–HCl, pH 7.5, 0.5 mM EDTA, 100 mM NaCl, 1 mM DTT, 1 mM PMSF, 100 nM s(hTopI).

2.3 Rolling Circle Amplification

1. Codelink Activated Slides (Surmodics).
2. Blocking buffer: 50 mM Tris, 50 mM Tris–HCl, 32 mM ethanolamine, pH 9.
3. Wash buffer 1: 4× SSC, 0.1 % SDS.
4. Wash buffer 2: 100 mM Tris–HCl, 150 mM NaCl, and 0.3 % SDS.
5. Wash buffer 3: 100 mM Tris–HCl, 150 mM NaCl, and 0.05 % Tween-20.
6. 99 % ethanol (EtOH).
7. RCA mixture: 1× Phi29 bufferd .supplemented with 0.2 μg/μL BSA, 250 μM dNTP, and 6 U Phi29 DNA Polymerase.
8. 1× Hybridization buffer: 20 % formamide, 2× SSC, and 5 % glycerol.
9. Fluorescent probes (*see* **Note 2**).
10. Vectashield for mounting slides.
11. Pap pen.

2.4 Microfluidic Devices

1. Photoresist SU8-3025.
2. Polydimethylsiloxane (PDMS) kit.
3. Weighing boat.
4. Vacuum desiccator.
5. Hole punchers.
6. PTFE tubes 30TF TBG Assy, Hamilton.

2.5 Microscope

1. Solid support slides.
2. Fluorescence microscope equipped with appropriate filters for the fluorescent probes chosen.
3. Camera for image analysis.
4. Software ImageJ.

3 Methods

3.1 Cell Culture

1. Culture cells in suitable medium and supplements (*see* **Note 3**).
2. Remove medium from a sub-confluent culture flask.
3. Wash the cells with 1× PBS, harvest with 0.25 % trypsin–EDTA, and transfer to an Eppendorf tube.
4. Spin at 600 ×*g* for 2 min and remove the supernatant.
5. Adjust the cell densities to 0.5 million cells/mL, and resuspend the cell pellet in cell resuspension buffer. From this step onwards, all samples are kept on ice and buffers are prepared ice cold, to prevent degradation and denaturation of proteins.
6. For REAAD in bulk (Subheading 3.3), cells are lysed in lysis buffer for 10 min and subjected to nuclear extraction buffer to analyze the activity of hTopI present in cell nuclei. For REAAD-on-a-chip this extraction happens directly in the microfluidic channel (*see* **Note 4**).

3.2 Preparation of the Primer-Coated Slides

1. Cut 5 × 5 mm pieces from the Codelink slide, and glue them to a glass slide. Let them dry for at least 10 min.
2. Draw the border of the pieces with a pap pen to have a water-repelling edge.
3. Covalently couple the 5′-amine-labeled primer to the slides (Fig. 1b) according to the manufacturer's protocol.
4. Block the slides by immersing the slides in the blocking buffer for 30 min at 50 °C.
5. Rinse twice in ion-exchanged H_2O for 1 min.
6. Wash for 30 min at 50 °C in the wash buffer 1.
7. Rinse twice in ion-exchanged H_2O for 1 min.
8. Air-dry the slides and keep them in a dry space for a maximum of 1 month.

3.3 REEAD: Bulk Measurement—Cell Nuclear Extraction

1. Lyse the cells in lysis buffer for 10 min in ice.
2. Spin 600 ×*g* for 5 min at 4 °C and discard the supernatant.
3. Add 100 μL nuclear extraction buffer without disturbing the pellet.
4. Wrap with parafilm around the tube and gently rotate for 1 h at 4 °C avoiding pellet resuspension.
5. Spin for 10 min at 8500 ×*g*, 4 °C. Transfer supernatant to a new tube and use ice-cold nuclear extracts for hTopI circularization reactions.

3.4 REEAD: Bulk Measurement—Rolling Circle Amplification

Purchase all the substrates, primers, and probes with high purity or perform gel purification (*see* **Note 1**). The substrate for hTopI folds into a dumbbell shape containing a primer hybridization sequence (Fig. 1a labeled *p*), a double-stranded region with the hTopI preferred cleavage sequence [9], and a sequence for hybridization with fluorescent probes (Fig. 1a labeled *i*).

1. Prepare the circularized reactions by incubating fresh nuclear extracts in 30 μL reaction volumes containing hTopI reaction buffer.

   The enzymatic reaction will join the nick of the s(hTopI) forming a circularized product as shown in Fig. 1a. Incubate the reaction at 37 °C for 30 min. Stop the circularization reaction by heat inactivation at 95 °C for 5 min. Flash freeze the reaction mixture in dry ice and keep at −80 °C until use or use fresh.
2. Transfer 5 μL of the reaction mixture onto the primer-coated slide (Fig. 1c).
3. Let the circularized products hybridize to the primers for 60 min at room temperature (20–25 °C) in a humidity chamber.
4. Wash the slides for 1 min in wash buffer 2, and 1 min in wash buffer 3, and dehydrate for 1 min in 99 % EtOH.
5. Let the slides air-dry.
6. Prepare the RCA mixture and add 3 μL to each slide.
7. Incubate for 1 h at 37 °C in a humidity chamber.
8. Wash the slide for 1 min in wash buffer 2, and 1 min in wash buffer 3, and dehydrate for 1 min in 99 % EtOH.
9. Let the slides air-dry.
10. Prepare a mixture containing 1× hybridization buffer and 0.2 μM fluorescent probe and add 5 μL of this mixture on each slides in the dark.
11. Incubate for 30 min at 37 °C in a humidity chamber in the dark.
12. Wash the slide for 10–15 min in wash buffer 2, and 5 min in wash buffer 3. Then transfer the slide to 99 % EtOH, dehydrate for 1 min, and air-dry.

3.5 Fabrication of Microfluidic Devices

The microfluidic setup consists of two devices: a flow-focusing droplet generator and a drop trap. The flow-focusing droplet generator consists of one inlet for the carrier fluid (continuous phase) and three inlets for the aqueous solution (discrete phase). Generation of droplets is achieved through the competition between the continuous and discrete phases (Fig. 2a) [7]. The generated droplets are then collected from the outlet. The function of the drop trap (*see* **Note 5**) is to hold the droplets in given locations (Fig. 2b).

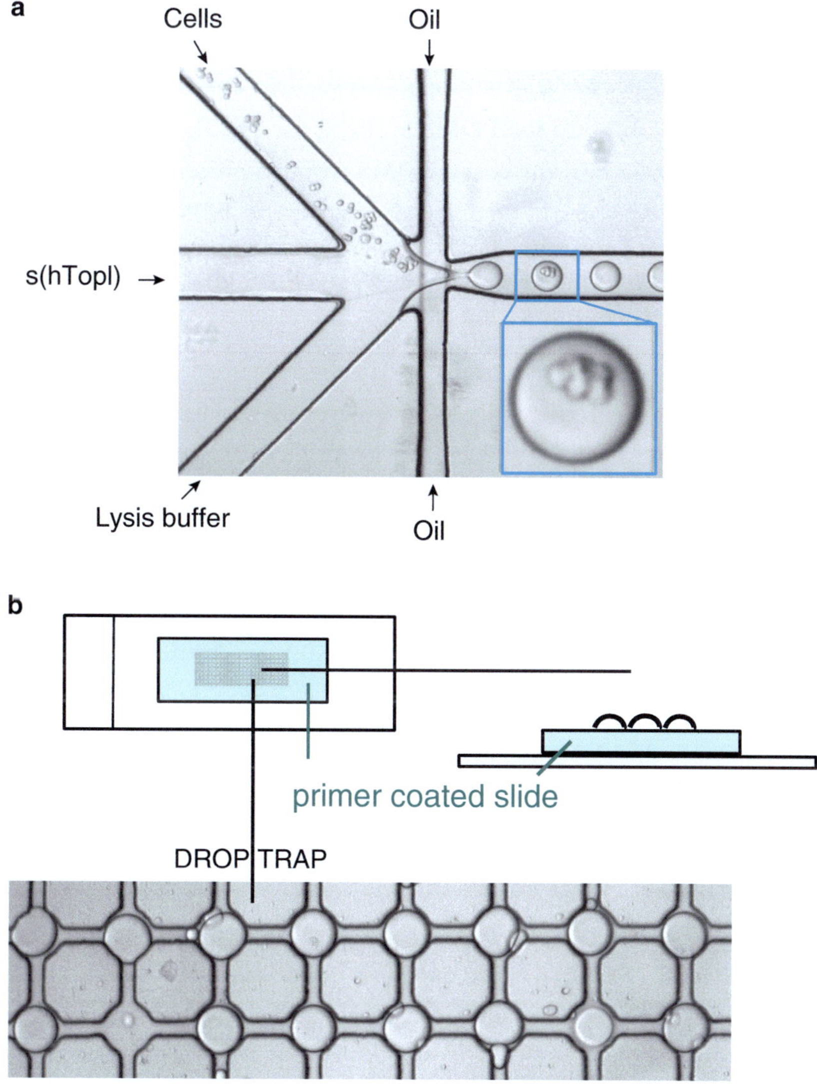

Fig. 2 Microfluidic setup: (**a**) Cells, s(hTopI), and lysis buffer are injected in the inlets and, by competition with oil, confined in droplets of picoliter dimension in which the lysis of the cells and circularization of s(hTopI) take place. (**b**) The droplets are confined in a drop trap placed on top of a primer-coated glass slide on which RCA takes place

Therefore, the geometry of the drop trap is designed according to the size of generated droplets. Both devices are fabricated by conventional soft lithography techniques [10].

1. Produce the master mold with SU8-3025 through a photolithography process with a transparency mask. Produce the mold at a height of around 35 μm following the manufacturer's protocol.
2. Prepare polydimethylsiloxane (PDMS) prepolymer in a 10:1 (base:curing agent) ratio.

3. Place the SU-8 master in a weighing boat.
4. Pour the PDMS prepolymer onto the SU-8 master and leave the weighing boat in a vacuum desiccator to remove bubbles.
5. Cure the PDMS at 65 °C for 1–2 h.
6. Peel the cured PDMS strip from the Si master mold.
7. For the droplet generator, the PDMS strip is punched with through holes (hole puncher) as fluidic connections and bonded with a cover glass through a thin layer of PDMS. The bonded chip is then left in an oven at 95 °C overnight to enhance bonding strength.

3.6 Preparation for the Droplet Generation

1. Highly volatile fluorinated carrier fluid, such as HFE-7500, is selected for fast evaporation at room temperature, higher density than aqueous solutions, and easy collection of final products (*see* **Note 6**). To maintain the stability of the generated droplet, fluorosurfactants (surfactants with fluorinated tails) are required.
2. Pre-dissolve the synthesized or purchased fluorosurfactant into fluorocarbon oil HFE-7500 in a 2 % weight percentage.

3.7 REEAD-on-a-Chip (Enzyme Extraction and Reaction in Drops)

Enzyme extraction in droplets generated by the flow-focusing droplet generator.

1. Connect four PTFE tubes to four syringes, respectively. Individually load the carrier fluid (oil), cell suspension (0.5 million/mL to ensure single-cell encapsulation in drops), lysis buffer, and 100 nM s(hTopI) into the four syringes (*see* Fig. 2a).
2. Insert the PTFE tube connected to the carrier fluid syringe into the oil inlet. Prior to the droplet generation, wet the channels of the droplet generator with the carrier fluid for at least 30 min, to render the PDMS surface hydrophobic.
3. Introduce the reagents, i.e., cells, s(hTopI) and hypotonic lysis buffer to the three corresponding inlets for aqueous solutions (Fig. 2a). Connect a collection tube to the outlet to introduce the generated droplets in an Eppendorf tube.
4. Two syringe pumps are used to control the flow rates of the carrier fluid and reagents independently, forming monodisperse water-in-oil droplets at a frequency of 0.8–1.5 kHz (*see* **Note 7**). The droplet volume and generation frequency are controlled by the flow rate ratio, determined by the competition between the continuous phase (carrier fluid: the oil/surfactant, flow rate 22.5 μL/min) and the disperse phase (aqueous reagents: cells, lysis buffer, and substrates, flow rate 2.5 μL/min). The droplet generation frequency is ~kHz. When the droplet generation is stable, harvest around 50 μL of the generated droplets.

Single-cell enzymatic reaction in droplets and located with a drop trap (Fig. 2b).

5. Place 5 μL of the collected droplets on a primer-printed glass slide prepared as described in Subheading. 3.2.
6. Gently place the drop trap on top of primer-coated glass slide containing droplets.
7. Leave the droplets to exsiccate and the REEAD substrates to hybridize to the primers before removing the drop trap. Proceed to RCA and hybridization of probes as stated above in Subheading 3.4.

3.8 Microscopic Observation and Analysis

1. Mount slides with 2.5 μL Vectorshield, add a cover glass, and analyze using a fluorescence microscope (using appropriate filters) equipped with a 63× (or 100×) oil immersion objective and a camera.
2. Find the printed area by focusing first on the pap pen-delimited square area.
3. Take several (at least 12) pictures per slide.
4. Import the pictures using the software ImageJ [11] and count the fluorescent spots. Representative picture is reported in Fig. 1c.

4 Notes

1. Primer and substrate designs: The DNA substrates consist of a single oligonucleotide, which is converted to a closed DNA circle by a single-enzyme cleavage-ligation event mediated by hTopI. HTopI does not recognize a specific sequence but has preferences [9]. For each circularized substrate subsequent RCA results in the creation of one RCP consisting of multiple (up to 10^3) tandem copies of the circularized substrate. Each RCP is optically detected at the single-molecule level by hybridization of fluorescently labeled probes to the RCP followed by microscopic analysis. For all the above mentioned reasons substrates need to be designed with a "p" sequence specific for a DNA primer and an "i" sequence that will be recognized by a fluorescent-labeled probe.
2. Fluorescent probes are oligonucleotides that are purchased with a fluorophore at 5′ end to allow the detection of the amplified product at the microscope.
3. In this chapter, human embryonic kidney (HEK293) cells were chosen as model and cultured in Dulbecco's modified Eagle medium (DMEM) supplemented with 10 % fetal bovine serum, 100 U/mL penicillin, and 100 g/mL streptomycin in a humidified incubator (5 % CO_2/95 % air atmosphere at 37 °C).

4. Lysis and nuclear extraction buffer are supplemented with 1 mM DTT and 2 μL of saturated PMSF added fresh to avoid degradation of the proteins.
5. The drop trap shown in Fig. 2b "traps" the droplets passively by having similar geometry as the generated droplets, in this case around 75 μm in diameter. For the droplets to break up rapidly on the drop trap, highly volatile fluorinated carrier fluid, such as HFE-7500 and HFE-7100, is selected over the FC series, such as FC-40 and FC-72. For the audience interested in the design of drop trap, many innovations [12, 13] are published and can also be adapted for the same purpose.
6. Fluorosurfactants are commercially available from RainDance Technologies and Bio-Rad along with their digital PCR system, whereas fluorosurfactants pre-dissolved in fluorinated oil may be purchased independently from Dolomite. If resources are available, published protocols are also available for synthesizing fluorosurfactants in house [14].
7. Be careful not to introduce any air bubbles in the syringe-tubing assembly; otherwise, the droplet generation may not remain stable throughout the process.

References

1. Leppard JB, Champoux JJ (2005) Human DNA topoisomerase I: relaxation, roles, and damage control. Chromosoma 114:75–85, http://www.ncbi.nlm.nih.gov/pubmed/15830206. Accessed 3 Oct 2013
2. Pommier Y (2006) Topoisomerase I inhibitors: camptothecins and beyond. Nat Rev Cancer 6:789–802, http://www.ncbi.nlm.nih.gov/pubmed/16990856. Accessed 3 Oct 2013
3. Pfister TD, Reinhold WC, Agama K, Gupta S, Khin S a, Kinders RJ et al (2009) Topoisomerase I levels in the NCI-60 cancer cell line panel determined by validated ELISA and microarray analysis and correlation with indenoisoquinoline sensitivity. Mol Cancer Ther 8:1878–84, Available from: http://www.pubmedcentral.nih.gov/articlerender.fcgi?artid=2728499&tool=pmcentrez&rendertype=abstract. Accessed 3 Oct 2013
4. Humana Press DNA Topoisomerase and cancer 2011. Press H, editor, Pommier Y
5. Stougaard M, Lohmann J, Mancino A, Celik S, Andersen FF, Koch J et al (2008) Single-molecule detection of human topoisomerase I cleavage- ligation activity. ACS Nano 3:223–33, http://pubs.acs.org/doi/abs/10.1021/nn800509b. Accessed Sep 18 2014
6. Ho Y, Grigsby C, Zhao F, Leong K (2011) Tuning physical properties of nanocomplexes through microfluidics-assisted confinement. Nano Lett 11:2178–82, http://pubs.acs.org/doi/abs/10.1021/nl200862n. Accessed 6 May 2014
7. Juul S, Nielsen CJF, Labouriau R, Roy A, Tesauro C, Jensen PW et al (2012) Droplet micro fluidics platform for highly sensitive and quantitative detection of malaria-causing plasmodium parasites based on enzyme activity measurement. ACS Nano 6:10676–83. http://pubs.acs.org/doi/abs/10.1021/nn3038594
8. Tesauro C, Juul S, Arnò B, Nielsen CJF, Fiorani P, Frøhlich RF (2012) Specific detection of topoisomerase i from the malaria causing P. falciparum parasite using isothermal rolling circle amplification. Conf. IEEE EMBS San Diego, CA, USA, 28 Aug–1 Sept 2012, pp 2416–2419. http://ieeexplore.ieee.org/xpls/abs_all.jsp?arnumber=6346451. Accessed 6 May 2014
9. Bonven BJ, Gocke E, Westergaard O (1985) A high affinity topoisomerase I binding sequence is clustered at DNAase I hypersensitive sites in Tetrahymena R-chromatin. Cell 41:541–51,

http://www.ncbi.nlm.nih.gov/pubmed/2985282

10. Qin D, Xia Y, Whitesides GM (2010) Soft lithography for micro- and nanoscale patterning. Nat Protoc 5:491–502, http://www.ncbi.nlm.nih.gov/pubmed/20203666. Accessed 29 Apr 2014
11. Schneider CA, Rasband WS, Eliceiri KW (2012) NIH Image to ImageJ: 25 years of image analysis. Nat Methods 9:671–675, http://www.nature.com/doifinder/10.1038/nmeth.2089. Accessed 28 Apr 2014
12. Bai Y, He X, Liu D, Patil SN, Bratton D, Huebner A et al (2010) A double droplet trap system for studying mass transport across a droplet-droplet interface. Lab Chip 10:1281–5, http://www.ncbi.nlm.nih.gov/pubmed/20445881. Accessed 17 Nov 2014
13. Huebner A, Bratton D, Whyte G, Yang M, Demello AJ, Abell C et al (2009) Static microdroplet arrays: a microfluidic device for droplet trapping, incubation and release for enzymatic and cell-based assays. Lab Chip 9:692–698, http://www.ncbi.nlm.nih.gov/pubmed/19224019. Accessed 31 Oct 2014
14. Chiu Y-L, Chan HF, Phua KKL, Zhang Y, Juul S, Knudsen BR et al (2014) Synthesis of fluorosurfactants for emulsion-based biological applications. ACS Nano 8:3913–20, http://www.ncbi.nlm.nih.gov/pubmed/24646088

Chapter 15

Microfluidic Chemical Cytometry for Enzyme Assays of Single Cells

Livia Shehaj, Lorena Lazo de la Vega, and Michelle L. Kovarik

Abstract

Cellular heterogeneity occurs, and should be probed, at multiple levels of cellular structure and physiology from the genome to enzyme activity. In particular, single-cell measures of protein levels are complemented by single-cell measurements of the activity of these proteins. Microfluidic assays of enzyme activity at the single-cell level combine moderate to high throughput with low dead volumes and the potential for automation. Herein, we describe the steps required to fabricate and operate a microfluidic device for chemical cytometry of fluorescent or fluorogenic reporters of enzyme activity in individual cells.

Key words Microfluidic, Enzyme assay, Chemical cytometry, Microchannel, Reporter substrate, Electrophoresis

1 Introduction

The origins of cellular heterogeneity can be traced through the central dogma of biology from DNA (through mutations) to RNA (through transcriptional regulation) to proteins (through translational, posttranslational, and degradation-based regulation). Cellular heterogeneity can arise at each of these levels of protein coding and expression [1]. However, after a protein is expressed, heterogeneity can also arise in its level of activity, which is controlled by posttranslational modifications, competition between or for substrates, formation of protein complexes, and localization [2]. As a result, a comprehensive understanding of heterogeneity in the proteome should include assays that directly evaluate the activity of enzymes within intact single cells.

One methodology to assay enzyme activity in single cells uses a combination of reporter substrates and chemical cytometry. Reporter substrates are fluorescent or fluorogenic substrates for an enzyme of interest. These substrates may be small molecules, lipids, or peptides, depending on the enzyme to be assayed. Fluorogenic and fluorescence resonance energy transfer (FRET)-based substrates

Anup K. Singh and Aarthi Chandrasekaran (eds.), *Single Cell Protein Analysis: Methods and Protocols*, Methods in Molecular Biology, vol. 1346, DOI 10.1007/978-1-4939-2987-0_15, © Springer Science+Business Media New York 2015

allow enzyme activity to be measured in intact cells as a function of space and time using microscopy or image cytometry. However, quantitative results from these substrates are limited by uncertainty about initial concentrations of substrate in each cell. In contrast, fluorescently labeled substrates permit accurate quantitation of the total amount of substrate loaded but require a separation step to separate the unmodified form of the reporter from any enzymatic products [1]. Chemical cytometry involves the lysis of a single cell, followed by electrophoretic separation and detection of its chemical contents. In concert with reporter substrates, this technique provides quantitative information about enzyme activity in individual cells. Chemical cytometry can separate unmodified and modified forms of a fluorescent reporter or multiple fluorogenic reporters that emit at the same wavelength. The combination of chemical cytometry with reporter substrates complements single-cell methodologies based on antibodies by eliminating the need to raise antibodies against a molecules of interest, reducing issues of nonspecific binding, and measuring enzyme activity toward a substrate directly.

Both capillary and microchannel systems can be used for chemical cytometry. Microfluidic systems have the advantages of low dead volumes and the potential for automation of upstream tasks, including reporter loading. Several related microchip designs have been developed for chemical cytometry (Fig. 1). Each design includes a cell channel to introduce cells from the sample inlet, a waste channel to remove cellular debris after lysis, and an electrophoresis channel for separation and detection of cell contents. Figure 1a, b also includes a focusing channel to direct cells to the lysis point. Herein, we describe experiments using the device shown in Fig. 1b, which includes empirically optimized channel widths for cell transport and fluid flow in the device. This design has provided throughputs up to 30 cells/min for the fluorogenic reporters described below [3] and up to 37 cells/h for peptide reporters for peptidase activity in drug-treated leukemia cells [4]. Devices based on this design are extremely robust; experiments on up to 600 individual single cells have been performed on a single device [3], and a 95-day longitudinal study has been completed that includes data for 678 cells collected on multiple devices over the course of several months [5]. This combination of high throughput and robust performance indicates that chemical cytometry is becoming mature enough for routine use. Below, we describe a protocol for fabricating and coating microfluidic devices for single-cell measurements; loading membrane-permeant, fluorogenic substrates into cells; and conducting single-cell measurements of enzyme activity using laser-based lysis, electrophoretic separation, and laser-induced fluorescence detection. As an example, in this chapter we demonstrate measurement of the activity of esterase enzymes in single cells. We include notes on adapting this protocol to alternative substrates, lysis schemes, and detection methods.

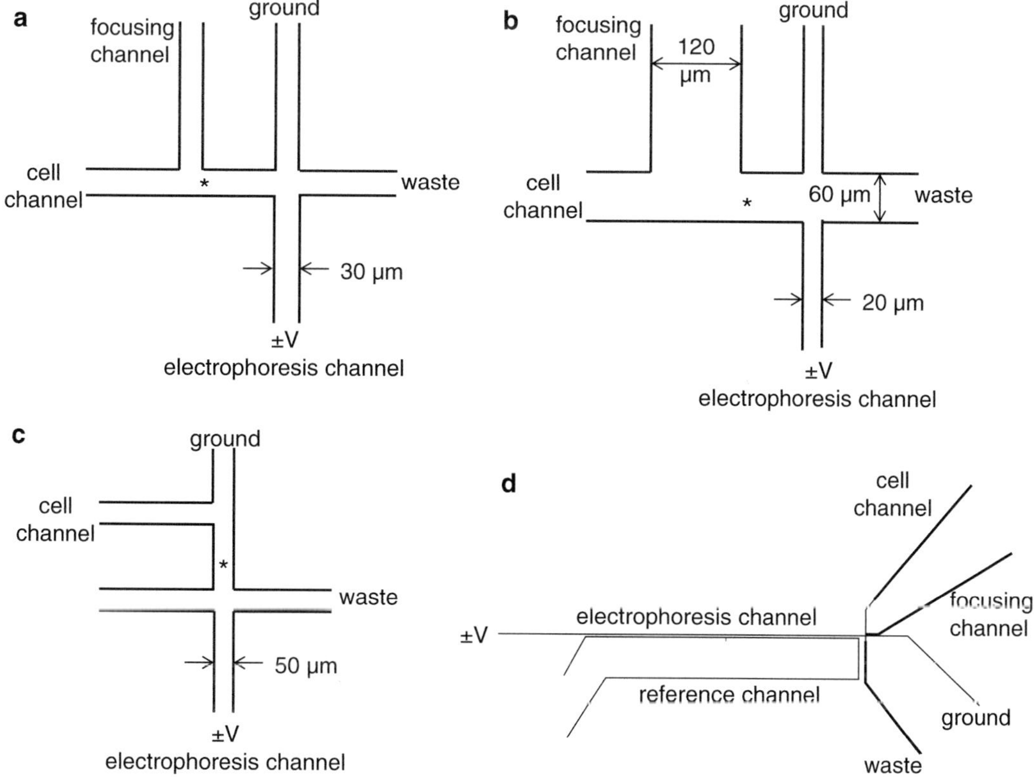

Fig. 1 (**a–c**) Schematics of microchip designs for chemical cytometry from Refs. 3, 9, 10, respectively. The approximate location of cell lysis in each device is indicated by an *asterisk*. In general, devices should have a cell channel, a waste channel, and a separation channel. Designs (**a**) and (**b**) include an additional focusing channel to direct cells to a specific location on the device. (**d**) Full photomask design for the device in (**b**). The length of the channel from ground to ±V is 3 cm, and all other channel lengths are to scale

2 Materials

2.1 Device Fabrication

1. Small unilamellar vesicle (SUV) buffer: 10 mM Tris base, 150 mM NaCl, pH 7.4.
2. 4″ silicon wafers (*see* **Note 1**).
3. SU-8 10 negative tone photoresist.
4. Propylene glycol monomethyl ether acetate (PGMEA).
5. Spin coater.
6. UV exposure system or mask aligner.
7. Photomask (*see* **Note 2**).
8. Poly(dimethylsiloxane) (PDMS, Sylgard 184, consists of silicone base and curing agent, Dow Corning).
9. Platinum-cured siloxane tubing.

10. #1 thickness cover glass (at least as large as device footprint).
11. Plasma cleaner.
12. Lipid solution: 25 mg/mL egg phosphatidylcholine in chloroform.
13. Sonifier or cell disruptor.
14. Microcentrifuge.
15. Vacuum desiccator.
16. Cutting mat.
17. 1 mm biopsy punch or cork borer.
18. Oven.

2.2 Chip Alignment and Substrate Loading

1. 0.22 μm filters.
2. Bath sonicator.
3. Separation buffer: 25 mM Tris–HCl, pH 8.4.
4. Extracellular buffer (ECB)-glucose: 10 mM 4-(2-hydroxyethyl)-1-piperazineethanesulfonic acid (HEPES), 135 mM NaCl, 5 mM KCl, 1 mM $MgCl_2$, 1 mM $CaCl_2$ and 10 mM glucose, pH 7.4.
5. Hemacytometer.
6. Esterase substrates: Fluorescein diacetate and carboxyfluorescein diacetate (20 mM stock solutions in DMSO, stored at −20 °C for no more than 2 weeks).

2.3 Chemical Cytometry with Laser Lysis (See Note 3)

1. Epifluorescence microscope with 532 nm laser filter and 20× air objective.
2. Fiber optic laser-induced fluorescence system:
   - 488 nm continuous wave laser (output ~2 mW; e.g., argon ion or solid state).
   - Single-mode fiber-optic patch cord to carry 488 nm excitation light with collimating connectors.
   - FITC filter cube.
   - 60× air objective.
   - XYZ translation stage.
   - Multimode visible light fiber patch cord with collimating connectors.
   - Photomultiplier tube.
   - Hardware filter (optional).
   - Data acquisition system.
3. Sub-nanosecond pulsed 532 nm laser.
4. 532 nm laser filter cube.
5. 5× beam expander.

6. Iris.
7. High-voltage power supply (max. voltage output ≥ 1500 V, max. current output ≈ 50 μA).
8. Pt wire electrodes.
9. Fluorescent bead solution: 1 mL of 1 μM fluorescein in water plus 1 drop of concentrated 5 μm green fluorescent microsphere suspension.

3 Methods

3.1 Preparing Small Unilamellar Vesicles for Chip Coating

1. Add 160 μL of 25 mg/mL lipid solution in chloroform to a 1.5 dram glass vial (*see* **Note 4**).
2. Dry the lipid solution carefully with a stream of nitrogen. Tip the vial to the side and slowly rotate it to create an even lipid film. Do not let the solution run higher than ¼ from the bottom of the vial. When fully dry, the lipid film appears whitish and opaque.
3. Add 2 mL of SUV buffer to the vial for a final lipid concentration of 2 mg/mL.
4. Using a sonifier or cell disruptor, apply 40 W at a duty cycle of 30 % for approximately 20 min, until the solution appears clear rather than cloudy (*see* **Note 5**).
5. Centrifuge the resulting solution for 5 min at 12,000 × *g*. Depending on the age of the sonifier tip, centrifugation may result in a small pellet of metal debris from the tip. Carefully remove the supernatant, and store at 4 °C for at least 2 h before use and up to 2 weeks.

3.2 Fabricating the Microfluidic Device

1. Descum a new, clean silicon wafer in air (oxygen) plasma on high RF power for 20 min.
2. Dispense ~10–15 mL of SU-8 10 photoresist on the wafer and spin-coat at 1500 rpm for 30 s (*see* **Note 6**).
3. Soft-bake the wafer for 10 min in a 95 °C oven.
4. Using a mask aligner or UV exposure system, expose the wafer through the photomask to a total exposure energy of 300 μJ (*see* **Note 7**).
5. Place the exposed wafer in a 95 °C oven for a 5-min post-exposure bake before developing for ~2 min in a recrystallization dish filled to a depth of ~1 cm with propylene glycol monomethyl ether acetate (PGMEA). Agitate the developer over the surface of the wafer during development manually or using an orbital shaker.
6. Using wash bottles of solvent, rinse the developed wafer gently with PGMEA and then with isopropyl alcohol. If any part of the

wafer turns white on exposure to isopropyl alcohol, development is incomplete. Rinse the wafer again with PGMEA, and return the wafer to the developing dish. Repeat the isopropyl alcohol rinse and blow the wafer dry with compressed air or nitrogen.

7. Hard-bake the wafer for 10 min on a 95 °C hotplate and then for 70 min on a 120 °C hotplate (*see* **Note 8**).
8. Mix 5 g of silicone elastomer base with 0.5 g of curing agent and carefully coat the silicon wafer with the mixture. This mass of PDMS will cover a 4″ silicon wafer and be held in place by surface tension. For larger or smaller substrates, adjust the total mass and keep the mass ratio between the base and curing agent at 10:1, respectively. Place the wafer under vacuum for about 20 min to remove any bubbles. Then cure the wafer on a hot plate at ~100 °C for about an hour. Longer curing times will not damage the PDMS (for example, devices may be left to cure overnight).
9. Using a razor blade to detach the edges of the PDMS replica from the wafer, carefully peel the PDMS from the master mold (*see* **Note 9**). Place the PDMS on a clean cutting mat with the channels facing up. Using a blade, cut the PDMS into separate devices and punch outlet holes at the end of each channel with a 1 mm biopsy punch or cork borer.
10. Place one PDMS device and a #1 cover glass in a Petri dish, making sure that there is no contact between the two pieces.
11. Expose the pieces to air (oxygen) plasma on high for 2 min (*see* **Note 10**), then isolate the chamber from the vacuum pump, and turn off the RF power. Slowly vent the chamber and turn off the pump (*see* **Note 11**).
12. Use a pair of tweezers to place the PDMS on the cover glass so that the sides that were facing up during the plasma treatment are now in contact with each other. Press out any air bubbles. This step should be completed in a single attempt within 60 s of plasma exposure as the reactive surface will be passivated with contact and over time (*see* **Note 12**).
13. Cut six small (4 mm) sections of siloxane tubing and place them into the Petri dish with the bonded chip. Repeat **step 11**.
14. After removing the Petri dish from the plasma cleaner, position the oxidized tubing around the punched out holes at the end of each channel, making sure that the sides exposed to the plasma are in contact. Press down firmly on each reservoir to ensure that they are bonded to the chip (*see* **Note 13**). It is not necessary to place a reservoir at the reference channel outlet, only at the reference inlet.
15. Once the device is bonded, add 30 μL of SUV suspension (prepared as described in Subheading 3.1) to each reservoir.

Channels should wet readily as PDMS is hydrophilic after exposure to oxygen plasma. A supported bilayer membrane coating will form spontaneously on the channel walls (*see* **Note 14**). The assembled and coated device should be stored at 4 °C for at least 1 h and up to 2 days before use.

3.3 Aligning the Microfluidic Chip with the Detection System

1. Syringe-filter ~1 mL of separation buffer through a 0.22 μm filter, and degas the filtered buffer for 15 min in a bath sonicator (*see* **Note 15**).
2. Remove the vesicle solution from the reservoirs of the microfluidic device and rinse each reservoir once with 30 μL filtered degassed separation buffer solution. Add 30 μL of separation buffer to the cell, focusing, ground, and waste reservoirs. Add ~10 μL of separation buffer to the high-voltage reservoir. Put 30 μL of fluorescent bead solution in the reference channel inlet, and leave the reference channel outlet empty. Prop the chip up vertically with the cell, focusing, ground, and waste reservoirs above the high voltage and reference outlets, and allow solutions to flush through the chip by gravity flow for 10 min. During this time, turn on and warm up electronic systems (i.e., laser, imaging system, detector, and power supply).
3. Replace the contents of each reservoir (except the reference reservoir) with 30 μL fresh filtered degassed separation buffer in each. Leave the fluorescent bead solution in the reference channel.
4. Align the chip on the microscope stage so that the focal point of the lysis laser is near the intersection of the cell channel and the focusing channel (see asterisk, Fig. 1b). Secure the chip on the stage with tape or clamps.
5. Position the high-voltage and ground electrodes in their respective reservoirs.
6. Turn on 488 nm laser and turn off room lights and/or enclose the system in a light-tight box.
7. Set the PMT to a low voltage and begin data collection. Align detection optics over the reference channel using the micrometers of the XYZ translation stage to move forward and back across the device and to adjust the detector focus. Once the focus (Z-setting) has been adjusted the first time, it should not need much adjustment for subsequent experiments. With the detection optics over the reference channel, the PMT should show a variable signal from beads flowing past the detector superimposed on a steady DC signal from the fluorescein solution (*see* **Note 16**).
8. Once you have maximized the signal on the reference channel, check the device to be sure that the objective is in line with the high-voltage reservoir (*see* **Note 17**).

9. Apply 1500 V to the high-voltage electrode and set the inlet electrode to ground (*see* **Note 18**). For the buffers suggested here, the current between the two electrodes should be ~7–12 μA. Current above 19 μA indicates chip failure.
10. Confirm that the signal for the reference channel remains stable while high voltage is applied (*see* **Note 19**).

3.4 Conducting the Chemical Cytometry Experiment

1. Pre-warm complete cell growth media and ECB-glucose in water bath to 37 °C (or the normal growth temperature of your cells). Put 0.4 mL warm complete media in a 1.5 mL microcentrifuge tube, and then add 100 μL cell culture (*see* **Note 20**). Mix well and add 10 μL of diluted cell culture to one side of the hemacytometer. Count the four outside corners of the hemacytometer. Record count. Cell density is COUNT × 5 × 2500 cells/mL.
2. Bring fluorescein diacetate and carboxyfluorescein diacetate stock solutions to room temperature, and prepare a dilute solution in ECB-glucose containing both substrates at a final concentration of 20 μM each (*see* **Note 21**).
3. Spin down 10^6 cells and resuspend in 1 mL of the solution containing the diacetate substrates (*see* **Note 22**). Incubate for 30 min at room temperature in the dark, then resuspend in 1 mL of plain ECB-glucose solution, and incubate for 30 min at 37 °C (or the appropriate growth temperature for the cell line used). Wash cells 2–3 times in 1 mL of ECB-glucose, and resuspend in 0.5 mL of ECB-glucose (final cell concentration of 2×10^6 cells/mL) (*see* **Note 23**).
4. Replace the contents of the cell reservoir with 30 μL of cell suspension in ECB-glucose (*see* **Note 24**).
5. Remove a small volume (5–10 μL) of buffer from the waste outlet, and add a small volume (~5 μL) of separation buffer to the focusing inlet to generate a flow of cells with a linear velocity of 100–200 μm/s. At this flow rate, the cells should require less than 1 s to traverse the intersection region (*see* **Note 25**).
6. Move the laser lysis focal point into the focusing channel (*see* **Note 26**). Fire the pulsed laser and adjust the focus until laser pulses do not damage the chip. Move the pulsed laser back to the lysis point and confirm that the chip is not damaged (*see* **Note 27**).
7. Re-enable the high voltage at 1500 V.
8. Confirm that the detector is still in place and focused on the reference channel.
9. Using a micrometer, move the detection system 200 μm back (from the center of the reference channel to the center of the separation channel), and turn up the PMT voltage. Wait for the baseline to stabilize. It may drop steadily for 2–3 min.

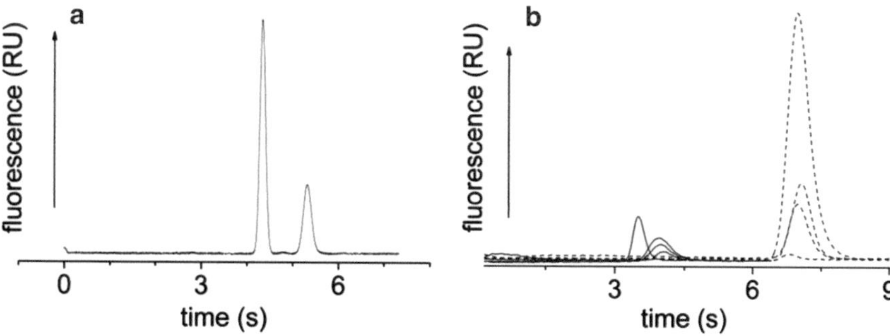

Fig. 2 Sample single-cell electropherograms obtained using the device shown in Fig. 1b at an electric field strength of 235 V/cm. In (**a**), the cell was loaded with both carboxyfluorescein and fluorescein diacetate substrates. In (**b**), cells were loaded with only one or the other substrate. *Solid traces* are for cells loaded with carboxyfluorescein diacetate; *dashed traces* are for cells loaded with fluorescein diacetate. Note that it is common for migration times to vary between devices and for peak areas to vary widely from cell to cell. The cell line used was the mouse leukemia line Ba/F3. Reproduced from Ref. 3 with permission from The Royal Society of Chemistry

10. Start data collection and when a cell intersects the pulsed laser focal point, fire the laser (*see* **Note 28**). Wait several seconds. At this electric field strength, carboxyfluorescein should reach the detector in 2–5 s on a phosphatidylcholine-coated device, depending on the detector location. The fluorescein peak should follow the carboxyfluorescein peak as shown in Fig. 2 (*see* **Note 29**). Adjust the PMT voltage as needed to obtain reasonable peak heights.
11. After each peak envelope is detected, fire the pulsed laser at the next cell entering the intersection (*see* **Note 30**).
12. At the end of each experiment, the electrodes and stir bar (if used, *see* **Note 28**) should be sterilized with ethanol. If needed for the cell line used, the device can be sterilized by soaking in 10 % bleach solution for several hours before disposal.
13. Electropherograms can be analyzed in Cutter 7 [6] or other peak analysis software (*see* **Note 31**).

4 Notes

1. Glass slides may be used in place of silicon wafers, but poorer adhesion between SU-8 and glass results in the need for an adhesion layer of solid SU-8 or an adhesion promoter.
2. Chrome photomasks are needed for designs with features <10 μm and recommended for designs with features <30 μm. Transparency photomasks are acceptable for patterns with larger features.
3. This protocol describes lysis by a pulsed laser, which has several advantages; cell debris, such as the membrane, does not typi-

cally enter the separation channel and lysis is extremely rapid (<1 ms) [7], so activation of stress pathways should not influence results for enzymes in these pathways. The major disadvantage is that this method requires a fairly expensive pulsed laser and the construction of a second optical path (Fig. 3) to accommodate the fluorescence detection and the pulsed laser for lysis. In this arrangement, the pulsed laser for lysis travels through free space, enters the epifluorescence microscope chassis, passes through the cover glass bottom of the microfluidic

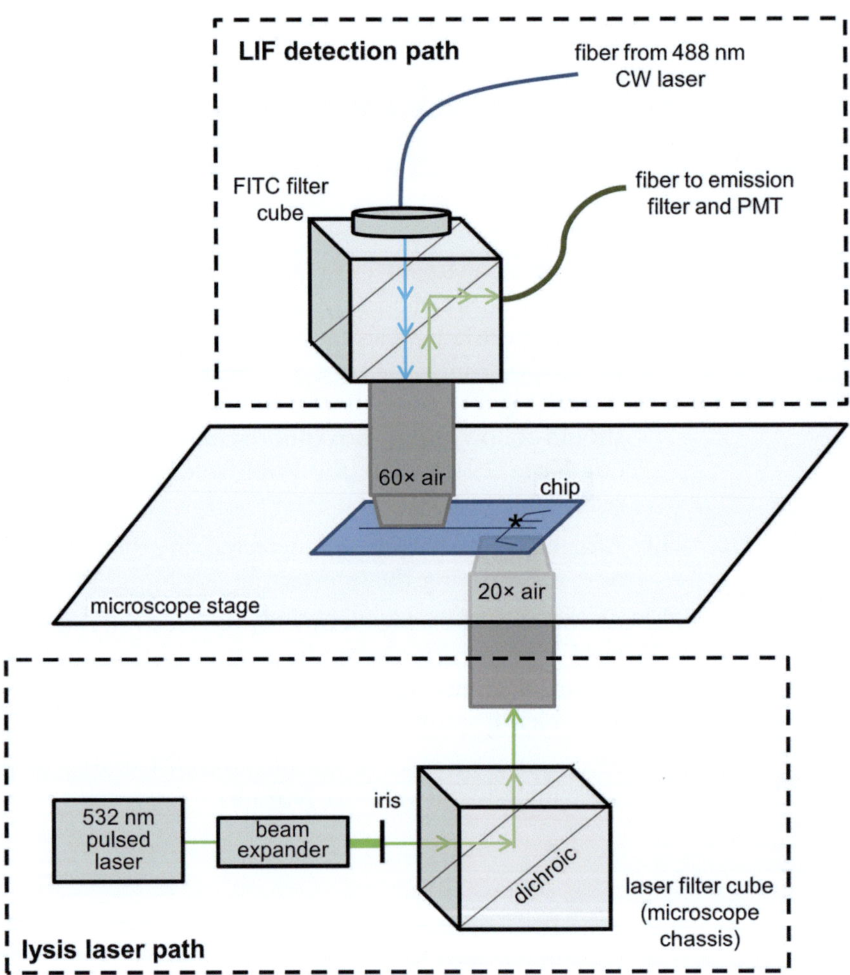

Fig. 3 Optical paths for lysis of cells by a pulsed laser and laser-induced fluorescence detection of the reporter substrate. The pulsed laser path includes beam expansion (e.g., 5×) and an iris for beam cleanup, and then enters the chassis of an epifluorescence microscope. An asterisk indicates the location of cell lysis (i.e., the focal point of the pulsed laser). The laser-induced fluorescence optical path must be constructed and mounted on the microscope stage. The filter cube can be mounted on an XYZ translation stage for alignment with the reference channel on the chip. Fiber optics carry 488 nm excitation light from a continuous wave laser and the resulting fluorescence signal. Recommended components are described in Subheading 2.3 and **Note 3**

device, and produces a plasma in the channel which lyses the cell. Depending on the quality of the alignment and focus, pulse energies from 4 to 10 μJ are required to produce a plasma for lysis; if needed, a polarizer can be placed in the path of the pulsed laser to adjust the pulse energy. The second optical path is used for laser-induced fluorescence detection of the cell contents after electrophoretic detection. This path uses fiber optics to carry excitation light from a 488 nm continuous wave laser to a FITC filter cube and 60× air objective mounted on an XYZ translation stage on the microscope stage above the chip. (Accommodating this apparatus may require moving or removing the bright-field condenser from the microscope.) This detection system is aligned to the reference channel of the device, which is 200 μm from the electrophoresis channel. Fluorescence from the reference or electrophoresis channel is collected by the 60× objective, reflected by the dichroic mirror in the filter cube, collimated by a lens in the fiber coupler, and carried by a multimode fiber optic to second coupler, emission filter, and PMT. The collimating coupler at the end of this fiber should be chosen so that the beam diameter (BD) and divergence angle (DA) across the distance from the lens to the PMT yields a spot size smaller than the active area of the PMT.

BD (mm) = 2 × f (mm) × NA, where f is the focal length of the lens and NA is the numerical aperture of the fiber.

DA (mrad) = a (μm) × f (mm), where a is the fiber diameter (*see* Ref. 8).

A less expensive and simpler alternative is to use the electric field both for lysis and electrophoretic separation. This method effects lysis in <33 ms [9, 10] and eliminates the need for the fiber optics (~$2500), pulsed laser (~$10 k), beam expander ($500), iris (~$50), 532 nm laser filter cube (~$1500), XYZ translation stage (~$900), and reference channel (since the microscope optical path can produce an image of the detection point for alignment of the 488 nm laser focal point with the electrophoresis channel). However, lysis is slower and debris is more likely to enter the electrophoresis channel. For electric field-based lysis, the device design shown in Fig. 1c and described in Ref. 10 was designed to move debris to the waste channel by pressure-based flow.

4. Lipid solutions in chloroform should only contact glass, stainless steel, or Teflon. Do not use plastic containers to store or transfer lipids in chloroform since these materials may introduce plasticizers into the lipid solutions. We recommend glass V-vials with Teflon closures for storage and glass syringes for transferring solutions. Lipid solutions can be stored for up to 6 months at −20 °C. To reduce oxidation, vial headspace can be backfilled with nitrogen or argon gas after each use.

5. The sonifier tip should not contact the container in any way as this can break the vial and seriously damage the tip. The sonifier tip should be at least 0.5 cm from the bottom of the vial to reduce tip damage from reflected sonic energy. Lipid solutions can be placed in an ice bath during sonication to reduce heating and oxidation, although we have found that this does not significantly change the electroosmotic flow obtained from phosphatidylcholine coatings. If the solution does not appear clear after 30 min of sonication, proceed to the next step, as cloudiness may arise from excessive debris from the tip rather than from remaining multilamellar vesicles.
6. This spin speed and SU-8 photoresist formulation should produce a film thickness of 25–30 μm, which is a suitable channel depth for mammalian cells 10–15 μm in diameter.
7. This protocol is written assuming a photomask design like the one in Fig. 1b will be used. Alternative chip designs are shown in Fig. 1a, c. The design recommended here (Fig. 1b) is an adaptation of an earlier design (Fig. 1a) originally published in 2003 [9]. A recent paper describes the development of the device in Fig. 1c to improve the reproducibility of flow rates and injections [10]. Alternative chip designs include simple cross chips [11–14], double tee designs [15, 16], and simple crosses supplemented with focusing channels [17]. The Fig. 1b design is recommended for preliminary experiments but can be adapted for specific cell and assay types. Modifications to the device design may change the required exposure energies and development times.
8. After hard-baking, a profilometer can be used to check the height of the pattern, and the master should be reusable indefinitely. We have fabricated 200+ devices from a single master without noticeable degradation of the pattern on the wafer.
9. Peeling the PDMS piece up in a direction parallel to the longest channels in the pattern makes damage to the master less likely.
10. When the plasma cleaner is turned on, the plasma should appear bright pink. Turn the needle valve to intensify the color. Adjust until the plasma appears maximally bright. This step should be completed slowly as there is often a lag between adjustment of the needle valve and the plasma color. After the needle valve setting has been optimized once, the setting should not need further adjustment.
11. If the pump is turned off without first starting to vent the chamber, the vacuum formed inside the chamber can draw in oil from the pump, which fouls the plasma chamber over time and interferes with bonding. Venting should be performed slowly because rapid venting can disrupt the items within the chamber.

12. If a permanent bond does not form between the PDMS and the cover glass, the plasma cleaning steps of Subheading 3.2 may be repeated several times.

13. From the underside of the chip, the tubing surface should look homogeneous in brightness. Due to a difference in refractive index between air and PDMS, the appearance of brighter spots indicates an air gap between the PDMS device and the siloxane tubing reservoirs, which may compromise the seal. Air gaps can be remedied by placing the assembled device in the plasma cleaner and repeating the plasma cleaning and bonding process. Press down firmly on the reservoirs that were not fully sealed as soon as the device is removed from the plasma cleaner.

14. Channels filled with solution should be difficult to see relative to channels filled with air, which appear bright due to the diffractive index difference between PDMS and air. If the solution does not fully fill the device, gentle vacuum may be applied to one outlet to draw solution through the device.

    The supported bilayer membrane coating has been shown to minimize protein adsorption during electrophoretic separations and allows many separations to be conducted in a physiological buffer [18]. Without the coating, separations in such high-ionic-strength buffers have poor efficiency

    The coating described here is a zwitterionic phosphatidylcholine bilayer, and as a result, electroosmotic flow in the device is minimal. This means that analyte molecules must be drawn electrophoretically into the separation channel, and lower electrophoretic mobility analytes will be injected into the separation channel less efficiently. For the device dimensions described and electric field strength (+300 V/cm) used in this procedure, analytes with an electrophoretic mobility of -2.83×10^{-4} cm^2V^{-1} s^{-1} (similar to these dyes) will be injected with close to 100 % efficiency. Analytes with lower mobilities will be injected with lower efficiency. The relationships between injection efficiency, electrophoretic mobility, and electric field strength are explored in Ref. 19, and the electrophoretic mobility of an analyte can be determined as described therein.

15. Degassing the buffer minimizes the volume of cavitation bubbles, which form when the pulsed laser is fired in the channel.

16. If you have trouble finding a signal, confirm that there are beads flowing through the reference channel. If the reference channel inlet contains 30 μL of reference solution, and the reference outlet is empty (hence, no need for a tubing reservoir), flow of beads through the channel should be rapid. Absence of flow is typically due to a bubble in the channel.

17. Because the reference channel is U-shaped, it is possible to align the detection system over the branch of the reference

channel that is further from the separation channel. As a result, it is important to visually confirm that the detector is aligned correctly on the device.

18. High voltage is hazardous. We strongly recommend safety settings that shut off the voltage at currents > 50–100 μA. The "let-go" current, above which a person may not be able to let go of an electrified lead, is 6–16 mA [20], but for the channel dimensions and buffer conductivities used here much lower currents (e.g., 5–20 μA) will be carried in a well-sealed device.
19. If a chip is poorly sealed, applying voltage across the separation channel can affect flow in the reference channel. In a sealed chip, these channels should be electrically isolated. Poor chip sealing is typically due weak plasma formation (*see* **Note 10**) or dirty PDMS pieces, cover glass, or plasma cleaner chamber.
20. Common microfluidic systems for chemical cytometry, including the one described here, are designed for flow-through assays. As a result, adherent cells must be trypsinized or EDTA-treated to obtain a suspension of single cells. Capillary systems, such as the one described in reference [21], are an alternative for assays of adherent cells on a surface.
21. Stock solutions of diacetate substrates in DMSO should be stored at −20 °C, protected from light, desiccated, and discarded after 2–4 weeks. Over time, DMSO will degrade, and the diacetate dyes will be hydrolyzed to yield the cell-impermeant fluorescent forms, which decreases loading efficiency and can increase background signal.
22. Fluorescein diacetate and carboxyfluorescein diacetate are membrane-permeant, fluorogenic reporters of esterase activity. The diacetate bonds of these molecules are cleaved by esterases within the cells to yield anionic, membrane-impermeant, fluorescent molecules. The simple and efficient loading of the diacetate forms, high quantum yields of the resulting dye molecules, and their straightforward electrophoretic separation makes these analytes an excellent choice for initial chemical cytometry experiments. Lipid-based reporters can also be loaded by simple incubation of cells with reporter, sometimes in the presence of histone [22]. Peptide reporters are often less membrane permeant and require other loading methods.

    Potential methods for the transport of non-lipophilic molecules across the membrane into the cell include pinocytosis, electroporation, myristoylation, and cell-penetrating peptides (CPPs). In pinocytosis, the cells in a hypertonic loading medium containing reporter substrate form pinosomes (vesicles similar to endosomes) that release peptides into the cytosol after addition of a hypotonic solution [23]. This method is effective for many mammalian cell types but exposes cells to osmotic pressure and can lead to inhomogeneous loading.

Electroporation is effective for loading most, if not all, reporters. When exposed to an electric field the lipids in the phospholipid bilayer are reoriented, creating a hydrophilic pore that allows molecules to diffuse into the cell [24]. The electric field stresses cells, however, and this process is likely to activate stress pathways and may lead to cell death in a significant fraction of cells. Myristoylation involves the addition of a hydrophobic myristoyl group to the reporter peptide through a disulfide bond. This chemical modification makes the peptide more membrane permeant and facilitates transport into the cell by insertion of the fatty functional group into the membrane. Once the peptide flips into the inner leaflet of the bilayer, the reducing environment in the cell cleaves the disulfide connector and allows the peptide to be released [25]. Alternatively, for assays of membrane-bound enzymes, the reporter can be covalently bound to the myristoyl group to sequester the reporter to membranes and potentially improve access to the enzyme of interest. Finally, CPPs are composed of amino acid sequences that help to transport cargo into the cell by either endocytosis or direct penetration through inverted micelle formation, pore formation, the carpeting, or thinning of the cell membrane [26]. Myristoylation and CPPs are gentler than pinocytosis or electroporation but make substrate synthesis more demanding.

23. Although some researchers have performed separations in cell media, ECB-glucose is generally preferable because complete media typically contains autofluorescent components that raise the background signal.

24. Separation buffer should be left in all other reservoirs. In the absence of lower ionic strength buffer in the focusing and separation channels, the device may clog more rapidly. Other low-ionic-strength buffers may be substituted for the Tris separation buffer described here to optimize separations of other reporter substrates.

25. This experiment uses gravity-driven flow based on a difference in liquid height in the reservoirs; however, in these small reservoirs, surface tension can also contribute to flow rate, resulting in unexpected changes in flow rate as solution is added to or removed from reservoirs. Patience may be required to obtain optimum flow rates in each channel. Other chip designs (e.g., Fig. 1c) include a syringe pump to provide flow.

26. The 532 nm pulsed laser presents substantial beam hazards, particularly after being focused through the microscope objective. All researchers in the nominal hazard zone should be fully trained and take all necessary safety precautions. In particular, proper eye wear should be used at all times during laser operation and a viewing screen, such as a cathode ray tube or computer

monitor, should be used to view the pulse. Do not observe the laser pulse through the eyepiece!

27. If the laser is focused above the channel, the laser pulse will burn the PDMS, resulting in a diffuse dark spot. If the laser focus is below the channel, the laser will crack the cover glass, resulting in a starburst-shaped damage site. If the pulsed laser cracks the cover glass in the channel, move slightly up, down, or to the right of the original lysis position. Lysis is not as effective over a damaged surface, and cells tend to adhere to the damaged region. Once damage has occurred, it will worsen each time the laser is fired in that location.

28. If no cells appear, fire the pulsed laser to create bubbles and confirm that flow is moving from the cell reservoir to the intersection and not vice versa. Use a pipettor to mix the cell reservoir contents to dislodge any settled cells. If still no cells appear, confirm that the cell channel is not blocked by debris or a bubble. Adjust reservoir levels to obtain flow of cells before realigning the chip. When an appropriate cell density has been identified for your application, cells should flow readily through the channel, and clogging is rarely an issue. For some cell types or treatments, however, cells may settle rapidly and block the channel inlet. In this case, it is possible to use a cuvette stir bar (or "flea") to keep cells in suspension. This is achieved by machining out a small (~1 × 1 cm) square in the stage insert so that the magnetic stir piece for a cuvette stirrer can be placed under the cell reservoir of the device. A 2 mm stir bar is then placed in the cell reservoir at a suitable stir speed to keep cells from settling.

    If intact cells enter the separation channel, add 5–10 μL of buffer to the high-voltage reservoir. You may need to disable the high voltage for 5–10 min to allow whole cells to be pushed out the separation channel and back into the waste channel by pressure-driven flow.

29. Over time, peak heights and areas will decrease as cells metabolize or expel the dyes. Carboxyfluorescein peak areas decrease more rapidly than fluorescein peak areas, so a commonly observed peak pattern is a small peak followed rapidly by a larger peak.

30. Although it is not necessary to wait for peaks from the first cell to reach the detector before firing on a second cell, data interpretation is more straightforward when peak envelopes for individual cells are well resolved, so a delay between cells is recommended for initial experiments until migration times are well characterized.

31. For the dyes used here, peak areas correspond to the quantity of compound in each cell. For other substrates, the peak area

may need to be corrected to account for injection efficiency (*see* **Note 14**). Peak identities can be characterized using electrophoretic separations of bulk lysate in a simple cross chip with spiking of known standards. High frequency peaks that are much narrower than peaks from dyes or substrates are due to scatter from cellular debris in the separation channel. Large data sets (100+ cells) can be further analyzed using regression modeling [5].

Acknowledgment

The authors thank Trinity College for financial support.

References

1. Kovarik ML, Allbritton NL (2011) Measuring enzyme activity in single cells. Trends Biotechnol 29:222–230
2. Ubersax JA, Ferrell JE Jr (2007) Mechanisms of specificity in protein phosphorylation. Nat Rev Mol Cell Biol 8:530–541
3. Phillips KS, Lai HH, Johnson E, Sims CE, Allbritton NL (2011) Continuous analysis of dye-loaded, single cells on a microfluidic chip. Lab Chip 11:1333–1341
4. Kovarik ML, Shah PK, Armistead PM, Allbritton NL (2013) Microfluidic chemical cytometry of peptide degradation in single drug-treated acute myeloid leukemia cells. Anal Chem 85:4991–4997
5. Kovarik ML, Dickinson AJ, Roy P, Poonnen RA, Fine JP, Allbritton NL (2014) Response of single leukemic cells to peptidase inhibitor therapy across time and dose using a microfluidic device. Integr Biol. doi:10.1039/C3IB40249E
6. Shackman JG, Watson CJ, Kennedy RT (2004) High-throughput automated post-processing of separation data. J Chromatogr A 1040: 273–282
7. Lai HH, Quinto-Su PA, Sims CE, Bachman M, Li GP, Venugopalan V, Allbritton NL (2008) Characterization and use of laser-based lysis for cell analysis on-chip. J R Soc Interface 5: S113–S121
8. OZ Optics collimators and focusers – receptacle style. http://www.ozoptics.com/ALLNEW_PDF/DTS0094.pdf. Accessed 17 Jan 2014
9. McClain MA, Culbertson CT, Jacobson SC, Allbritton NL, Sims CE, Ramsey JM (2003) Microfluidic devices for the high-throughput chemical analysis of cells. Anal Chem 75: 5646–5655
10. Metto EC, Evans K, Barney P, Culbertson AH, Gunasekara DB, Caruso G, Hulvey MK, Fracassi da Silva JA, Lunte SM, Culbertson CT (2013) An integrated microfluidic device for monitoring changes in nitric oxide production in single T-lymphocyte (Jurkat) cells. Anal Chem 85:10188–10195
11. Sun Y, Yin X (2006) Novel multi-depth microfluidic chip for single cell analysis. J Chromatogr A 1117:228–233
12. Wang HY, Lu C (2006) Microfluidic chemical cytometry based on modulation of local field strength. Chem Commun 33:3528–3530
13. Greif D, Galla L, Ros A, Anselmetti D (2008) Single cell analysis in full body quartz glass chips with native UV laser-induced fluorescence detection. J Chromatogr A 1206:83–88
14. Mellors JS, Jorabchi K, Smith LM, Ramsey JM (2010) Integrated microfluidic device for automated single cell analysis using electrophoretic separation and electrospray ionization mass spectrometry. Anal Chem 82:967–973
15. Xia F, Jin W, Yin X, Fang Z (2005) Single-cell analysis by electrochemical detection with a microfluidic device. J Chromatogr A 1063: 227–233
16. Ye F, Huang Y, Xu Q, Shi M, Zhao S (2010) Quantification of taurine and amino acids in mice single fibrosarcoma cell by microchip electrophoresis coupled with chemiluminescence detection. Electrophoresis 31:1630–1636
17. Xu C, Wang M, Yin X (2011) Three-dimensional (3D) hydrodynamic focusing for continuous sampling and analysis of adherent cells. Analyst 136:3877–3883
18. Phillips KS, Kottegoda S, Kang KM, Sims CE, Allbritton NL (2008) Separations in

poly(dimethylsiloxane) microchips coated with supported bilayer membranes. Anal Chem 80: 9756–9762

19. Kovarik ML, Lai HH, Xiong JC, Allbritton NL (2011) Sample transport and electrokinetic injection in a microchip device for chemical cytometry. Electrophoresis 32: 3180–3187
20. OSHA Directorate of Training and Education. How electrical current affects the human body. https://www.osha.gov/SLTC/etools/construction/electrical_incidents/eleccurrent.html. Accessed 6 Jan 2014
21. Dickinson AJ, Armistead PM, Allbritton NL (2013) Automated capillary electrophoresis system for fast single-cell analysis. Anal Chem 85:4797–4804
22. Jiang D, Sims CE, Allbritton NL (2010) Microelectrophoresis platform for fast serial analysis of single cells. Electrophoresis 31: 2558–2565
23. Okada CY, Rechsteiner M (1982) Introduction of macromolecules into cultured mammalian cells by osmotic lysis of pinocytic vesicles. Cell 29:33–41
24. Tsong TY (1991) Electroporation of cell membranes. Biophys J 60:297–306
25. Nelson AR, Borland L, Allbritton NL, Sims CE (2007) Myristoyl-based transport of peptides into living cells. Biochemistry 46:14771–14781
26. Madani F, Lindberg S, Langel U, Futaki S, Graslund A (2011) Mechanisms of cellular uptake of cell-penetrating peptides. J Biophys 2011:414729

Chapter 16

Quantitative Detection of Nucleocytoplasmic Transport of Native Proteins in Single Cells

Zhenning Cao and Chang Lu

Abstract

The detection of protein translocation (i.e., the movement of intracellular proteins among various subcellular compartments) conventionally relies on imaging and subcellular-fractionation-based techniques that do not generate information on a large cell population with single-cell resolution. Although special flow cytometric tools such as imaging flow cytometry may generate single-cell data on processes such as nucleocytoplasmic transport, such equipment is expensive (thus has limited accessibility) and has low throughput for examining cells due to the reliance on high-speed imaging. Here we describe a protocol for detecting translocation of native proteins using a common flow cytometer which detects fluorescence intensity without imaging. We conduct chemical release of cytosolic proteins and fluorescence immunostaining of a targeted protein. The detected fluorescence intensity is quantitatively correlated to the cytosolic/nuclear localization of the protein at the single cell level. Our technique provides a simple route for studying nucleocytoplasmic transport with single-cell resolution using common flow cytometers.

Key words Translocation, NF-κB, Flow cytometry, Selective release, Immunostaining

1 Introduction

The subcellular localization of intracellular proteins provides critical information about the activation state of the molecules. During cell signaling events, a variety of mechanisms function to localize signaling molecules to the appropriate subcellular compartments where they can carry out their intended functions. The translocation of a protein from the cytosol to the plasma membrane, for example, can result from the receptor-triggered binding of signaling proteins with PH (pleckstrin homology), C1, C2, and related domains to membrane phospholipids [1, 2]. Nucleocytoplasmic transport (the translocation between the cytosol and the nucleus) occurs via nuclear pore complexes (NPCs) with the participation of a large family of transport receptors called karyopherins [3, 4]. The transport receptors bind to nuclear localization signals (NLSs) or nuclear export signals (NESs) in their cargos and mediate their

Anup K. Singh and Aarthi Chandrasekaran (eds.), *Single Cell Protein Analysis: Methods and Protocols*, Methods in Molecular Biology, vol. 1346, DOI 10.1007/978-1-4939-2987-0_16, © Springer Science+Business Media New York 2015

transport through the NPCs. Dysfunctions in the regulatory steps may lead to disrupted translocation or mislocalization. Not surprisingly, the subcellular mislocalization of proteins is linked to diseases ranging from metabolic disorders to cancer [5, 6]. For example, mislocalizations of Akt, NF-κB, FOXO, p27, and p53 have been well documented as key features in a variety of cancers [6, 7]. Thus, modulation of protein localization has been proposed and practiced as an important therapeutic approach for cancer treatment [5, 6]. The subcellular location of a target protein can serve as a useful readout for high-content screening of cancer drugs [6]. Furthermore, protein translocation and activation may occur in response to therapeutic measures such as chemotherapies and radiation [8–10]. Understanding the mechanism is critical for therapy regulation and chemoresistance reversal.

Protein translocations such as nucleocytoplasmic transport have been studied by various techniques. Subcellular fractionation followed by immunoblotting has been widely used for quantifying protein subcellular localization [11]. However, due to the homogenization of materials from a cell population, the result only reflects the population average without information on cell subsets and population heterogeneity. Fluorescence microscopy readily recognizes nuclear/cytosolic localization of a labeled protein at the single cell level. However, confocal microscopy with high spatial resolution is required for the observation of localization on the membrane. In addition, all imaging techniques are restricted to interrogating only a small number of cells due to the limitation of the frame size. Thus, the data do not accurately represent a large population of cells and the distribution of a property in the cell population is difficult to establish. Classical flow cytometry possesses extremely high throughput owing to high sensitivity and the ultrafast response times of single element photodetectors such as photomultiplier tubes (PMTs) [12, 13]. However, a classical flow cytometer detects fluorescence from the entire cell and is insensitive to the subcellular localization of a protein (Fig. 2a, b) [14]. Laser scanning cytometry (LSC) [15, 16] or imaging flow cytometry [17, 18] based on array detectors (e.g., CCD cameras) does permit quantification of the nuclear/cytosolic localization through automated image acquisition and analysis of solid-phase or flowing cell samples. These imaging-based instruments require sufficient exposure time in order to generate enough spatial resolution for the analysis. This restricts their throughputs to no more than several hundred cells per second. Additionally, the complex algorithms used for image analysis can often lead to errors and bias [19–21]. Furthermore, the requirement of high-end cameras and image storage/analysis capabilities makes the price of LSC and imaging flow cytometers prohibitively high (>$200 K). Thus these tools are only available at a small number of centralized facilities. We explored total internal reflection fluorescence (TIRF)-based flow cytometry for probing protein translocations [22]. However, it required special setup and might not yield very high throughput.

We recently demonstrated a simple method that combines selective chemical release of cytosolic proteins and standard fluorescence immunostaining for detecting the translocation of native proteins at the single-cell level [23]. We demonstrated the proof of principle by detecting nucleocytoplasmic transport of an important transcription factor NF-κB [23]. NF-κB undergoes nucleocytoplasmic transport from the cytosol to the nucleus in order to regulate transcription and gene expression, upon extracellular stimuli (e.g., by TNFα) [24, 25]. Briefly, we used saponin (i.e., a class of amphipathic glycosides) to selectively release cytosolic fraction of intracellular proteins by dissolving the cholesterol content and permeabilizing the plasma membrane [26]. Such treatment was followed by fluorescence immunostaining of residual NF-κB. The cell population was then screened by a common flow cytometer for fluorescence intensity of each cell. We showed that the fluorescence intensity of a cell could be correlated to the subcellular localization of the protein. Taking advantage of common flow cytometry which is widely accessible, our approach detects the translocation of native proteins without imaging and with high throughput. Our approach is readily compatible with analysis of primary samples from animals and patients.

We outline the procedure and working principle of our protocol in Fig. 1, in comparison to those of the standard immunostaining protocol, and how these procedures generate different results for single-cell screening with standard flow cytometry.

In our experiments, TNFα was applied to stimulate cells and produce NF-κB translocation from the cytosol to the nucleus. With standard fluorescence immunostaining (Fig. 1a), HeLa cells with NF-κB translocation (i.e., +TNFα) and those without NF-κB translocation (i.e., untreated) are cross-linked by formaldehyde and permeabilized by Triton X-100 before all intracellular NF-κB is fluorescently labeled by antibodies that specifically target NF-κB. The crosslinking by formaldehyde conserves all proteins in the cells and the permeabilization ensures the full access of the targeted protein (i.e., NF-κB in here) by the antibodies for labeling. When these cells are subsequently screened by flow cytometry, which detects fluorescence from the entire cell, the translocation does not create difference in the fluorescence intensity detected because it only varies the subcellular localization of the protein, not the overall expression level. In comparison, in our new protocol, an extra selective release step (**step 1** of Fig. 1b) is involved before immunostaining. During this step, saponin is used to treat the cells to dissolve cholesterol in the plasma membrane and make the membrane

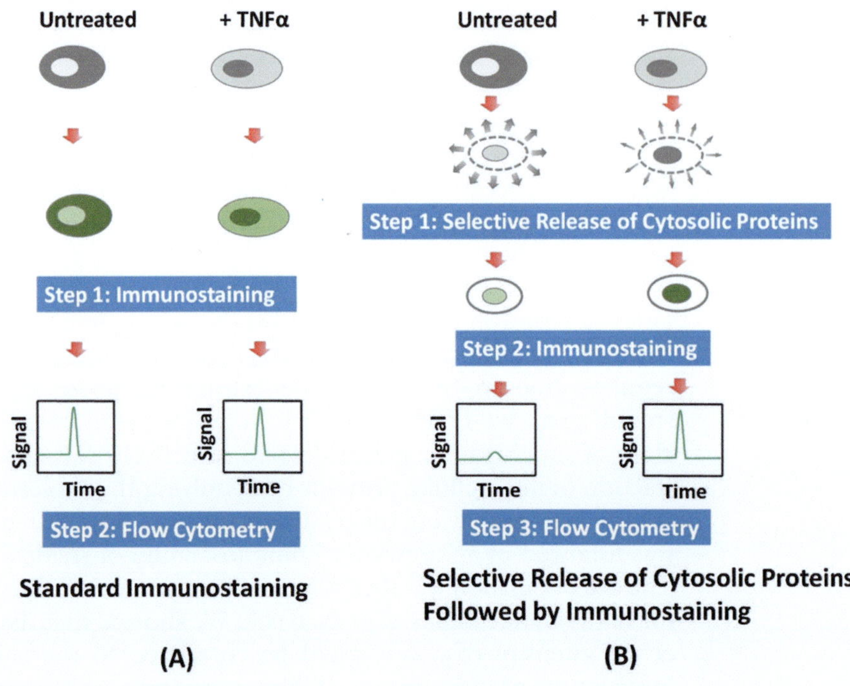

Fig. 1 The comparison between flow cytometric screening of cells after (**a**) standard fluorescence immunostaining and (**b**) selective release of cytosolic proteins followed by immunostaining. There is no difference in the fluorescence intensity between cells with and without nucleocytoplasmic transport with standard fluorescence immunostaining, whereas the fluorescence intensity of single cells is correlated with the protein subcellular localization (or the activation state) with our selective release combined with immunostaining protocol (reproduced from Ref. 23)

leaky [26–28]. Saponin is known to primarily permeabilize the plasma membrane while keeping the cholesterol-poor internal membranes (e.g., the mitochondrial membrane and nuclear envelope) intact [26, 29]. Cytosolic proteins are preferentially released out of the cells and the nucleic proteins are largely unaffected in their amounts. Thus, cells with NF-κB translocation (with NF-κB mostly in the nucleus) lost far less total amount of NF-κB due to the release, compared to the cells without NF-κB translocation (with NF-κB primarily in the cytosol). The selective release step is immediately followed by immunostaining that fixes the cells and labels all the remaining intracellular NF-κB (Fig. 1b). The flow cytometry results (**step 3** of Fig. 1b) now are significantly different for cells with translocation and those without translocation, with the latter showing much smaller fluorescence intensity (Fig. 2c, d). Thus, we are able to link the subcellular localization of the protein with the detected fluorescence intensity of a cell treated by our protocol.

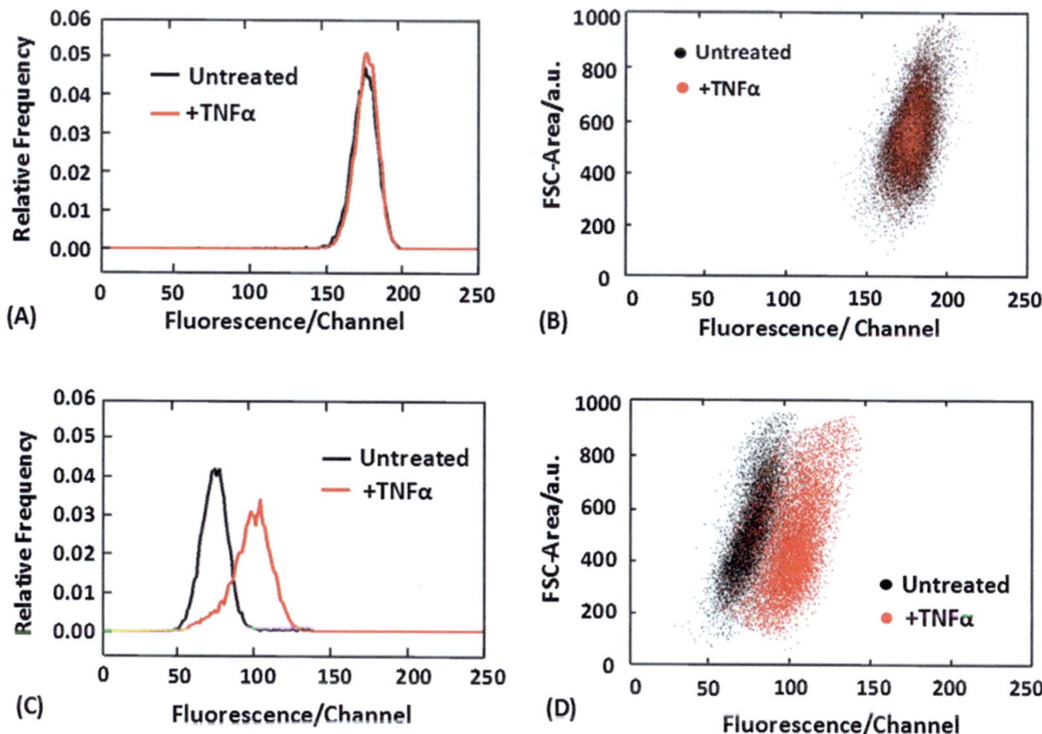

Fig. 2 The detection of NF-κB translocation using the selective release/immunostaining protocol and conventional flow cytometry. The cell populations with and without TNFα stimulation were examined. The difference between the two populations was not revealed in either the fluorescence intensity histograms (**a**) or 2D dot plots (**b**) involving both fluorescence intensity and forward scatter (i.e., FSC), when standard immunostaining was used. In comparison, there was pronounced difference between the two populations in both the fluorescence intensity histograms (**c**) and 2D dot plots (**d**) when selective release followed by immunostaining was conducted. TNFα stimulation was conducted at 37 °C with 50 ng/ml TNFα for 30 min. Selective release was performed using 0.05 % saponin for 10 min at room temperature (reproduced from Ref. 23)

2 Materials

2.1 Cell Culture and Sample Preparation

1. HeLa cells (ATCC, CCL-2).
2. Cell culture medium: Dulbecco's modified Eagle's medium (DMEM) supplemented with 10 % (vol/vol) FBS, 2 mM L-glutamine, 100 U ml^{-1} penicillin, and 100 mg ml^{-1} streptomycin. Store at 4 °C for up to 1 year.
3. Fetal bovine serum (FBS).
4. Penicillin–streptomycin solution.
5. L-Glutamine solution.
6. Trypsin–EDTA (0.05 % (wt/vol)).

7. Sterile phosphate-buffered saline (PBS): Dissolve one PBS buffer tablet in 100 ml DI water. Autoclave for 20 min to sterilize. Store at 4 °C for up to 1 year.
8. Recombinant human TNF alpha: Aliquot into smaller volume (e.g., ~10 μl) and store at −20 °C to avoid extra freeze-thaw cycles.
9. Laminar flow hood.
10. Cell culture incubator set at 37 °C with humidified, 5 % CO_2 atm.
11. Cell culture flasks and well plates.
12. Pipettes/tips.
13. Centrifuge tubes (0.6, 1.5, 15, and 50 ml).
14. Centrifuge.
15. Spray bottles.
16. Hemocytometer.
17. Ice maker.
18. Mini lab rotator (24 rpm).
19. 30 μm Nylon-mesh filters.

2.2 Immunofluorescence Labeling

1. Blocking solution: Prepare a solution of 0.5 % bovine serum albumin (BSA) (wt/vol) in sterile PBS buffer. Filter-sterilize with a 0.2 μm membrane and store at 4 °C for up to 2 months.
2. Fixation solution: Dilute 16 % formaldehyde (FA) in sterile blocking buffer to a final concentration of 4 %. Filter-sterilize with a 0.2 μm membrane and store at 4 °C for up to 2 months.
3. Permeabilization solution: Dilute 1 % Triton X-100 in blocking buffer to a final concentration of 0.2 %. Filter-sterilize with a 0.2 μm membrane and store at 4 °C for up to 2 months.
4. Selective releasing solution: Dissolve 0.05 % (w/v) Saponin in DMEM. Filter-sterilize with a 0.2 μm membrane and store at 4 °C for up to 1 month.
5. Primary antibody: NF-κB p65 (Santa Cruz Biotechnology).
6. Secondary antibody: DyLight™ 488 Goat anti-mouse IgG1 antibody (Biolegend).
7. Sodium azide.

2.3 Solution Preparation

1. Ultrapure DI water.
2. Glass syringe with Lure-Lock tip (50 ml).
3. Sterile syringe filters with 0.2 μm pore diameter.

2.4 Microscope

1. Inverted phase-contrast and epifluorescence microscope system with 100 W mercury lamp (IX-71; Olympus) equipped with 10× and 20× objectives (0.5 NA).
2. High-resolution CCD camera (Hamamatsu, ORCA-285).
3. Fluorescence filter cubes (exciter HQ480/40×, emitter HQ535/50 m; Chroma Technology).
4. Image acquisition software (QImaging QCapture Pro).
5. ImageJ software (or similar).

2.5 Flow Cytometer

1. BD FACSCanto II workstation (BD Bioscience).
2. Sapphire 488–20 laser (488 nm, 20 mW, Coherent).
3. BP Filter or LP mirror (530/30, intended dye: FITC).
4. Flow cytometer controlling and analysis software (BD FACSDiva).

3 Methods

3.1 Cell Sample Preparation

1. Use cells of 80 % confluency for the analysis. For HeLa cells, this corresponds to about 3.0×10^6 cells in a 75 cm^2 flask. This total amount of cells is sufficient for 5~6 samples (given that ideally 5×10^5 cells are needed for one flow cytometric screening) (*see* **Note 1**).
2. Aspirate the medium and briefly rinse the adherent cells with 10 ml of ice-cold PBS.
3. Aspirate PBS and add 3 ml of the trypsin–EDTA. Disperse the solution by gently shaking the plate and incubate it in the incubator for 5 min at 37 °C (*see* **Note 2**) (Table 1).
4. Gently shake the plate until cells detach. Detached cells can be easily identified under a microscope, as they come off in round shape.
5. Immediately add 10 ml of cell culture medium to inactivate trypsin. Disperse suspended cells in medium by gently pipetting three or four times to produce a single-cell suspension (*see* **Note 3**).
6. Transfer the suspension to a 15 ml conical tube and pellet the cells by centrifugation at $300 \times g$ for 5 min at 4 °C (*see* **Note 4**).
7. Discard the supernatant and resuspend the cell pellet in 1.5 ml of fresh cell culture medium by gently pipetting with 1 ml pipet tip several times. Put the cells on ice and use a small aliquot to estimate the number of cells using a hemocytometer (*see* **Note 5**).
8. Aliquot around one-sixth of the cell suspension (250 μl and $\sim 5 \times 10^5$ cells) to a new 0.6 ml centrifuge tube (*see* **Note 6**).

Table 1
Troubleshooting

Step	Problem	Possible reason	Solution
Step 3 of Subheading 3.1	Cell clumping	Insufficient trypsinization	Increase the amount of trypsin–EDTA solution Increase the time of trypsin–EDTA incubation
Step 3 of Subheading 3.1	Reduced cell viability	Excessive trypsinization	Decrease the amount of trypsin–EDTA solution Decrease the time of trypsin–EDTA incubation
Step 1 of Subheading 3.2	No translocation	Reagent out of date Insufficient mixing	The shelf life of human recombinant TNFα is 2 months at −20 °C Agitate the cell solution several times during stimulation
Step 4 of Subheading 3.3	Excessive loss of cells after selective releasing	Inefficient pelleting	Increase the centrifugal force above 600 × *g*
Step 4 of Subheading 3.4	Loss of fixed cells	Inefficient pelleting	Increase the centrifugal force above 600 × *g*
Step 14 of Subheading 3.5	Loss of cells during immunostaining	Cell absorption on the tube surface Cell rupture due to excessive mechanical force	Use clear, low-retention tubes Mix by gentle shaking of the tube instead of pipetting
Step 5 of Subheading 3.7	Excessive amount of outliers	Too few cells are examined	Eliminate subsets in the FSC vs. SSC plot which contain <100 cells Acquire a minimum amount of 20,000 exponentially growing cells for analysis

3.2 Cell Stimulation

1. Add TNF-α into the cell suspension and mix the solution by gently pipetting up and down several times (Table 1).
2. Stimulate cells by incubation with TNF-α for a period of time at 37 °C under constant, mild agitation (>30 min for complete NF-κB translocation from the cytosol to nucleus).
3. Centrifuge the cell suspension at 300 × *g* for 5 min at 4 °C.
4. Slowly aspirate and discard the supernatant. Add 200 μl of the fresh DMEM, and gently disperse the pellet by pipetting up and down until the cells are resuspended.
5. Repeat **steps 3** and **4** once.

3.3 Selective Release of Cytosolic Proteins

1. Centrifuge the cell suspension at 300 × *g* for 5 min at 4 °C.
2. Slowly aspirate and discard the supernatant and resuspend cells with 200 μl of the selective releasing solution by gently pipetting several times.
3. Incubate cells with the selective releasing solution for 5 min on a rotator mixer (24 rpm) at the room temperature (*see* **Notes 7** and **8**).
4. Centrifuge the cell suspension at 600 × *g* for 5 min at 4 °C (*see* **Note 9**) (Table 1).
5. Slowly aspirate and discard the supernatant and resuspend cells with 200 μl of the blocking solution by gently pipetting several times.
6. Repeat **steps 4** and **5** once.

3.4 Fixation and Permeabilization of Treated Cells

1. Centrifuge the cell suspension at 600 × *g* for 2 min at 4 °C.
2. Slowly aspirate and discard the supernatant and resuspend cells in 200 μl pre-warmed fixation solution by gently pipetting several times (*see* **Note 10**).
3. Incubate cells with the fixation solution for 40 min at room temperature (*see* **Note 11**).
4. Centrifuge the cell suspension at 600 × *g* for 5 min at 4 °C (Table 1).
5. Slowly aspirate and discard the supernatant and wash the sample with 200 μl blocking solution by gently pipetting several times.
6. Repeat **steps 4** and **5** once.
7. Centrifuge the cell suspension at 600 × *g* for 5 min at 4 °C.
8. Slowly aspirate and discard the supernatant and resuspend the cells in 200 μl of the permeabilization solution by gently pipetting several times.
9. Incubate cells in the permeabilization solution for 5 min at room temperature (*see* **Note 12**).
10. Centrifuge the cell suspension at 600 × *g* for 5 min at 4 °C.
11. Slowly aspirate and discard the supernatant and wash the cells with 200 μl of the blocking solution by gently pipetting several times.
12. Repeat **steps 10** and **11** once.

3.5 Fluorescence Immunostaining

1. Add 0.75 μl of the primary antibody solution (diluted 1:200 in blocking solution, containing ~100 ng antibody) to the solution and mix by pipetting several times (*see* **Note 13**).
2. Incubate cells with the primary antibody for 1 h at room temperature.
3. Centrifuge the cell suspension at 600 × *g* for 5 min at 4 °C.

4. Slowly aspirate and discard the supernatant and wash the cell pellet with 200 μl of the blocking solution by gently pipetting several times.
5. Repeat **steps 3** and **4** once.
6. Centrifuge the cell suspension at 600 × *g* for 5 min at 4 °C.
7. Slowly aspirate and discard the supernatant and resuspend the cells in 200 μl blocking solution by gently pipetting several times.
8. Add 0.5 μl (containing ~50 ng) of fluorophore-conjugated secondary antibody (1:300 diluted) and mix the solution by pipetting several times (*see* **Note 14**).
9. Incubate cells with the secondary antibody for 30 min at the room temperature in the dark.
10. Centrifuge the cell suspension at 600 × *g* for 5 min at 4 °C.
11. Slowly aspirate and discard the supernatant and wash the cell pellets with 200 μl blocking solution by gently pipetting several times.
12. Repeat **steps 10** and **11** once.
13. Centrifuge the cell suspension at 600 × *g* for 5 min at 4 °C.
14. Slowly aspirate and discard the supernatant and resuspend the pellet with 200 μl PBS solution containing 0.1 % sodium azide by gently pipetting several times (Table 1).
15. The cells can be stored at 4 °C for up to 24 h, if not immediately analyzed by flow cytometry.

3.6 Observation by Fluorescence Microscopy (Optional)

At this stage of the protocol, it is optional to use an aliquot of the prepared cell sample for analysis by fluorescence microscopy. The images (Fig. 3) confirm the proposed mechanism in Fig. 1. Our technique renders the fluorescence intensity different for cells with translocation and those without translocation, whereas standard immunostaining reveals the different localizations of the protein via imaging with the overall fluorescence intensity from whole cells being the same for the two types of cells.

1. Transfer immunostained cell samples to 96-well plate. Make sure that each well has around 1,000 cells and 200 μl PBS solution.
2. Mix the cell solution by pipetting several times in each well.
3. Centrifuge the sample at 50 × *g* for 15 min at 4 °C so that the floating cells settle at the bottom of wells (*see* **Note 15**).
4. Capture both phase-contrast and fluorescent images of cells at five different locations in each sample using a fluorescence microscope equipped with a 20× dry objective and a CCD camera. The excitation and emission were filtered by a fluorescence filter cube for observing green fluorescence.

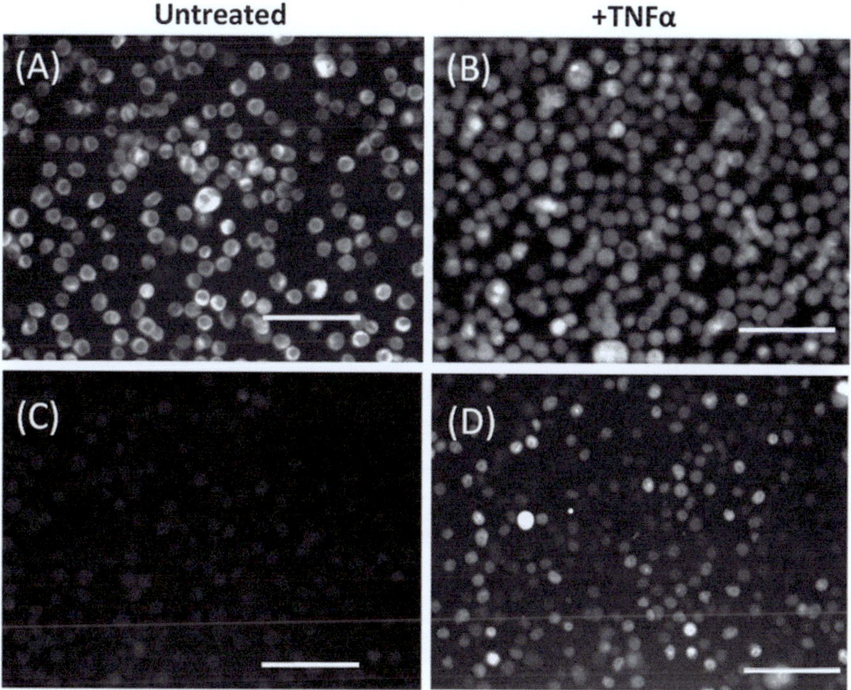

Fig. 3 Fluorescence images of untreated cells (**a**, **c**) and TNFα-stimulated cells (**b**, **d**) after standard immunostaining (**a**, **b**) and our selective-release-based immunostaining (**c**, **d**). TNFα stimulation was performed by adding 50 ng/ml TNFα for 30 min at 37 °C. Selective release was performed by 0.05 % saponin for 10 min at room temperature. Scale bar = 100 μm (reproduced from Ref. 23)

3.7 Cell Screening by a Common Flow Cytometer

1. Start the flow cytometer and the computer, perform the usual fluidics maintenance and launch FACSDiva 6.0.
2. Open the appropriate acquisition document from the storage folder or create a new one.
3. Remove aggregated cells from the sample using a nylon-mesh filter (*see* **Note 16**).
4. Run samples on the flow cytometer with 488 nm excitation.
5. When acquiring data on the flow cytometer, create a forward scatter (FSC) versus side scatter (SSC) plot. To exclude cell debris, air bubbles, and laser noise from analysis, ensure that the expected cell population is visible in FSC/SSC plot by adjusting the individual FSC (voltage 200–400, linear mode) and SSC (voltage 200–400, linear mode) settings (Table 1).
6. A gate is applied to the region around your cell population of interest on this FSC/SSC plot.
7. Necessary fluorophore analysis histogram/plots can now be created as needed (i.e., FSC vs. counts or FSC vs. FITC).
8. Collect at least 20,000 events in each sample.
9. After completing data acquisition, perform the usual fluidics maintenance and proceed to data analysis with Flow Jo (version 7.6.1).

4 Notes

1. Each cell line has a characteristic optimal confluency and should be regularly subcultured. HeLa cells are passaged every 2–3 days at a ratio of 1:10 by trypsinization.
2. The trypsinization time may need to be optimized for a particular cell type. Insufficient trypsinization may lead to cell aggregates. Excessive trypsinization may cause cell damage or death.
3. Single-cell suspension is required for immunostaining and flow cytometry.
4. If multiple culture flasks are used, mix all cell suspensions in one tube for cell counting to avoid variability among different cultures.
5. High shear stress induced by vigorous pipetting may damage cells and cause nucleocytoplasmic translocation to create artifacts in the final result.
6. Single-cell suspension must be obtained at this point for optimal TNF-α stimulation and selective releasing.
7. The concentration of saponin in the selective releasing solution should be optimized for different cell/target protein combinations (Fig. 4a). Low saponin concentration may lead to incomplete releasing of cytosolic proteins. High concentration of saponin may cause releasing of nuclear proteins. The cytosolic proteins NF-κB of HeLa cells can be completely released by selective releasing solution with 0.05 % saponin in 5 min. The saponin concentration and the treatment time need to be optimized together.

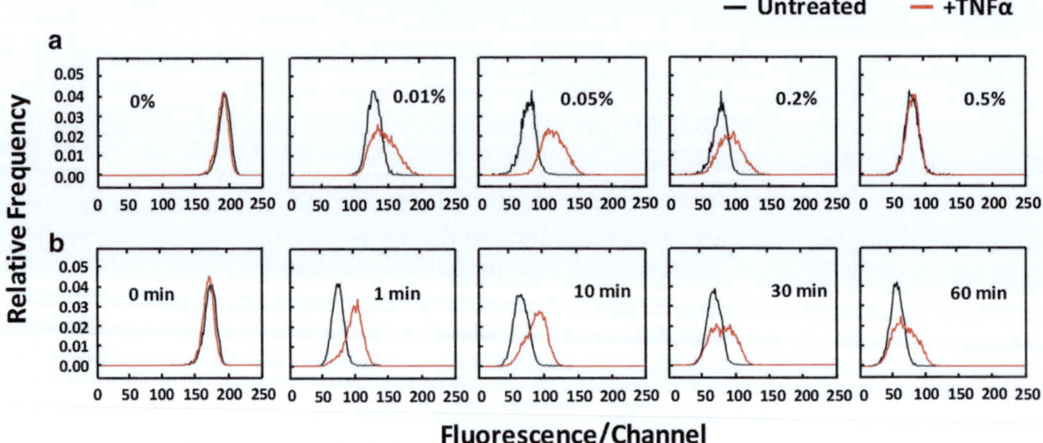

Fig. 4 The optimization of conditions (saponin concentration and treatment duration) for maximal differentiation of untreated and TNFα-stimulated cells using selective-release-based immunostaining and flow cytometry. TNFα stimulation was conducted by incubating cell with 50 ng/ml TNFα at 37 °C for 30 min. Fluorescence histograms were obtained after selective-release-based immunostaining and flow cytometric screening using NF-κB antibody. (**a**) Various saponin concentrations used when the treatment duration was 10 min. (**b**) Various treatment durations used while the saponin concentration was 0.05 % (reproduced from Ref. 23)

8. The treatment time for the selective releasing should be optimized for different cell/target protein combinations (Fig. 4b). Insufficient incubation time may lead to incomplete releasing of cytosolic proteins. Prolonged incubation time may cause rupture of nuclei and releasing of nuclear proteins. In our experiment, the cytosolic NF-κB of HeLa cells can be completely released in 5 min by the selective releasing solution (containing 0.05% saponin).
9. The density of cells decreases after the selective releasing. Increased centrifugation speed is critical to pellet most of the cells.
10. Pre-warming the fixation solution at 37 °C is critical for maintaining cell integrity.
11. Cells after selective release are easy to rupture. A prolonged fixation time of >30 min instead of the standard 10-min treatment is critical for the integrity of these cells.
12. The permeabilization provides access to intracellular target proteins. Instead of the standard 15-min permeabilization time recommended in most conventional immunostaining protocols, HeLa cells after the selective release can be completely permeabilized by Triton X-100 within 5 min. The permeabilization time needs to be optimized for a particular cell type. Insufficient permeabilization leads to inadequate labeling. Prolonged permeabilization may cause loss of the target protein and decreased cell integrity.
13. The amount of the primary antibody needs to be optimized. Insufficient antibody leads to incomplete labeling. Excessive antibody leads to increased nonspecific binding. In our experiment, 200 ng primary antibody per 1×10^6 cells yields the best result. It is also crucial to keep staining conditions (i.e., concentrations of cells and antibodies) consistent among samples in order to generate good reproducibility.
14. We used DyLight 488 conjugates, with excitation and emission properties nearly identical to those of FITC. Compared to FITC, DyLight 488 is brighter for flow cytometry and fluorescence microscopy.
15. Use the centrifugation speed of $50 \times g$ to avoid concentration of cells at the edge of the well bottom.
16. Cell aggregates potentially clog the tubing of the flow cytometer.

Acknowledgments

This research was supported in part by NSF Grant 0967069.

References

1. Pawson T (1995) Protein modules and signalling networks. Nature 373:573–580
2. Lemmon MA, Ferguson KM (2000) Signal-dependent membrane targeting by pleckstrin homology (PH) domains. Biochem J 350(Pt 1): 1–18
3. Nakielny S, Dreyfuss G (1999) Transport of proteins and RNAs in and out of the nucleus. Cell 99:677–690
4. Nigg EA (1997) Nucleocytoplasmic transport: signals, mechanisms and regulation. Nature 386:779–787
5. Davis JR, Kakar M, Lim CS (2007) Controlling protein compartmentalization to overcome disease. Pharm Res 24:17–27
6. Kau TR, Way JC, Silver PA (2004) Nuclear transport and cancer: from mechanism to intervention. Nat Rev Cancer 4:106–117
7. Bellacosa A, Kumar C, Di Cristofano A, Testa J (2005) Activation of AKT kinases in cancer: implications for therapeutic targeting. Adv Cancer Res 94:29
8. Baldwin AS (2001) Control of oncogenesis and cancer therapy resistance by the transcription factor NF-kappaB. J Clin Invest 107:241–246
9. Kim HJ, Hawke N, Baldwin AS (2006) NF-kappa B and IKK as therapeutic targets in cancer. Cell Death Differ 13:738–747
10. Yamamoto Y, Gaynor RB (2001) Therapeutic potential of inhibition of the NF-kappaB pathway in the treatment of inflammation and cancer. J Clin Invest 107:135–142
11. Huber LA, Pfaller K, Vietor I (2003) Organelle proteomics – implications for subcellular fractionation in proteomics. Circ Res 92:962–968
12. Perez O, Nolan G (2006) Phosphoproteomic immune analysis by flow cytometry: from mechanism to translational medicine at the single-cell level. Immunol Rev 210: 208–228
13. Krutzik P, Irish J, Nolan G, Perez O (2004) Analysis of protein phosphorylation and cellular signaling events by flow cytometry: techniques and clinical applications. Clin Immunol 110:206–221
14. Wang J et al (2008) Detection of kinase translocation using microfluidic electroporative flow cytometry. Anal Chem 80:1087–1093
15. Deptala A, Bedner E, Gorczyca W, Darzynkiewicz Z (1998) Activation of nuclear factor kappa B (NF-kappa B) assayed by laser scanning cytometry (LSC). Cytometry 33:376–382
16. Harnett MM (2007) Laser scanning cytometry: understanding the immune system in situ. Nat Rev Immunol 7:897–904
17. Arechiga AF et al (2005) Cutting edge: FADD is not required for antigen receptor-mediated NF-kappaB activation. J Immunol 175:7800–7804
18. Fanning SL et al (2006) Receptor cross-linking on human plasmacytoid dendritic cells leads to the regulation of IFN-alpha production. J Immunol 177:5829–5839
19. Basiji DA, Ortyn WE, Liang L, Venkatachalam V, Morrissey P (2007) Cellular image analysis and imaging by flow cytometry. Clin Lab Med 27:653–670
20. Vigo J, Salmon J, Lahmy S, Viallet P (1991) Fluorescent image cytometry: from qualitative to quantitative measurements. Anal Cell Pathol 3:145
21. Falkmer UG (1992) Methodologic sources of errors in image and flow cytometric DNA assessments of the malignancy potential of prostatic carcinoma. Hum Pathol 23:360–367
22. Wang J et al (2010) Embellishment of microfluidic devices via femtosecond laser micronanofabrication for chip functionalization. Lab Chip 10:1993–1996
23. Cao ZN, Geng S, Li LW, Lu C (2014) Detecting intracellular translocation of native proteins quantitatively at the single cell level. Chem Sci 5:2530–2535
24. Hoffmann A, Baltimore D (2006) Circuitry of nuclear factor κB signaling. Immunol Rev 210: 171–186
25. Ghosh S, May MJ, Kopp EB (1998) NF-{kappa} B and Rel proteins: evolutionarily conserved mediators of immune responses. Sci Signal 16:225
26. Wassler M, Jonasson I, Persson R, Fries E (1987) Differential permeabilization of membranes by saponin treatment of isolated rat hepatocytes. Release of secretory proteins. Biochem J 247:407–415
27. Jamur MC, Oliver C (2010) Permeablization of cell membrances. Immunocytochemical methods and protocols. pp. 63–66
28. Goldenthal KL, Hedman K, Chen JW, August JT, Willingham MC (1985) Postfixation detergent treatment for immunofluorescence suppresses localization of some integral membrane proteins. J Histochem Cytochem 33:813–820
29. Negrutskii BS, Deutscher MP (1992) A sequestered pool of aminoacyl-tRNA in mammalian cells. Proc Natl Acad Sci U S A 89:3601–3604

INDEX

A

Antibody conjugation ... 50, 116
Antigen-specific T cell ... 115, 126, 130

B

Bacteria analysis ... 11–24, 55–67
Basic fibroblast growth factor (bFGF) ... 88, 91–94, 97
B cell assays ... 39

C

CDM. *See* Chemically defined media (CDM)
Cell culture ... 53, 57, 70–71, 74, 75, 77, 80, 81, 85, 88, 90–95, 102, 103, 107, 108, 170, 172, 173, 175–179, 182, 195, 213, 228, 243–245
Cell signaling ... 53, 71, 106, 109, 169, 170, 173, 175, 239
Cell surface receptors ... 71, 78
Chemical cytometry ... 221–237
Chemically defined media (CDM) ... 88, 93, 94
Combinatorial tetramer staining ... 122
CyTOF. *See* Mass cytometry (CyTOF)
Cytokine ... 28–31, 33, 35, 36, 38, 39, 41–43, 71, 78, 82, 111, 115, 154, 165

D

DNA barcoding ... 47–54

E

Electrophoresis ... 2, 3, 5, 6, 8, 222, 231
ELISpot. *See* Enzyme-linked immunospot (ELISpot)
Enzyme activity ... 209–218, 221, 222
Enzyme assay ... 221–237
Enzyme-linked immunospot (ELISpot) ... 27–44
Escherichia coli ... 11–24, 55–67, 161, 173

F

FC. *See* Flow cytometry (FC)
Fine needle aspirates ... 48
Flow cytometry (FC) ... 1, 53, 55, 69–82, 100, 115, 133, 134, 218, 240–243, 245, 248–251
Fluorescence ... 20, 56, 62–67, 72, 75–77, 79, 96, 97, 100, 134, 136, 137, 141, 142, 146, 147, 152, 153, 157, 164, 166, 167, 170, 172, 179, 180, 185, 187, 196, 222, 224, 230, 231, 240–243, 245, 247–250
Fluorescence microscopy ... 6, 55, 56, 89, 95, 98, 136, 162, 170, 172, 186, 198, 210, 212, 217, 224, 230, 240, 245, 248–249, 251
Fluorescent protein fusion library ... 56
Fluorescent reporter ... 152, 222
Fluorospot ... 31, 38–41

H

High-throughput platform ... 56
HPSC. *See* Human pluripotent stem cells (hPSC)
HTopI. *See* Human topoisomerase I (hTopI)
Human ... 31, 33, 39, 40, 87, 88, 90, 99, 101–110, 112, 113, 117, 123, 146, 173, 185, 244, 246
Human pluripotent stem cells (hPSC) ... 85–98
Human topoisomerase I (hTopI) ... 209–217

I

Imaging ... 3–5, 8, 20, 36, 56, 59–60, 62, 63, 100, 133–146, 157, 162–164, 179, 186, 187, 190, 191, 196, 199, 200, 202, 204, 205, 227, 240, 242, 245, 248
Immune monitoring ... 27
Immunoassay ... 11–24
Immunoblot ... 3, 6, 240
Immunofluorescence ... 96, 244
Immunophenotyping ... 102, 104–108, 110, 112
Immunostaining ... 71, 72, 78, 108, 134, 241–243, 246–251
In situ protein analysis ... 135, 169

L

Laser scanning cytometry (LSC) ... 133–148, 240

M

Mass cytometry (CyTOF) ... 99–113, 115–130
Matrigel ... 88, 90, 92–94, 97
Microchannel ... 72, 73, 77, 78, 91, 170–172, 222

Anup Singh and Aarthi Chandrasekaran (eds.), *Single Cell Protein Analysis: Methods and Protocols*, Methods in Molecular Biology, vol. 1346, DOI 10.1007/978-1-4939-2987-0, © Springer Science+Business Media New York 2015

Microfluidic.. 11–24, 48, 56–62, 66, 69–82, 85–98, 169–184, 209–218, 221–237
Microfluidic hPSC array 86–88, 90–97
Microfluidic image cytometry 85–98
Multiplexing.. 47, 116

N

NF-κB.. 240–244, 246, 250, 251

P

Pathway analysis...47
PDMS. *See* Polydimethylsiloxane (PDMS)
Peptide-MHC tetramer 115–130
Phosphoprotein 109–110
Phospho-specific flow cytometry (phospho-flow) 101, 108–110, 112
Polydimethylsiloxane (PDMS)............................ 13, 15–17, 22–24, 56–59, 81, 88, 90, 97, 170–172, 175, 176, 178, 212, 215, 216, 223, 226, 227, 232–234, 236
Protein complementation assay 210, 211
Protein-protein interaction............................. 151–167, 169
Proteomics.. 1, 47, 55, 169
Proximity ligation assay 169–183

R

Reporter substrate 221, 222, 230, 234, 235
Rolling circle amplification (RCA) 170, 173, 179, 210–212, 214, 215, 217

S

Selective release 241–243, 247, 249, 251
Signaling network profile 99–113, 169–183
Single-cell... 1–9, 11–24, 27–44, 47–67, 69–82, 85–113, 116, 130, 133–146, 151–167, 169–183, 185–205, 209–218, 221–237, 239–251
Single cell analysis 11–24, 27–44, 47–67, 69–82, 135, 209
Single cell biology... 99
Single cell measurement 88, 90, 100, 163, 209–218, 222
Single cell proteomic and transcriptomic analysis... 55
Single-cell resolution.................................. 69, 70, 169–183
Single cell segmentation................................. 180–181, 183
Single molecule detection................................ 56, 186, 187, 190, 191, 193, 194, 196, 202, 217
Single molecule methods.. 192–194
Single particle tracking.................................. 186, 194–196, 199, 200, 203, 205
Split-protein sensor 151, 152
Split-ubiquitin (split-Ub)... 152
Stochastic gene expression........................... 55, 56, 160, 241
Superresolution...185

T

T cell assays 38
Time-lapse analysis 158
Translocation... 239–243, 246, 248, 250

Y

Yeast 57, 152–159, 161–163, 166

MIX
Papier aus verantwortungsvollen Quellen
Paper from responsible sources
FSC® C105338

If you have any concerns about our products,
you can contact us on
ProductSafety@springernature.com

In case Publisher is established outside the EU,
the EU authorized representative is:
Springer Nature Customer Service Center GmbH
Europaplatz 3, 69115 Heidelberg, Germany

Printed by Libri Plureos GmbH
in Hamburg, Germany